A Dictionary of Aviation

Other books by David W. Wragg

World's Air Fleets
World's Air Forces
Flight Before Flying   (Frederick Fell, New York)
Speed in the Air

A Dictionary
of AVIATION

Compiled by
David W. Wragg

A World of Books That Fill a Need

FREDERICK FELL PUBLISHERS, INC. NEW YORK

Published by Frederick Fell Publishers, Inc.
New York 1974

Originally Published in 1973 by
Osprey Publishing Ltd, P.O. Box 25
707 Oxford Road, Reading, Berkshire

Copyright © 1973 by David W. Wragg.

For information address:
Frederick Fell Publishers, Inc.
386 Park Avenue South
New York, N.Y. 10016

Library of Congress Catalog Card No. 74–75382

Published simultaneously in Canada by
George J. McLeod, Limited, Toronto 2B, Ontario

MANUFACTURED IN THE UNITED STATES OF AMERICA

International Standard Book Number: 0–8119–0236–6

Introduction

The prime purpose of *A Dictionary of Aviation* is to provide a guide to the more important events and personalities in aviation history, as well as to the more commonly used terms and concepts, and to the major airlines, guided weapons, aircraft, and aircraft manufacturers. It is considered that the specialist involved in any particular field of aviation activity does not require such a book to deal with his own interest in depth since he will be either fully conversant with all aspects of his work, or at least know of suitably specialised publications which can assist. The title of the book reflects the non-technical approach, in that the term 'aviation' probably means more to the layman than 'aerospace', the more correct and appropriate word, since the book deals with lighter-than-air flight and spaceflight, albeit sparingly in the latter instance, as well as with aviation proper. Air forces are dealt with briefly and collectively; those requiring more detailed historical and current information on individual air forces and air arms are referred to *World's Air Forces*, published by Osprey in 1971, which deals with this aspect of aviation activity in considerably greater detail than is possible in a work of the extensive coverage of this dictionary.

DAVID W. WRAGG

A

A-: current United States Navy designation for attack aircraft, such as the A-1 Skyraider, A-4 Skyhawk, A-7 Corsair, etc. The designation was used originally by the then United States Army Air Corps and United States Army Air Force for attack aircraft, including light bombers and dive-bombers, and to some extent preceded wide-scale use of the B- bomber designation for such aircraft as the A-20 (q.v.) Havoc and A-30 (q.v.) Baltimore.

A-1 Skyraider, Douglas: probably one of the most long-lasting and versatile warplanes of modern times, the Skyraider originated from a 1944 United States Navy requirement, and more than 3,000 were built before production ended in 1957. The aircraft was used extensively in both the Korean and Vietnam Wars, and users in Vietnam have included the U.S.A.F., U.S.N., and Vietnamese Air Force. A few aircraft also survive in Armée de l'Air service, operating from former French colonial territory in Africa. Originally a carrier-borne attack-bomber, the main use of the A-1 today is on counter-insurgency (COIN) duties. A maximum speed of 300 mph, a range of up to 3,000 miles with wing tanks, and a load of up to 8,000 lb of bombs, rockets, napalm tanks, etc. is provided for, using a single Wright R-3350 piston-engine. Most of the surviving aircraft are single-seat A-1H/J and twin-seat A-1E versions, although production covered several versions from A-1A to A-1J. Amongst other notable users of the aircraft in the past was the Royal Navy, mainly on anti-submarine duties. Airborne-early-warning versions were also produced, the EA-1E and EA-1F.

A-4 Skyhawk, McDonnell Douglas: developed as a simple, light, and cheap jet replacement for the A-1 (q.v.) Skyraider by Douglas, starting in 1950, the aircraft made its first flight in 1954. The first production version was the A-4A, with basic avionics and a 7,700 lb thrust Wright engine, and deliveries of this version started in 1956, followed by the A-4B, with detail improvements, and the A-4C, with more sophisticated avionics. The A-4E, with an 8,500 lb Pratt and Whitney J52 engine, appeared in 1961, and still more power is available in the A-4F, with its 9,300 lb thrust turbojet. Although normally a single-seat aircraft, tandem twin-seat training versions are the TA-4E, which first flew in 1965, and the TA-4F. It is the only aircraft in the U.S.N.'s inventory which can be parked in the hanger of an aircraft carrier without any need for folding wings, and it was a logical choice for the Royal Australian Navy to order for operations from the small H.M.A.S. *Melbourne* (*see* aircraft carriers, Australian). The R.A.N. version is designated A-4G, while the Royal New Zealand Air Force uses A-4Ks, the Argentinian Air Force has ex-U.S.N. A-4Bs, and the Israeli Defence Force/Air Force uses A-4Es. A wide variety of weapons can be carried in the tactical strike role, including Bullpup (q.v.) air-to-surface missiles, from a fuselage and four wing strong points, to a total of some 8,000 lb of ordnance. A range of up to 3,000 miles with external tanks is available, with a maximum speed at sea level of 680 mph.

A-5 Vigilante, North American Rockwell: the A-5 Vigilante first flew in 1958, and was designed to meet an all-purpose carrier-borne bomber requirement for the United States Navy, with nuclear strike capability and a Mach 2 speed. The A-5A first became operational in 1962, aboard the U.S.S. *Enterprise* (*see* aircraft carriers, United States), and was followed by the extended range A-5B, which had extra fuel in a 'saddleback' fuselage fairing. The RA-5C reconnaissance version first flew in 1962, and in addition to new aircraft of this type, the A-5A and A-5B aircraft have now all been converted to RA-5C standard, with a ventral fairing holding cameras,

1

side-looking radar, and other reconnaissance equipment. All RA-5Cs retain their attack capability, and a feature of the aircraft is a bomb bay in the form of a tunnel running the full length of the fuselage, with bombs ejected at the aft end. There are also under-wing strongpoints for bombs or additional fuel tanks. The crew of two have tandem seating. The RA-5C has a range of between 2,000 and 3,000 miles, Mach 2 speed, and uses two re-heated General Electric J79-GE-8 turbojets of 17,000 lb thrust each. In U.S.N. service only.

A-6 Intruder, Grumman: the Intruder was designed to meet a United States Navy and Marine Corps need for a long-range, all-weather, low-level attack aircraft, arising from experience gained during the Korean War. Carrier-borne operation and a heavy warload were also part of the requirement. Grumman won the resulting design competition in 1957 with the Intruder, and the A-6 first flew in 1960. Basically the aircraft is a two-seat machine, using two 8,500 lb thrust Pratt and Whitney J52-P-6 turbofans which give speeds up to Mach 0·95, and carrying up to 18,000 lb of ordnance in a semi-recessed bomb bay and on under-wing strongpoints. The EA-6A, which first flew in 1963, is a U.S.M.C. reconnaissance version, with a four-seat cockpit, and radomes on the fin and on the wing pick-ups, which can operate in the passive reconnaissance or electronics countermeasures role. The A-6 uses a digital integrated attack navigation system (DIANE) which enables the aircraft to fly and deliver its weapons automatically. U.S.N. and U.S.M.C. service only.

A-7 Corsair II, Ling-Temco-Vought: not to be confused with earlier piston-engined aircraft of the same name by Vought, the Corsair II won a 1963 U.S.N. design contest for an A-4 (q.v.) Skyhawk replacement, and was a derivative of the F-8 (q.v.) Crusader fighter, to which it bears more than a superficial resemblance. The first flight of this single-seat, single-engined aircraft took place in 1965. Powerplants are Pratt and Whitney TF30 turbofans of 11,350 lb thrust (A-7A) or

12,200 lb thrust (A-7B), or Allison TF-41 (licence-built Rolls-Royce military Spey) of 14,240 lb thrust (A-7D for U.S.N.) or 15,000 lb thrust (A-7E for U.S.A.F.). Maximum range is 3,600 miles with external tanks for ferrying, and warload is 10,000 lb on two fuselage and six wing strongpoints. The U.S.N.'s versions are carrier-borne, and the aircraft has seen extensive use during the Vietnam War.

A-20 Havoc, Douglas: the Douglas A-20 Havoc twin-engined light bomber served with the then United States Army Air Force, the Royal Air Force, and the Soviet Air Force during World War II. The aircraft was based on the Douglas DB-7, which also served with the U.S.A.A.F. and the R.A.F., which named it the 'Boston'. A-20s were sometimes used as long-range fighters as well as ground-attack aircraft. In appearance, the aircraft was a mid-wing monoplane with tricycle undercarriage and a single-fin tailplane; powerplants were two 2,000 hp Pratt and Whitney radial engines.

A-30 Baltimore, Martin: the Baltimore, also known as the Martin 187, was one of a small series of successful light-medium bombers produced by this American company during World War II, stemming from the Martin XB-10 design of 1932. Two Pratt and Whitney 2,000 hp radials powered this mid-wing monoplane, which was used extensively in North Africa by the Royal Air Force.

A.106, Agusta: a lightweight single-seat anti-submarine warfare helicopter for operation from warships. A single tubomeca-Agusta TAA 230 turbine engine gives a maximum speed of 100 mph. The Italian Navy operates twenty-four of these helicopters, while the Italian Army uses a further twelve on A.O.P. duties.

A.109, Agusta: A twin-turbine helicopter, mainly for the executive market but with military applications if required. The A.109C uses two 400 shp Allison 250-C20 engines, enabling it to carry up to eight persons. First flight was in 1971.

A.300B, Airbus Industrie: the A.300B is the European airbus, denoting that the aircraft is a collaborative venture between the aircraft industries of France, Germany, the Netherlands, Great Britain, and Spain, although the manufacturer, Airbus Industrie, is French-registered. Consideration of the project started in 1965, as a Hawker Siddeley Trident (q.v.) development, and in 1966 three groups – Hawker Siddeley, Breguet, and Nord; Sud Aviation; and Dornier, Messershmitt-Bölkow-Blohm, and V.F.W. – were considering three separate airbus designs. The following year the British, West German, and French governments instructed H.S.A., Sud (which later became Aerospatiale on an amalgamation with Nord), and Dornier, V.F.W. and M.B.B., to formulate a common project, for which it was planned to use a new Rolls-Royce engine, the 50,000 lb thrust RB.207. A decision to use a General Electric CF-6 engine was taken in 1969, and early in that year the British Government withdrew official support for the project, leaving H.S.A. in the position officially of a subcontractor with responsibility for design and development of the wing, but unofficially as advisers for the entire project. In 1971 Dornier left the Deutsche Airbus consortium (of which V.F.W. and M.B.B. were the other two partners) rather than contribute substantially to the project's development costs. C.A.S.A. of Spain joined the project in 1971. France and Germany have almost half of the project each, excepting the 8 per cent share of the Netherlands and the 2 per cent of Spain. Entry into airline service of the 250–300-seat aircraft is scheduled for 1974. First orders include six of the 1,400 mile range A.300B-2 for Air France, with ten options, while Iberia has ordered four of the 2,100 mile range A.300B-4, with eight aircraft on option. Lufthansa has also ordered the aircraft. The B-1 version is the prototype. A special Rolls-Royce RB.211-powered version might be made available for B.E.A. and other interested airlines. Initial development costs were estimated at £260 million in 1972.

A.A.M.: *see* air-to-air missile.

A.B.M.: *see* anti-ballistic missile defences.

A-bombs: *see* Atom-bomb.

AC-: U.S.A.F. designation for gunship conversions of transport aircraft, often with machine guns firing through window openings in a modern equivalent of a broadside. This form of conversion was introduced during the Vietnam War. AC-47 is a Dakota conversion, AC-119 a Packet, and AC-130 a Hercules.

accident: *see* air safety.

ACE: Allied Command, Europe, a NATO (q.v.) command subordinate to the Supreme Headquarters, Allied Powers in Europe (SHAPE (q.v.)), which in turn is one of the two major NATO commands. ACE has some 3,000 combat aircraft and 150 aircraft put at its disposal by the NATO air forces in Europe.

Achgelis, Focke-Wulf: the Focke-Wulf Achgelis was the first practical helicopter and first flew in 1936, although it was not further developed and did not enter production. The fuselage still owed much to the conventional aircraft, but twin rotors were placed, one on either side, at the end of skeletal wings.

ACLANT: Allied Command Atlantic is one of NATO's (q.v.) two major commands. Headquarters is in the United States at Norfolk, Virginia, and ACLANT is responsible for maritime security from the North Pole to the Tropic of Cancer (the northern tropic). Priority is given to anti-submarine warfare, and in the event of war some 300 land-based aircraft and a part of the United States Navy's carrier-borne aircraft strength could be made available, plus some 300 warships from the British, Canadian, Danish, Dutch, Portuguese, and United States navies. Peacetime strength consists of a handful of frigates.

A.C.V.: *see* Air Cushion Vehicle.

Ader, Clement (1841–1925): a Frenchman, Clement Ader was one of the more important amongst those who hoped to succeed in developing a machine capable of heavier-than-air flight. Ader was certainly one of the more controversial as well. His first 'aircraft' was a steam-powered monoplane called the 'Éole' and this became the first machine in history to leave the ground through its own efforts when Ader attempted to fly it in 1890, although the distance covered only qualified to be counted as a powered leap. In many ways the design of the 'Éole' was already obsolete, since it used the even-then discredited bat-like wing and had only a rudimentary control system.

Undaunted, Ader produced another steam-powered monoplane, also with a bat-wing, in 1897. This was the 'Avion III' (q.v.), which was fitted with two steam engines, each driving a propeller, and still lacking an adequate control system. The 'Avion III' failed to leave the ground, in spite of two attempts by Ader, and a claim made by him in 1906 to have flown 300 metres in the machine – which was completely untrue since his failure had been observed by witnesses of impeccable character.

advanced technology engine: the term given to the modern generation of jet engines, used initially on the airbuses and 'jumbo' jets but also now finding other applications. The principle is that a turbine engine will function more efficiently if the various sets of turbines can revolve at their individual optimum speeds, instead of at the same speed. The true advanced technology engine has a triple spool, meaning that instead of having a single driving shaft, there are three, in order that the three sets of blades may revolve at different speeds. An interim state is the twin-spool engine, allowing only two different speeds for the turbines. Advantages of the advanced technology engines lie mainly in their low noise levels, although the use of new materials, such as carbon fibres, will in due course also lead to lighter and more economical power-plants.

A.E.G.: one of the many German aircraft manufacturers in existence immediately before and during World War I. Apart from the firm's own designs, such as the G.IV bomber biplane, the company was one of those producing the Rumpler Taube for the Military Air Service.

Aerfer-Aermacchi: a joint production arrangement between two Italian firms engaged on production of the AM.3 (q.v.) light liaison and A.O.P. aircraft for the Italian Army.

aerial reconnaissance: aerial reconnaissance was one of the first duties to fall to the military aeroplane. At the outbreak of World War I in 1914, it was envisaged that the main task of the Royal Flying Corps would be aerial reconnaissance duties, and other air arms at this time expected and prepared for the same kind of operation. A number of scout aircraft were developed with this in mind, including the Bristol, Nieuport, and Morane Scouts, and these became some of the first fighter aircraft. Later, Royal Aircraft Factory R.E.8 'Harry Tate' aircraft were used, although usually armed after the evolvement of the fighter.

Aerial reconnaissance was not restricted to the land, for during World War I Short seaplanes were carried by seaplane tenders, usually converted cruisers or steam packets, and after the war, many battleships and cruisers were equipped to carry seaplanes. During the 1920s and 1930s, aircraft such as the Fairey IIIF and its successor, the Seal, were used for reconnaissance work at sea.

World War II saw a number of fast reconnaissance aircraft developed, including versions of the Bristol Beaufighter (q.v.) and de Havilland D.H. 98 (q.v.) Mosquito, the North American F-51 (q.v.) Mustang, the Northrop P-61 Black Widow, the Focke-Wulf Fw.189, the Mitsubishi Ki-46, and the Petlyakov Pe-2 (q.v.). Maritime reconnaissance (q.v.) from warships frequently took the form of Supermarine Walrus seaplanes, plus carrier-borne fighter types such as the Grumman (q.v.) series of aircraft, Seafires and Sea Hurricanes. The German submarine menace, and the effects of surface raiders on Allied convoys already being hit hard by the submarines, made long range maritime reconnaissance important, and if anything this importance has increased since. Major maritime-reconnaissance types of World War II included the Short Sunderland (q.v.) and Convair Catalina (q.v.) flying-boats, and the Vickers Wellington (q.v.), Boeing B-17 (q.v.) Flying Fortress, Consolidated B-24 (q.v.) Liberator and Lockheed Ventura and Privateer landplanes. Although convoys were of less importance to the Germans as a means of supply, their attraction as targets entailed the use of aircraft such as the six-engined Blohm und Voss (q.v.) Bv 222 flying-boat, and the Focke-Wulf Fw.200 (q.v.) Condor, a military version of the four-engined airliner.

During the post-war period, the Western Allies used a variety of aircraft on reconnaissance duties, including special versions of the English Electric Canberra (q.v.) bomber, the Handley Page Victor (q.v.), the de Havilland Venom (q.v.), the Hawker Hunter (q.v.), the North American F-86 (q.v.) Sabre, and the Republic RF-84F Thunderflash, while maritime-reconnaissance aircraft included the Avro Shackleton (q.v.) and the Lockheed P-2 (q.v.) Neptune. Against this, the Soviet Bloc armed forces also used a variety of conversions of fighter and bomber aircraft, while placing a great deal of the emphasis for maritime-reconnaissance on flying-boats, such as the Beriev Be-6 (q.v.) 'Madge', Be-10 (q.v.) 'Mallow', and Be-12 (q.v.) 'Mail', although more recently the trend to landplanes has also been marked in the U.S.S.R. One of the most advanced reconnaissance aircraft ever built, the B.A.C. TSR-2, was cancelled in 1964, well after the first flight and while some thirty machines were actually in course of production.

Aerial reconnaissance today can be divided into three main types, strategic, tactical and maritime, although the latter is dealt with more thoroughly under its appropriate heading.

Strategic air reconnaissance requires the use of such aircraft as the Canberra, the McDonnell Douglas RF-4 Phantom, the Lockheed RF-104 version of the F-104 (q.v.) Starfighter, the Dassault Mirage (q.v.) III-R, the Lockheed U-2 (q.v.) and its successor, the SR-71 (q.v.), and the North American Rockwell RA-5C Vigilante. Such aircraft are always equipped with cameras, and in the most sophisticated versions may have side-looking radar. A new aircraft under development for this task, amongst others, is the Panavia 200 (q.v.) Panther. In the future, there is likely to be an increasing tendency for strategic reconnaissance functions to be taken over by space satellites.

Tactical reconnaissance includes at its upper level many of the jet fighters already mentioned as undertaking strategic reconnaissance work, plus aircraft such as the Northrop F-5A/B Freedom Fighter, and advanced jet training aircraft in the reconnaissance variant, such as the Aermacchi MB.326. At a lower level, and in support of ground forces within a very confined area, aircraft such as the Cessna 0-1 (q.v.) Bird Dog and 0-2 (q.v.) Super Skymaster, were in use in Vietnam. As well as light aircraft and helicopters, reconnaissance 'remotely-piloted' vehicles are nowadays being used for tactical reconnaissance.

The aim of reconnaissance over land targets is primarily to identify suitable targets, such as links in a communications network, strategic industry, airfields, and port areas; to obtain information on air defences for strategic use; or to provide information about the deployment of enemy land forces, and the conditions and terrain prevailing in their area, for tactical use.

Aeritalia: the main Italian airframe manufacturer, Aeritalia was created from the Fiat concern which entered aircraft production in 1918, shortly before the end of World War I, with the R-2 reconnaissance-bomber biplane. The Italian Air Force, the Regia Aeronautica, was maintained at a greater strength during the inter-war period than most European air arms, and Fiat was able to produce a series of fighters designed by Celestino Rosatelli (q.v.), including the C.R.20, C.R.30, and C.R.32 biplanes, in addition to producing some light aircraft designs, such as the A.S.1 biplane of 1929. At the outbreak of World War II, the company was producing the C.R.50 Freccia fighter monoplane, which supplemented C.R.40 and C.R.42 fighters in Regia Aeronautica service, as well as the B.R.20M Cicogna bomber and the G.12 trimotor transport. Other notable wartime Fiats were to include the G.52 and G.55 fighters.

After the war Fiat became principally engaged on licence production of British and American designs, including the de Havilland Vampire (q.v.) fighter and trainer, followed by the North American F-86 (q.v.) Sabre, and then a share in the European Lockheed F-104 (q.v.) Star-

Aer Lingus–Irish International Airlines

fighter production programme. Trainer versions of the wartime fighters were developed and produced in the G.46 and G.59. During the early 1950s, the Fiat G.91 (q.v.) won a NATO design competition for a strike fighter, and this aircraft entered production during the late 1950s, remaining in production throughout much of the 1960s with the twin-engined G.91Y development. Recently the company has been concentrating on development and production of the G.222 (q.v.) light transport, and is the Italian partner in the Panavia 200 (q.v.) Panther multi-role combat aircraft. For most of its existence Fiat has produced aero-engines, sometimes under licence, for many of its products. The present title was taken in 1972.

Aer Lingus–Irish International Airlines: a single integrated structure for what were two separate airlines, Aer Lingus Teoranta, formed in 1936 to operate services to the United Kingdom, and Aerlinte Eireann, formed in 1947 to operate North Atlantic services, Aer Lingus is today the Irish Republic's state-owned airline. Operations began with a de Havilland Dragon Rapide service from Dublin to Bristol, which was later extended to London, and by the outbreak of World War II in 1939 a service to Liverpool was also being operated. Post-war development included Aerlinte Eireann's services to North America with Lockheed Super Constellation; or services which long enjoyed a considerable degree of protection from competition since Aer Lingus could operate from Dublin across the Atlantic, while American airlines were restricted to operations into Shannon.

Currently, Aer Lingus operates a fleet of Boeing 747s, 707s, 720s and 737s, and B.A.C. One-Elevens, on a European and North Atlantic route network.

Aermacchi: an Italian aircraft manufacturer of long standing, Aermacchi's mainstay today is the highly successful MB.326 trainer and tactical strike aircraft, although the company is also working on the Aerfer-Aermacchi AM.3 S.T.O.L. A.O.P. aircraft for the Italian Army.

The company first came into prominence during World War I with the Macchi M.14 fighter and the Nieuport-Macchi M.7 and M.8 fighter-bombers, while the M.8 and an M.9 development were available in civil versions during the post-war period for airline and private use. However, along with the rest of the Italian aircraft industry, the company suffered from the neglect of the armed forces immediately after the end of the war and before Benito Mussolini assumed power in 1923. Aermacchi produced seaplanes for the Schneider Trophy (q.v.) races, notably the Macchi M.33 and the M.39, the latter being a mid-wing monoplane which won the 1926 race. An end product of the Schneider Trophy successes was the Macchi C.200 Saetta of the late 1930s, which was one of the better Italian fighters of World War II, in spite of engine shortcomings, although these were overcome by re-engining with Mercedes Benz engines to produce the C.202 and C.205.

After World War II the company produced de Havilland Vampire jet fighters under licence for the Aeronautica Militaire Italian, before introducing its own design, the MB.326 (q.v.), in 1960. The company also undertakes licence production of the Lockheed Starfighter for the A.M.I., and designed the LASA-60 utility plane for the civil market.

Aero: this Czechoslovak company was responsible for the design of a number of military aircraft after the end of World War I, when Czechoslovakia was established as an independent state. Amongst the first of these were the Aero A-18 fighter, which was followed by the A-24 bomber and A-11 and A-12 A.O.P. aircraft, all of which were operated by the Czech Army Air Force. During the 1930s, the Aero A-30 and A-100 reconnaissance aircraft and Aero A-32 A.O.P. aircraft were put into production. The Aero B-17 was a licence-built Tupolev SB-2 bomber, which entered Czech military service just before World War II started.

Civil work was not neglected during this period. One of the first Czech-produced aircraft was the Aero A-10, which could carry three passengers on the Prague-Kosice route, and this was followed by the ten-passenger A-22, operated by the manufacturer on a domestic air service. The Aero A-23 entered service in 1928, and in turn was followed by the A-35 and A-38 during the early 1930s.

Post-war products of the company, which is now state-owned since Czechoslovakia is a member of the Soviet Bloc, include the L-60, a high-wing single-engined monoplane used for a variety of duties, of which the most important is aerial crop-spraying, and the L-200, a four-person twin-engined air taxi. A helicopter project is the HC-3, a five-seat executive machine with a single piston engine. The firm is most famous for its L-29 (q.v.) Delfin, a tandem two-seat jet trainer which is used by most East European air forces, and which first flew in 1959. A successor to the Delfin, the L-39, first flew in 1968.

Aero 3: this is one of a small series of wooden Yugoslav light training aircraft, with no connection with the Czechoslovak Aero concern. Basically, the Aero 3 is a development of the Aero 2. A tandem dual-seat low-wing monoplane, with a Lycoming 0-435-A piston engine, the Aero 3 has a top speed of about 140 mph. About sixty of these aircraft are in Yugoslav military service.

aerobatics: a contraction of the term 'aerial acrobatics'. Aerobatics are of considerable importance in the evolvement of fighter manoeuvres, as well as having an entertainment and a training value.

The first aerobatic manoeuvre was the spin, recovery from which was first performed by an Englishman, Frederick Langham, after stalling in thick fog while flying an Avro biplane in 1911. The first pilot to enter a spin deliberately in order to demonstrate recovery was Lieutenant Wilfred Parke, R.N., in April 1912, while flying an Avro cabin-biplane. The next aerobatic manoeuvre, the loop, was first performed by Lieutenant Nesterov of the Imperial Russian Flying Corps, while flying a Nieuport in August 1913.

Aero Commander: now a division within the North American Rockwell Group,

Aero Commander has been producing light and executive aircraft for a number of years. These have been largely based on the twin-engined, high-wing, Aero Commander 500 (q.v.) series, including the Aero Commander 500 itself, and the 520 and 560 developments, but with the light Aero Commander 100 (q.v.) and Darter Commander single-engined models, which were also high-wing aircraft, to complete the range. The Darter Commander design was 'bought in' by the company, having started life as the Volaire 100, but is now out of production. A jet executive aircraft, the twin-engined Aero Commander 1121, or Jet Commander, was in production for some time before the production rights were sold to Israel Aircraft Industries (q.v.), who produce the aircraft as the 1123 Commodore Jet. The reason for Aero Commander leaving the important executive jet market was almost certainly to avoid conflict with the North American T-39 (q.v.) Sabreliner, or Sabre Commander, as it is now known.

Recently, a new range of low-wing, single-engined light aircraft has been introduced, including the Aero Commander 111A 112 (q.v.).

Aero Commander 100: a range of high-wing, single-engined monoplanes using Lycoming piston engines – usually in the 0-360 series – of which the most recent versions are the Darter Commander, or basic 100, which started life as the Volaire 100, and the higher-performance Lark Commander, which has superior interior finish and equipment in order to appeal to the lower end of the business aircraft market. Maximum cruising speed is in the region of 140 mph, with a range of 500 miles. The aircraft have four seats and non-retractable undercarriages. Now replaced by the Aero Commander 111A/ 112 (q.v.).

Aero Commander 111A/112: the latest addition to the Aero Commander range is the four-seat, low-wing, 111A/112, with a Lycoming engine. The basic version, the 111A, has a non-retractable undercarriage, but a retractable undercarriage and detail refinements are available on the 112. Maximum speed is in the order of 160 mph, with a range of around 500 miles.

Aero Commander 500: a successful range of high-wing, twin-engine, executive aircraft has been developed for civil and military use over the post-war period by Aero Commander, using their 500 aircraft as the starting point. Seating varies from five to eight seats in the 500/540/560 range, of which the 500 uses 290 hp Lycoming 10-540-EIA piston engines, and the 540 and 560 up-rated versions of this engine, while a 680F-P has a pressurised cabin, and the Grand Commander has a lengthened fuselage, but otherwise is of 560 standard; the Turbo Commander has the eleven-seat Grand Commander fuselage and 605 shp Air Research TPE-331 turboprop engines. Maximum speed and range vary according to model, but are generally over 250 mph and around 1,200–1,500 miles. Undercarriages are retractable.

aerodrome: *see* airport.

Aerodrome, Langley's: Samuel Langley (q.v.) was awarded a United States Army contract worth $50,000 in 1898 to build a heavier-than-air flying machine. However, the tandem wing 'Aerodrome', which was catapulted off the roof of a houseboat moored on the Potomac River in 1903, only succeeded in crashing twice into the river after fouling the catapult mechanism. A more successful American aviation pioneer, Glenn Curtiss (q.v.), altered the aircraft extensively, fitting one of his own engines, in 1914, and attempted to fly it, but only succeeded in producing a series of powered leaps.

aerodynamics: the science of flight in heavier-than-air machines can be loosely termed aerodynamics, although a stricter and more accurate description is that it is the science of the motion of air, or other gases, over a body.

The two pioneers who are credited with contributing extensively to the development of the science were the Englishmen Sir George Cayley (q.v.) and Horatio Phillips (q.v.). Cayley's model glider of 1804 was the first real aeroplane with main plane (or wings) and adjustable tail surfaces with a fin, giving both control and stability. In a paper, 'On Aerial Navigation, published in 1809, Cayley laid the basis for all subsequent studies of aerodynamics, breaking away from the ornithopter concept and pointing towards the idea of a mainplane, tailplane, fuselage and undercarriage.

Phillips's work on high aspect cambered aerofoils (q.v.) was the basis for all subsequent successful wing design, as well as being an extension of Cayley's work. To Phillips must go the credit for discovering that if an aerofoil is made with a deeply-curved upper surface and a shallowly curved lower surface, there will result a high lifting power, the bulk of which is provided by a suction effect on the upper surface rather than by a pressure effect on the lower surface.

Aeroflot: Aeroflot is the world's largest airline, although this is due to its being operated on a semi-military basis and undertaking all manner of aviation activity, air taxi services and crop spraying as well as the more usual air transport. The exact size of the airline can only be guessed at, although it is the only airline in the Soviet Union, and the proportion of its fleet out of service at any one time is likely to be high compared to a western airline, due to the shortcomings of Soviet civil equipment.

The airline's immediate predecessor was Dobroflot, formed in 1923 at the start of the first five-year plan from the amalgamation of several small pioneer airlines. In 1932 Aeroflot was formed, and expanded rapidly until the German invasion of Russia in 1941 halted many of its services. In spite of the fact that Dobroflot started life with the then high route mileage of 6,000 miles, by 1941 Aeroflot was still only carrying 0·16 per cent of all Soviet domestic traffic.

The Douglas DC-3 Dakota was introduced to Soviet service during World War II, along with other Allied designs, and the aircraft was produced under licence in the U.S.S.R. as the Lisunov Li-2, which formed the backbone of the Aeroflot fleet throughout the late 1940s and the 1950s, even after such Russian designs as the Ilyushin Il-12 (q.v.) and Il-14 (q.v.)

entered service. At this time, almost all of Aeroflot's aircraft were single- or twin-engined because air travel was virtually restricted to government officials, and it was not considered worth wasting scant resources on developing aircraft of DC-4 or DC-6 type. After World War II ended, Aeroflot was behind the formation of airlines in Russian-occupied Eastern Europe, using standard packages of Li-2s.

Aeroflot's first jet, the twin-engined Tupolev Tu-104 (q.v.), was introduced in 1956, and soon followed by a shorter range version, the Tu-124, and turboprop airliners, including the Ilyushin Il-18 (q.v.), Antonov An-24 (q.v.), and Tupolev Tu-114 (q.v.). Modern jets include the Tupolev Tu-134 (q.v.), Tu-144 (q.v.), and Tu-154 (q.v.), and the Ilyushin Il-62 (q.v.), approximating to the B.A.C. One-Eleven/Douglas DC-9, Concorde, Boeing 727, and Vickers VC-10 respectively; and the Yakovlev Yak-40. It is unlikely that Antonov An-22 heavy transports are in Aeroflot service, but many types of light aircraft and a full range of helicopters are likely to be in use by the airline.

aerofoil: a surface designed to obtain a reaction from the air through which it is moved. It is usually an alternative expression for a wing, but it can be a lifting aerofoil fuselage (q.v.), a propeller or rotor, or indeed any control surface of an aircraft.

Aerolineas Argentinas: although Aerolineas Argentinas was formed in 1949 on the nationalisation of all Argentinian airlines, with the exception of the air force-operated L.A.D.E., the airline had its origins in Aeroposta Argentina, which dated from 1927, when it had been formed as a subsidiary of the French airline, Cie Générale Aeropostale. Three other airlines included in the nationalisation were A.L.F.A., or Aviacion del Litoral Fleuva Argentina, Z.O.N.D.A., or Zonas Oeste y Norte de Aerolineas, and F.A.M.A., or Flota Aero Mercante Argentina. All four airlines had a Government minority shareholding at the time of nationalisation.

The new airline had a very mixed fleet, including Douglas DC-3s, DC-4s, and DC-6s, Avro Yorks, Lancastrians and Ansons, Vickers Vikings, and Junkers Ju.52/3M trimotors, which gave rise to difficulties during the early years. However, by 1959 Argentina was able to become the first South American nation to operate jet airliners, with the delivery of six de Havilland Comet 4 airliners to Aerolineas Argentinas. Further jet equipment, Sud Caravelles, was put into service in 1962, and Boeing 707s followed in 1966.

Currently, Aerolineas Argentinas operates a fleet of Boeing 707s, 737s, Sud Caravelles, and Hawker Siddeley 748s on an extensive domestic network, and on services throughout South America, and to the United States and Europe.

Aeromexico: today a state-owned airline, Aeromexico was formed as a private enterprise company, Aeronaves de Mexico, in 1934, using a small fleet of Beechcraft aeroplanes operating a Mexico City to Acapulco service; at this time the resort of Acapulco was only a small fishing village. Pan American World Airways acquired a 40 per cent interest in Aeronaves in 1940, enabling further expansion of the airline to take place, including the acquisition of Transportes Aereas de Pacifico in 1941. A number of other small airlines have been acquired during the history of Aeromexico.

After the war the Aeronaves fleet consisted of Douglas DC-3 and DC-4 aircraft, and rapid expansion of the internal air service network operated by the airline took place, largely through the acquisition of the existing operators. The Pan American shareholding was sold in 1957 to Mexican interests, and later that year the first international service, to New York, was introduced. The airline was acquired by the Mexican Government in 1959, after a pilots' strike severely damaged the finances of Aeronaves. Further airline acquisitions followed the nationalisation of Aeronaves, including Aerovias Guest, with services to Miami and to South America.

A Government-sponsored reorganisation of Mexican air transport in 1970 has resulted in Aeronaves de Mexico controlling an air transport system which includes the airline's own network and that of eight smaller carriers – although the smaller airlines have not actually been acquired. The present title was adopted in 1972. Currently, Aeromexico operates an extensive domestic network, with international services to the United States, Montreal, Paris, Madrid, and Caracas, using a fleet of McDonnell Douglas DC-8s, including the 'Super Sixty' series, and DC-9s.

aeronaut: traveller by air, usually in a balloon or airship, either as crew or passenger.

aeronautics: the all-embracing term for field of balloons, airships, and heavier-than-air flight.

Aeronca: an American light aircraft manufacturer now no longer trading, but prominent during the pre-World War II period with a number of single- and dual-seat high-wing monoplanes, including the C-2. Wartime and early post-war production included aircraft for both civil and military use, amongst which were the four-seat 'Secan', and the twin-seat 'Champion' and its military counterpart, the L-16A. The manufacturing rights for the 'Champion' were later sold to a manufacturer of that name.

aeroplane: *see* aircraft.

aerospace: the modern term which groups aviation and space activities together, e.g. 'Society of British Aerospace Companies' has taken over from 'Society of British Aircraft Constructors'.

Aerospatiale: this is the nationalised part of the French aircraft industry, originating from the amalgamation of Nord and Sud Aviation during the late 1960s.

Nord was already noted as the manufacturer of the Noratlas transport aircraft of the early post-war period, although during the war the company had produced the Messerschmitt Bf. 108 trainer, and continued this after the war as the Nord 1100 and 1101 Noralpha. The company also produced the Nord 262 (q.v.) turboprop feeder-liner and the Nord 500 experimental tilt-wing V.I.O.L. air-

craft, while acting as the French partner in the Franco-German C-160 (q.v.) Transall military transport project and producing a range of tactical guided missiles, including the SS.11 (q.v.). The guided missile work fitted in well with Sud's practice and sounding rocket work on the formation of Aerospatiale.

Sud Aviation dated only from 1957, on the amalgamation of Ouest-Aviation and Sud-Est Aviation, which both dated from 1936 and the pre-World War II nationalisation of many small French aircraft manufacturers, which proved tragic for French military aircraft production at the outset of the war. Apart from the successful Caravelle (q.v.) jet airliner, which was one of the first jet aircraft to cater for short distance operations, and the Ouragan light jet bomber, Sud had considerable experience of helicopter and light aircraft manufacture. Sud helicopters have included the Djinn, a light helicopter for aerial crop-dusting, the famous SE. 3130 Alouette (q.v.) II, and SE. 3160 Alouette (q.v.) III and more recently the SA. 330 (q.v.) Puma and SA. 341 (q.v.) Gazelle, which are now part of a package collaborative deal with Westland (q.v.). The largest Sud helicopter is the SA. 321 (q.v.) Super Frelon, while the latest is the W.G.13 (q.v.) Lynx, on which Aerospatiale is Westland's junior partner. Light aircraft, including the M.S. Rallye and Gardan GY-80 Horizon, were originally produced by the Centre Est subsidiary, but these are now produced, with their successors, by the light aircraft division.

Apart from the several helicopter projects, Aerospatiale is the French partner and design leader on two European projects of considerable importance, the Anglo-French Concorde (q.v.) supersonic airliner and the Franco-German–Dutch–Spanish A.300B (q.v.) airbus, for which Sud formed Airbus Industrie (q.v.).

Aerostar: *see* Mooney Aircraft.

aerostat: *see* aerostation.

aerostation: the art of flying balloons (q.v.) and airships (q.v.), for which the technical name is aerostat.

Aerovan, Miles: one of the many post-

World War II attempts to produce a light aircraft with potential for exploiting freight traffic as well as carrying passengers, the Miles Aerovan could be considered in many ways to be the precursor of today's Short Skyvan (q.v.). Aptly nicknamed 'The Flying Tadpole', the Aerovan was a high-wing monoplane, with a twin-fin tailplane, a non-retractable undercarriage, two piston engines, and accommodation for freight, or twelve passengers. The aircraft saw service with a number of small airlines in Britain and Europe, including East Anglian Flying Services, the predecessor of the ill-fated Channel Airways. A four-engined development, the Miles M. 68, was never able to realise its potential, but the Aerovan itself was in small-scale production throughout the late 1940s.

A.E.W.: *see* airborne-early-warning.

AFCENT: Allied Forces Central Europe, a subordinate command within SHAPE (q.v.), which is one of the two major NATO commands. AFCENT has its headquarters at Brussum in the Netherlands, and is divided into two sub-commands, NORTHAG (Northern Army Group) with its own air support from the Second Allied Tactical Air Force, and CENTAG (Central Army Group), with air support from the Fourth Allied Tactical Air Force.

AFNORTH: Allied Forces Northern Europe, a subordinate command within SHAPE (q.v.), with its headquarters at Kolsaas in Norway. The area covered by AFNORTH includes Norway, Denmark, the Baltic, and Northern Germany. The Royal Norwegian Air Force, the Royal Danish Air Force, and two Luftwaffe wings are available for air support.

AFSOUTH: Allied Forces Southern Europe, another of SHAPE's (q.v.) subordinate commands, and based on Naples. Its area of responsibility includes Italy, Greece, and Turkey, the Mediterranean and Black Sea, and air support is provided by the Italian Air Force, the Royal Hellenic Air Force, and the Turkish Air Force, with United States Navy aircraft from Sicily and the Sixth Fleet, and Royal

Air Force aircraft from Malta and Cyprus. A surveillance force, MARAIRMED, Maritime Air Forces, Mediterranean, has been formed to counter Soviet Naval expansion in the area, and this uses Italian, United States, and British maritime-reconnaissance aircraft.

A.G.M.: air-to-ground, or more accurately air-to-surface (q.v.), missile or stand-off bomb.

Agusta/Agusta-Bell: an Italian helicopter manufacturer most renowned for its licence-production of Bell designs, but also including production of Sikorsky and Boeing helicopters, as well as development and production of its own projects.

Agusta was established in 1907 by Giovanni Agusta to manufacture biplanes. Although the founder died between the wars, the company survived and expanded, even though at one time diversification into motor-cycle production was necessary. A basic training aircraft, the AP-111, was produced after World War II ended, and in 1954 licence-production of the Bell 47 (q.v.) helicopter started. This has been followed by the Bell 204 and 205 (q.v.) Iroquois, the Bell 206 (q.v.) JetRanger, and the civil Bell 212 (q.v.) and it is from these aircraft, and from the development of the Bell 47 with cabin accommodation, that the Agusta-Bell name has become famous.

Other Agusta helicopter work has entailed licence-production of the Sikorsky SH-3D and the Boeing-Vertol Bv. 114 (q.v.) Chinook for the Italian and Iranian armed forces. Agusta's own helicopter designs now include the A.106 (q.v.) for the Italian Navy and Army, and the A.109 (q.v.) executive helicopter. Work on V.T.O.L. includes research on the design for the A.120B Helibus and the A.123 compound helicopter. The Agusta-designed EMA-124 is under production at the Meridionali factory. The company holds a 30 per cent interest in Savoia-Marchetti (q.v.), the Italian light aircraft manufacturer.

AH-1G Huey Cobra, Bell: the Huey Cobra is a combat helicopter development of the successful Bell 204 UH-1 Iroquois. A

11

cockpit with tandem twin-seating is fitted in a narrow fuselage, with stub wings for stability and as weapons attachment points. The Huey Cobra is faster than its utility counterpart, with a maximum speed of 200 mph, while retaining the standard Bell two-blade rotor and using a 1,100 shp Lycoming T53 turbine. Although its utility counterpart is in widespread service and manufacture, the Huey Cobra is purely a United States Army machine for ground attack and close support missions, including those in Vietnam.

AH-56A Cheyenne, Lockheed: the Cheyenne is another in the United States Army's combat helicopter series, intended to replace conventional utility helicopters operating in the gunship role on attack duties. A cockpit with tandem seating for two and stub wings for increasing stability and as weapons attachment points are standard features for this type of machine, although the Cheyenne has the unusual feature of a pusher propeller for which power from the 3,435 shp Lycoming engine can be switched as speed increases, so avoiding the drag problems of high-speed flight dependent on the main rotor. The Cheyenne holds the world speed record for a rotorcraft, and is capable of a maximum speed of 250 mph, and a range of 875 miles. The United States Army contract was terminated in 1969, due to technical problems, even though these have since been overcome.

aileron: a movable surface on the outer trailing edge of an aircraft's wings controlling the rolling or banking movements of the aircraft. Early biplanes had the ailerons sometimes mounted separately between the upper and lower wings, and gyroplanes (q.v.) or autogiros (q.v.) retained the aileron, even though the conventional wing was replaced by a rotor leaving the ailerons protruding from the fuselage sides. The earliest aircraft frequently used a system of wing warping (q.v.), or flexing, in order to achieve the control now provided by the ailerons.

Airacobra, Bell P-39: see P-39 Airacobra, Bell.

Airacomet, Bell: although the correct title is Bell P-59A Airacomet, the aircraft never became operational and only fourteen were built, largely as an evaluation exercise for the then U.S.A.A.F. The Airacomet was the first American jet aircraft, and used two General Electric engines based on those of the British experimental Gloster E.28/39 Whittle (q.v.), which gave a top speed of only 400 mph, marking the aircraft as inferior to both the British Gloster Meteor (q.v.) and the German Messerschmitt Me. 262 (q.v.). First flight was late in 1944.

Air Afrique – Société Aerienne Africaine Multinationale: formed in 1961 by twelve former French colonies in Africa–Cameroon, Central African Republic, Chad, Congo–Brazzaville, Dahomey, Gabon, Ivory Coast, Mauretania, Niger, Senegal, Togo, and Upper Volta – with each state contributing 6 per cent of the capital, and the remainder being provided by the Société pour le Développement du Transport Aerien en Afrique, or Sodentraf. Operations started soon after the airline's formation, and included domestic routes and an African network, as well as services to Europe. Cameroon and Chad withdrew in 1971 and 1972 respectively, leaving the remaining states to hold 7 per cent of the capital each. The airline is based in the Central African Republic, although it was originally based in Cameroon.

Air Afrique operates a fleet of McDonnell Douglas DC-10s and DC-8s, including the 'Super Sixty' series, Aerospatiale Caravelles, N.A.M.C. YS-11s, and Douglas DC-4s and DC-3s.

Air Algérie–Algerian Airways: formed in 1953 from a merger of an airline of the same name with Compagnie Air Transport, Air Algérie developed during the years prior to independence with the assistance of Air France (q.v.), which today is a minority shareholder with approximately 15 per cent, while the Algerian Government holds the remainder of the share capital. A network of services is maintained throughout North Africa, and to major points in Europe and the Soviet Union, with a fleet of Boeing 707s and 737s, Aerospatiale Caravelles, Convair 640 Metropolitans, and Douglas DC-3s.

air arm: strictly speaking, any military aircraft operator is an air arm, but in general use the term is confined to army, navy, or marine corps aircraft used as a part of the concept known today as organic airpower (q.v.). The first real air arm was that of the British Army, formed in 1898 with a Royal Engineers company operating balloons at Woolwich. The United States Air Force did not formally become independent of the United States Army until 1947, although for most purposes a degree of independence was achieved earlier, but fought through World War II as the United States Army Air Force, and so was officially still an air arm. The term may see a more general use in the future, with the tendency for some countries, such as Canada, to move towards a unified defence force concept.

airborne: in flight, but usually meaning that an aircraft has just left the ground.

airborne assault: usually refers to the dropping of paratroops from troop-carrying aircraft. Gliders have also been used on assaults coming into this category, and today the helicopter is the main vehicle for airborne assault, with World War II-type paratroop drops discredited in the eyes of some. Glider-borne forces assisted in the invasion of Normandy by the Allies towards the end of World War II, while amongst several paratroop assaults can be counted the invasion of Crete by German forces earlier in the war. The advantages of using helicopters include the less-exacting standard of training required for the troops involved, the ability to pick troops up, as well as set them down, and the reduced time for making an assault, with less risk of a strong counter-attack. Paratroopers are not only vulnerable while in the air, but also immediately on landing.

airborne-early-warning: today, an airborne-early-warning system is considered an essential part of any air defence system (q.v.), since it both extends the range of surface-based radar to give an earlier warning of attack by fast enemy aircraft and, even more important, is the only real way of detecting low-flying attacking aircraft seeking to intrude below the coverage of surface-based radar.

Amongst the first A.E.W. aircraft, shortly after the value of the measure was appreciated during the closing stages of World War II, was the Douglas EA-1 Skyraider (*see* A-1), used by both the United States Navy and the Royal Navy, and followed by the Grumman E-1B (q.v.) Tracer and Fairey Gannet (q.v.) for the American and British navies respectively. All of these aircraft were carrier-borne, as is the Tracer's successor, the E-2A Hawkeye. The U.S.A.F. has used Lockheed EC-121 Constellations (q.v.) on this duty, while the R.A.F. has a force of converted Avro Shackletons, which originally served in the maritime reconnaissance role. A new American concept is the airborne warning and control system, or A.W.A.C.S., which will use extensively modified Boeing C-135 (q.v.) aircraft.

airbrake: a device used to create extra drag, helping to steady or slow an aircraft. Usually airbrakes are fitted to the wings, especially on aircraft such as the Junkers Ju. 87 (q.v.) dive-bomber, but as devices for reducing the speed of aircraft after landing, and as an economical alternative to thrust reversers whenever possible, airbrakes are fitted to the tail of an aircraft; a user of this type of airbrake is the Fokker F-28 (q.v.) Fellowship.

airbus: the term given to large short-range aircraft, such as the Airbus Industrie A.300B (q.v.), the Lockheed L-1011 TriStar (q.v.), and the McDonnell Douglas DC-10 (q.v.).

Airbus Industrie: a company registered in France, and concerned with the production and sales of the European airbus, the A.300B (q.v.). The original partners were Aerospatiale of France, Hawker Siddeley of Great Britain, and Dornier, V.F.W. and Messerschmitt-Bolkow-Blohm of West Germany. Hawker Siddeley is now a sub-contractor and design consultant following the withdrawal of British Government support for the aircraft, and Dornier has left the project. Fokker is now a member of the consortium, following both a merger with V.F.W. and Netherlands Government support for the project, and C.A.S.A. of Spain has also joined.

Air Canada: a subsidiary of the state-owned Canadian National Railways, Air Canada was formed in 1937 to purchase the Canadian Airways Company and its fleet of two Lockheed Electras and a Stearman biplane, after a plan for a mixed enterprise Canadian National and Canadian Pacific Airways failed. The first service operated between Vancouver and Seattle, with the first of an extensive domestic service network being introduced in 1938; although the new airline's first service was international, the title from its formation until 1964, when the present title was adopted to reflect the widening of the route network, was Trans Canada Airlines, or T.C.A. Additional Electras and the larger Lockheed 14s were rapidly acquired, and by 1940 the airline was operating Canada's first trans-continental air service.

Unusually, T.C.A. was one airline which managed some expansion during World War II, and after 1943 gained invaluable transatlantic experience, operating Avro Lancastrians on the Canadian Government's Trans-Atlantic Air Service for V.I.P. passengers and forces air mail.

Late in 1945 the airline received the first of thirty Douglas DC-3 airliners, for domestic services and routes to the United States, followed in 1947 by the first Canadair North Stars – Canadian-built Douglas DC-4s with Rolls-Royce Merlin engines – for scheduled transatlantic flights. New services to Europe were opened during the 1950s, largely using Lockheed Super Constellations, while in 1955 the airline became the first North American turboprop airliner operator, with Vickers Viscounts for domestic and United States services. T.C.A.'s first jet airliners, Douglas DC-8-40s, were introduced in 1960. Vickers Vanguards entered service in 1961.

Currently, Air Canada operates a fleet of Boeing 747s, McDonnell Douglas DC-8s (including the 'Super Sixty' series) and DC-9s, Lockheed TriStars, and Vickers Viscounts on an extensive domestic and United States network, with services to Europe and the Caribbean.

Air Ceylon: formed in 1947 by the Ministry of Communications and works, with a fleet of three Douglas DC-3s for domestic services, Air Ceylon introduced international services to London and Sydney in 1949, using Australian National Airways aircraft. The airline became a corporation in 1951, with the Ceylon Government holding 51 per cent of the share capital, and Australian National Airways the remainder, until 1955, when A.N.A.'s share was purchased by K.L.M., and a service to Amsterdam was introduced. The agreement with K.L.M. ended in 1961, and only domestic services and some international services to India were operated until an agreement was concluded with B.O.A.C. in 1962. Ownership of the airline since 1961 has been entirely in the hands of the Ceylon Government.

Today, management assistance is received from the French airline, U.T.A. (q.v.), and in addition to the domestic network, international services are operated to Western Europe, Australia, India, and the Far East, with a fleet which includes Douglas DC-8, Hawker Siddeley Trident and 748, and Douglas DC-3 aircraft.

Airco: the name for the Aircraft Manufacturing Company, which was extant during World War I. Much of the company's fame can be credited to its chief designer, Sir Geoffrey de Havilland, who formed the de Havilland company after the war, but designed for Airco the D.H.2 and the D.H.4 bomber.

air corps: the air arm of an army or a marine corps.

aircraft: the term aircraft, or aeroplane, relates to heavier-than-air flying machines, to the exclusion of balloons, airships, gliders, kites, and various types of projectiles, in its contemporary sense.

Much of the early work on heavier-than-air flight involved man-powered flight, which is still some way off in practical terms even today. Attempts which entailed imitating birds or bats also had to be abandoned before man could get off the ground. Even after balloons and airships had achieved a fair degree of acceptance, heavier-than-air flight was pursued because of the poor degree of control available from lighter-than-air craft, the need to be able to fly into a headwind, and the greater speed and load-carrying potential of the aeroplane.

Amongst those who laid the foundations on which more famous pioneers were ultimately able to build were the Englishmen Sir George Cayley (q.v.), Horatio Phillips (q.v.), W. S. Henson (q.v.), and John Stringfellow (q.v.). The first aircraft to leave the ground was a steam-powered monoplane built by a Frenchman, Félix du Temple (q.v.), in 1874, although the aircraft only made a powered leap (i.e. the distance covered was too short for control to be exercised, and the force which counted was that of the take-off or down-ramp run and not any sustaining power of the engine). Ten years later a Russian steam-powered monoplane, built by Mozhaiski (q.v.), repeated this performance. The first plane to leave the ground through its own efforts, but again only for a powered leap, was Clement Ader's (q.v.) 'Éole' of 1890. A test-rig designed by the American-born Sir Hiram Maxim (q.v.) showed considerable potential in 1894, but was not pursued, while at the time of the Wright brothers' (q.v.), successful efforts in 1903, another American, Samuel Langley (q.v.), built his unsuccessful 'Aerodrome' (q.v.) for the United States Army.

The first true aeroplane in history was that built and flown by the Wright brothers in 1903. The achievement was not repeated in Europe until Alberto Santos-Dumont's (q.v.) '14-bis' of 1906, and Samuel F. Cody's (q.v.) flight of 1908 in British Army Aeroplane No. 1. However, by 1909 Blériot (q.v.) was able to fly across the English Channel, and the first international aviation meeting (q.v.) was held at Reims, in France.

A fair degree of reliability had been achieved by the outbreak of World War I (q.v.) in 1914, and the aircraft had become easily recognisable as such to modern eyes.

In 1913 (q.v.) the Russian, Sikorsky (q.v.), built the world's first four-engined aircraft, the Grand, and followed this in 1914 with his Ilya Mourametz. Subsequent development after the war tended to be along specialised lines, with airliners (q.v.), bombers (q.v.), fighters (q.v.), reconnaissance aircraft, etc.

aircraft carrier: the aircraft carrier is a warship specifically designed to carry fixed-wing (q.v.) aircraft which are not necessarily capable of vertical take-off, and for which it must have a through flight deck running from stem to stern.

The first flights to and from a warship were by the American, Lieutenant Eugene Ely, U.S.N., who flew a Curtiss biplane from a platform constructed over the forward gun turret of the cruiser U.S.S. *Birmingham* in 1910. The following year, Ely landed an aircraft on a platform constructed over a stern turret of the U.S.S. *Pennsylvania*. In January 1912 Lieutenant Charles Samson, (q.v.), R.N., flew a Short S.27 biplane from the battleship H.M.S. *Africa*, which had a platform constructed over one of the forward gun turrets for this purpose, and in May that year, Lieutenant Samson, R.N., then made the first take-off from a ship under way when he flew from another battleship, H.M.S. *Hibernia*, also in an S.27.

Before World War I a converted cruiser, H.M.S. *Hermes*, served as a seaplane tender, and a number of cross-Channel ferries were taken over by the Admiralty and converted for this purpose too. In 1917 a cruiser under construction and nearing completion, H.M.S. *Furious*, was modified to have a flight-deck built over the foredeck, on which Commander E. H. Dunning, R.N., made the first landing on a ship under way. H.M.S. *Furious* was later modified to have a second flight-deck aft, with connecting trackways through to the foredeck, and further modified during the 1920s to have a through flight deck and a side superstructure. An Italian cruiser under construction in the United Kingdom was taken over by the Admiralty in 1917, and completed as the world's first through-deck warship, H.M.S. *Argus*, although without any real superstructure.

15

Other World War I efforts in this field included aircraft take-offs from barges towed behind destroyers.

The first aircraft carrier to be completed with an 'island' superstructure was H.M.S. *Eagle*, in 1920, although the first ship to be laid down as an aircraft carrier was H.M.S. *Hermes*, laid down in 1918 and completed in 1923.

The first American aircraft carrier was the U.S.S. *Langley*, converted from a collier in 1922, while Japan's first aircraft carrier also entered service in 1922, having been laid down in 1919. France's first aircraft carrier, the *Bearn*, converted from a battleship, entered service in 1925. By 1930 the Royal Navy had five aircraft carriers, and the United States Navy had three.

The aircraft carrier achieved its full potential as a weapon in World War II, although H.M.S. *Hermes* had the unwanted distinction of being the first such ship to be sunk by an opposing vessel. The Japanese attack on the United States Fleet at Pearl Harbor was carrier-borne, as was the Royal Navy's attack on the Italian Fleet at Taranto. Carrier-borne aircraft played a significant part in the Battles of Midway and the Coral Sea, and were also present in the Battle of Cape Matapan. The *Bearn* acted as an aircraft transport prior to the fall of France, and an American carrier, the *Wasp*, carried aircraft to Malta. Merchant vessels were converted to act as escort carriers by the Americans, and many of these entered Royal Navy service, and went far beyond their original duty of convoy escorts.

After the war the aircraft carrier proved itself to be of value to the French when land bases in French Indo-China became untenable due to intensive guerrilla activity. Later, carriers were to prove useful in the Korean War, in the confrontation between British and Indonesian forces following the formation of the Federation of Malaysia, and in the Vietnam War. The Indian Navy used its one carrier to good effect during the 1972 war with Pakistan.

Meanwhile, a number of developments had taken place. In 1945, a de Havilland Vampire made the first landing by a jet on an aircraft carrier, H.M.S. *Ocean*. To help bring the aircraft carrier into the jet age, the Royal Navy developed the angled flight deck (q.v.), mirror landing system, and steam catapult. These developments were quickly adopted by other navies.

After the war the aircraft carrier found a number of new users, including the Royal Canadian Navy and the Royal Netherlands Navy (although neither of these has aircraft carriers at the present time), the Royal Australian Navy, the Argentinian Navy, the Brazilian Navy, and the Indian Navy. Aircraft carriers of about 30,000 tons are under construction for the Soviet Navy.

The aircraft carrier concept has come in for some criticism in recent years, both from those who merely see defence expenditure as too costly anyway, and from the supporters of other types of weapon, struggling for a larger share of severely limited funds. However, the carrier is a highly mobile and self-contained airfield, ready for action at almost any time, and in this it is more economical and less open to attack than a comparable level of shore-based forces moved at great cost and over a period of time into a forward battle area. A variety of strategic and tactical uses also make the carrier a versatile weapon.

An attempt to cut costs and make the most of V.T.O.L. aircraft has led to the mini-aircraft carrier concept, known as a 'through-deck cruiser' in Britain and a 'sea control ship' in the United States. Basically, these have the appearance of old-fashioned aircraft carriers, with protruding foredecks beyond the flight deck, and a slight degree of angling is provided on the flight deck. Deck parking space is limited, superstructure is over-sized, and lifts are badly sited. Entry into service should be around 1978.

aircraft carriers, Australian: the Royal Australian Navy operates one aircraft carrier, the 16,000 ton H.M.S. *Melbourne*, from which a squadron of McDonnell Douglas A-4 Skyhawks are operated, with Grumman Trackers and Westland Wessex helicopters. Australian naval aviation using aircraft carriers dates from after World War II, and an older vessel, H.M.A.S. *Sydney*, is used as a fast troop transport. Both ships are ex-Royal Navy.

aircraft carriers, British: the Royal Navy's aircraft carrier strength has declined from the five vessels, plus two commando carriers, of the mid-1960s, the *Eagle*, *Ark Royal*, *Victorious*, *Hermes*, and *Centaur*, and the *Albion* and *Bulwark*, to one aircraft carrier, H.M.S. *Ark Royal*, and the two commando carriers, although one or both of these will be replaced by H.M.S. *Hermes* after conversion. H.M.S. *Ark Royal* has been extensively modernised, although lacking the 3-D radar of H.M.S. *Eagle* (withdrawn in 1972) or *Hermes*, and is equipped with all modern aids. *Ark Royal* has a displacement of 53,000 tons with a complement of 3,000 men, including aircrew, and is equipped with McDonnell Douglas F-4K Phantom interceptors, Hawker Siddeley Buccaneer S.2 bombers, Westland Gannet airborne-early-warning aircraft, and Westland S-61 Sea King helicopters for anti-submarine warfare. The *Ark Royal* is due to be replaced during the late 1970s by the first of a number of through-deck cruisers for deployment of Sea Kings, and possibly Harrier V.T.O.L. fighters as well.

aircraft carriers, French: the Maritime National operates two light fleet aircraft carriers, the *Clemenceau* and the *Foch*, both of 22,000 tons, dating from 1961 and 1963 respectively, while the *Arromanches*, formerly H.M.S. *Colossus*, of 14,000 tons, acts as a training and helicopter carrier. L.T.V. F-8 Crusader fighters, Dassault Étendard fighter-bombers, Breguet Br. 1050 Alize A.S.W. aircraft, and helicopters are operated from both the larger vessels. There is also a helicopter carrier, *La Résolue*.

aircraft carriers, Indian: the Indian Navy has operated one aircraft carrier, the I.N.S. *Vikrant* of 16,000 tons, since the early 1960s. One squadron of Hawker Sea Hawk fighters can be carried at any one time, with a small number of Breguet Br.1050 Alize A.S.W. aircraft and two

Alouette helicopters. The ex-Royal Navy ship (formerly H.M.S. *Hercules*) was used to good effect in early 1972 in the war with Pakistan arising from the civil war in the former East Pakistan, or Bangladesh.

aircraft carriers, United States: the United States Navy now has the largest air arm of any navy, with sixteen attack carriers, some of which are nuclear-powered, seven anti-submarine carriers, and eight helicopter carriers. The largest carriers of this force are the U.S.S. *Nimitz* and *Enterprise*; of 85,000 tons displacement and capable of carrying 100 aircraft, these are the largest aircraft carriers ever. There are six Forrestal Class vessels, although another two ships, the U.S.S. *America* and *Kennedy*, could be described as improved Forrestal Class, and these vessels can carry up to ninety aircraft and have a displacement of 75–77,000 tons. The attack carrier force also includes two Midway Class vessels, of 62,000 tons displacement and able to carry eighty aircraft, and four Oriskany Class vessels, which are improved Essex Class and can carry seventy aircraft, having a displacement of 43,000 tons.

Seven Essex Class ships have been modernised and act as anti-submarine carriers, while three other Essex Class ships are used as helicopter carriers, along with some older vessels and a new class, of which the U.S.S. *Iwo Jima* is the first. A number of other older vessels remain in reserve, on training duties, or as aircraft ferries.

Aircraft used by the U.S.N. and U.S.M.C. include the McDonnell Douglas F-4 Phantom and A-4 Skyhawk, the Grumman A-7 Intruder, E-1 Tracer, S-2 Tracker, C-1 Trader, E-2 Hawkeye, and C-2 Greyhound, the L.T.V. A-7 Corsair and F-8 Crusader, and the Sikorsky S-61 and Kaman UH-2 Seasprite helicopters. The U.S.M.C. has Hawker Siddeley Harrier V.T.O.L. fighters. A new A.S.W. aircraft, the Lockheed S-3 Viking, and a new interceptor, the Grumman F-15 Tomcat, are under development at present.

Currently, design studies are being undertaken into the concept of the sea control ship, similar to the Royal Navy's through-deck cruiser, although larger.

aircraft carrying ships: almost without exception, modern destroyers and frigates are equipped to carry helicopters for anti-submarine duties. Although cruisers no longer carry seaplanes, and battleships no longer exist, most cruisers can also carry helicopters, and in the future may also carry V.T.O.L. fighters; cruisers of this type include the Italian Navy's *Andrea Doria, Caio Duilio,* and larger *Vittorio Veneto,* and the Soviet Navy's *Moskva* and *Leningrad.* The Royal Navy and the United States Navy both operate assault ships, with fairly extensive helicopter accommodation.

The use of helicopters is not confined to warships in the strict sense, and many fleet auxiliaries carry helicopters as flying cranes for ferrying supplies. Survey ships also carry helicopters in some cases.

Aircraft Transport and Travel Ltd: this was a British company, formed in 1916, which became the first airline to operate scheduled international services when it started a London to Paris service in 1919. (Prior to World War I, domestic Zeppelin airship services had operated in Germany.) The initial equipment for A.T.&T.'s services included converted D.H.4 and D.H.9 bombers. A London to Amsterdam route was operated on behalf of the newly-formed K.L.M. Royal Dutch Airlines later in 1919. The company also operated the first international airmail (q.v.) services. Aircraft Transport and Travel eventually led to Imperial Airways late in 1923, and through this step the company was in fact an ancestor of the present-day B.O.A.C. (q.v.).

aircrew: the crew of an aircraft, particularly military, as against ground crew. Civil aircraft usually distinguish between flight-deck crew: captain, first officer, and sometimes a navigator and flight engineer; and cabin crew: stewards and stewardesses.

air cushion vehicle: the correct term for what is usually known as a hovercraft, which is a misleading term since it can also apply to V.T.O.L. aircraft and helicopters which also have a hover characteristic.

The inventor of the air cushion vehicle was an Englishman, Sir Christopher Cockerell (q.v.), who designed the first practical A.C.V., the SR-N1 (q.v.), which first flew in 1959. As the term implies, an air cushion vehicle rides on a cushion of air, which is provided by ducted propellers. The provision of rubber skirts proved to be a major step forward, improving the operation of A.C.V.s by containing the air cushion. Modern A.C.V.s are either amphibious, or non-amphibious sidewall craft in which the sides of the craft extend down to the water. Amphibious hovercraft are ideal for operation in areas with water and sandbanks or mudflats and swamps, while the sidewall craft can dispense with the noisy and costly aircraft engines of the amphibious models, using diesels and water-immersed propellers instead, and are both more economical and more acceptable environmentally. Semi-amphibious hovercraft with beaching capability are now available and in use.

The first hovercraft public service in the world was initiated by British United Airways in 1962 across the Dee Estuary in North Wales, using a Vickers amphibious hovercraft, and since then hovercraft services have been inaugurated at several places in the British Isles, in Portugal, and in Scandinavia, while there are also freight services in the less-developed countries with inadequate surface communications. The largest hovercraft available today is the British Hovercraft Corporation's SR-N4 (q.v.) Mountbatten hovercraft, of which five are in service on cross-Channel routes between Great Britain and Europe. Probably the largest single hovercraft operator is the Imperial Iranian Navy, which uses a fleet of ten British Hovercraft Corporation SR-N5 (q.v.) Warden and BH-7 (q.v.) Wellington hovercraft on anti-piracy and anti-smuggling operations in the Persian Gulf.

The main hovercraft manufacturers today are the British Hovercraft Corporation (q.v.), Vosper-Thorneycroft (q.v.), Hovermarine (q.v.), Hover-Air, Bell (q.v.), and Mitsui.

air defence system: the object of an air defence system is to protect national airspace (q.v.) in the first instance, and ultimately the territory beneath the airspace. The fully integrated modern concept of aerial defence has come a long way from the visual observation of attacking enemy formations, often of Zeppelin airships, during World War I, via the listening posts and radar, with fighter control, of World War II, although many of the World War II practices are the basis for those in use today.

A modern air defences system relies extensively on the use of radar, including broad beam radar for detecting any intrusion by enemy aircraft, and narrow beam tracking radar for keeping the enemy aircraft or formations under observation, while airborne-early-warning aircraft, such as the Royal Air Force's Shackletons (q.v.), the United States Air Force's EC-121s, and the Royal Navy's Gannets (q.v.), extend the radar coverage beyond the horizon and also help to pick up low-flying enemy aircraft which might otherwise be able to sneak under the defending radar cover. Transponders (q.v.) fitted to aircraft assist in the elimination of friendly aircraft from any need for investigation, while airliners fly under air traffic control (q.v.) along airways (q.v.).

Computers are used to assist in assessing the likely targets for incoming aircraft, and to help in guiding interceptor (q.v.) aircraft or surface-to-air (q.v.) missiles to the enemy aircraft. An advantage in using interceptors instead of missiles whenever possible lies in the lower risk of mistaken attack of a friendly aircraft. Computer assistance in identification of enemy aircraft is also possible and, with an efficient communications system, also assists in obtaining the best deployment of the available defensive forces.

air force: although it is possible to use the term air force as an alternative to air arm (q.v.), and indeed the World War II United States air arm was the United States Army Air Force, it is almost always used to identify a separate armed service with the specific functions of air defence, aerial attack and reconnaissance, and air transport. The first air force within the usual meaning of the term was the Royal Air Force, formed from the Royal Flying Corps and Royal Naval Air Service in 1918. Most countries have a separate air force, of which the most recently formed, apart from those in the newly-independent nations, is the Royal Danish Air Force, dating only from 1950. Canada has reversed the process by forming a unified defence organisation, the Canadian Armed Forces, while Israel also has such an arrangement. Army origins are usually detectable in the retention of army rank for air force personnel, but the major exceptions to this rule include the Royal Air Force, Royal Australian Air Force, Royal New Zealand Air Force, and Indian Air Force, which have distinctive air force ranks, particularly for officers, owing as much to naval as to military practice; examples include the ranks of Flight-Lieutenant, Wing-Commander, and Group Captain instead of Captain, Lieutenant-Colonel, and Colonel respectively, while a purely air force rank is that of Squadron-Leader, equivalent to Major or Lieutenant-Commander.

The advantages of having a separate air force come from the realisation that airpower is a weapon to be used in its own right, and not just in support of naval or ground forces, although many air forces still retain these functions, even when the concept of organic airpower (q.v.) is practised. World War II (q.v.) was lost and won in the air, although this would have been wasted effort had not the surface forces been able to play their part successfully, and this is the true measure of the importance of airpower (*see also* air superiority).

Although the major airlines are included in this book, space has not permitted coverage of air forces and air arms, but this information is available in the author's *World's Air Forces* (Osprey, 1971), which is devoted entirely to the subject, giving brief histories and details of equipment for each air force and air arm in the world. However, the air forces and air arms of the world are:

Abu Dhabi: Air Wing, Abu Dhabi Defence Forces;

Afghanistan: Royal Afghan Air Force;

Albania: Albanian People's Army Air Force;

Algeria: Algerian Air Force–Force Aérienne Algérienne;

Argentina: Argentinian Air Force–Fuerza Aérea Argentina; Argentinian Naval Aviation–Comando de Aviación Naval; Argentinian Army Aviation Command;

Australia: Royal Australian Air Force; Royal Australian Navy-Fleet Air Arm; Australian Army Aviation Corps;

Austria: Austrian Air Force–Österreichische Luftstreitkräfte;

Belgium: Belgian Air Force–Force Aérienne Belge; Belgian Navy–Force Navale Belge; Belgian Army Aviation–Force Terrestre Belge;

Bolivia: Bolivian Air Force–Fuerza Aérea Boliviana;

Brazil: Brazilian Air Force–Forca Aérea Brasileira; Naval Air Arm–Forca Aeronavale;

Bulgaria: Bulgarian Air Force;

Burma: Union of Burma Air Force;

Cambodia: Cambodian Air Force;

Canada: Canadian Armed Forces–Forces Armées Canadiennes;

Central African Republic: Central African Air Force–Force Aérienne Centrafricaine;

Ceylon: *see* Sri Lanka;

Chad Republic: Chad Air Squadron–Escadrille Tchadienne;

Chile: Chilean Air Force–Fuerza Aérea de Chile; Chilean Navy;

China, Republic of (Taiwan): Chinese Nationalist Air Force; Chinese Nationalist Army;

Chinese People's Republic: Air Force of the People's Liberation Army. Chinese Navy;

Colombia: Colombian Air Force–Fuerza Aérea Colombiana;

Congo (Brazzaville): Congo Air Force;

Cuba: Cuban Air Force–Fuerza Aérea Revolucionaria;

Czechoslovakia: Czechoslovak Air Force–Ceskoslovenske Letectvo;

Dahomey: Dahomey Air Force–Force Aérienne du Dahomey;

Denmark: Royal Danish Air Force–Flyvevaabnet;

Dominican Republic: Dominican Air Corps–Aviación Militar Dominicana;

Ecuador: Ecuadorian Air Force–Fuerza Aérea Ecuatoriana;

Egypt (United Arab Republic): Egyptian Air Force;

Eire: Irish Army Air Corps;

El Salvador: Salvadorian Air Force–Fuerza Aérea Salvadurena;

Ethiopia: Imperial Ethiopian Air Force;

Finland: Finnish Air Force–Ilmavoimat;

France: French Air Force–L'Armée de l'Air; French Naval Aviation–Aéronautique Navale (L'Aéronavale); French Army Aviation–Aviation Légère de l'Armée de Terre;

Gabon: Gabon Air Force–Force Aérienne Gabonnaise;

German Democratic Republic: Luftstreitkräfte und Lufverteidigung (L.S.K.);

German Republic, Federal: West German Air Force–Luftwaffe; German Naval Air Arm–Marineflieger; Army Air Corps–Heeresflieger;

Ghana: Ghana Air Force;

Greece: Royal Hellenic Air Force;

Guatemala: Guatemalan Air Force–Fuerza Aérea de Guatemala;

Guinea Republic: Guinea Air Force–Force Aérienne de Guinea;

Haiti: Haitian Air Corps;

Honduras: Honduras Air Force–Fuerza Aérea Hondurena;

Hong Kong: Hong Kong Auxiliary Air Force;

Hungary: Hungarian Air Force;

India: Indian Air Force; Indian Naval Aviation;

Indonesia: Indonesian Republican Air Force–Angatan Udera Republik Indonesia (A.U.R.I.); Indonesian Naval Aviation–Angatan Laut Republik Indonesia;

Iran: Imperial Iranian Air Force;

Iraq: Iraqui Air Force;

Israel: Israel Defence Force/Air Force;

Italy: Italian Air Force–Aeronautica Militare Italaino; Italian Naval Aviation-Marinavia; Italian Army Aviation–C.A.A.L.E;

Ivory Coast: Ivory Coast Air Force–Force Aérienne de Côte d'Ivoire;

Jamaica: Air Wing, Jamaica Defence Force;
Japan: Japanese Air Self-Defence Force; Japanese Maritime Self-Defence Force; Japanese Ground Self-Defence Force;
Jordan: Royal Jordanian Air Force;
Kenya: Kenya Air Force;
Korea, Democratic People's Republic of (North): Korean People's Army Air Force;
Korea, Republic of (South): Republic of Korea Air Force (R.O.K.A.F.);
Kuwait: Kuwait Air Force;
Laos: Royal Lao Air Force;
Lebanon: Lebanese Air Force–Force Aérienne Libanaise;
Libya: Libyan Air Force;
Malagasy Republic: Malagasy Air Force Armée de l'Air Malgache;
Malaysia: Royal Malaysian Air Force;
Mali Republic: Mali Air Force–Force Aérienne du Mali;
Mauritania: Mauritanian Islamic Air Force–Force Aérienne de la Republique Islamique de Mauritanie;
Mexico: Mexican Air Force–Fuerza Aérea Mexicana; Naval Aviation–Armada da Mexico;
Mongolia: Air Force of the Mongolian People's Republic;
Morocco: Royal Moroccan Air Force–Aviation Royale Chérifienne;
Muscat and Oman: Sultan of Oman's Air Force;
Nepal: Royal Flight and Army Aviation;
Netherlands: Royal Netherlands Air Force–Koninklijke Luchtmacht; Royal Netherlands Naval Air Service–Marine Luchtvaartdienst;
New Zealand: Royal New Zealand Air Force; Royal New Zealand Navy;
Nicaragua: National Guard Air Corps–Fuerza Aérea, Guardia Nacional;
Niger: Niger Air Force–Force Aérienne de Niger;
Nigeria: Federal Nigerian Air Force;
Norway: Royal Norwegian Air Force–Kongelige Norske Luftsvaret;
Pakistan: Pakistan Air Force; Pakistan Navy; Pakistan Army Aviation;
Paraguay: Paraguayan Air Force–Fuerza Aérera del Paraguay;
Peru: Peruvian Air Force–Fuerza Aérea

Peruana; Peruvian Naval Aviation–Servicio Aéronavale;
Philippines: Philippine Air Force;
Poland: Polish Air Force–Polskie Lotnictwo Wojskowe;
Portugal: Portuguese Air Force–Forca Aérea Portuguesa;
Rhodesia: Rhodesian Air Force;
Rumania; Rumanian Air Force;
Saudi Arabia: Royal Saudi Air Force;
Singapore: Singapore Air Defence Command;
Somali: Somalian Aeronautical Corps–Cuerpo Aeronautica del Somalia;
South Africa: South African Air Force–Suid Afrikaanse Lugmag;
South Yemen Republic: Air Force of the South Yemen People's Republic;
Spain: Spanish Air Force–Ejercito del Aire; Naval Aviation–Marinha; Army Aviation–Arma Españá;
Sri Lanka: Sri Lanka Air Force;
Sudan: Sudanese Air Force;
Sweden: Royal Swedish Air Force–Flygvapnet; Royal Swedish Navy; Army Aviation–Armen;
Switzerland: Swiss Air Force and Anti-Aircraft Command–Kommando Flieger und Fliegerabwehrtuppen;
Syria: Syrian Air Force;
Tanzania: Tanzanian People's Defence Force Air Wing;
Thailand: Royal Thai Air Force; Royal Thai Navy; Royal Thai Border Police;
Togo: Togo Air Force–Force Aérienne Togolaise;
Tunisia: Tunisian Air Force;
Turkey: Turkish Air Force–Türk Hava Kuvvetleri; Turkish Army;
Uganda: Uganda Army Air Force/Police Air Wing;
Union of Soviet Socialist Republics: Soviet Military Aviation Forces–Sovietskaya Voenno-Vozdushnye Sily (S.V.V.S.); National Air Defence–Protivo-Vozdushniya Oborona Strany (P.V.O.S.); Independent Naval Air Fleet–Aviatsiya–Voenno Morskikh Flota (A.-V.M.F.);
United Kingdom of Great Britain and Northern Ireland: Royal Air Force; Fleet Air Arm; Army Aviation;
United States of America: United States Air Force; United States Navy and

Marine Corps; United States Army Aviation;

Uruguay: Uruguayan Air Force–Fuerza Aérea Uruguaya; Naval Aviation–Aviación Naval;

Venezuela: Venezuelan Air Force–Fuerzas Aéreas Venezulanas;

Vietnam (South): Vietnamese Air Force;

Vietnam, People's Republic of (North): Vietnamese People's Air Force;

Yemen: Yemen Republican Air Force;

Yugoslavia: Yugoslav Air Force–Jugoslovensko Ratno Vazduhoplovstvo;

Zaire: Zaire Air Force–Force Aérienne Zaire;

Zambia: Zambia Air Force;

airframe: the aircraft itself, minus engines and avionics (q.v.).

Air France: the history of Air France dates from 1933, and the formation of the airline from the merger of five airlines, Farman Airlines, Cie Internationale de Navigation Aérienne (C.I.D.N.A.), Air Union, Air Orient, and Aeropostale. The oldest of these was Farman Airlines, which dated from 1919, and by 1930 was operating to many of the major centres in Europe, while Air Union operated to the Mediterranean area of France, and Air Orient operated to the Far East, starting in 1931 with a Marseille–Damascus–Saigon service. Aeropostale, dating from the early 1920s, was owned by the French Government, which had saved the airline from bankruptcy in order to maintain its airmail services.

Air France started life with a French Government shareholding of 25 per cent, and expanded rapidly, having ninety aircraft by the outbreak of World War II in 1939, when normal operations ceased and the airline's aircraft and personnel were placed at the disposal of the military, except for a London–Paris diplomatic service. Services ceased with the fall of France in 1940, to be restarted in 1942 in North Africa.

After the war the airline was nationalised, although later provision was made for a 30 per cent private enterprise shareholding. Services restarted in 1946, with flights to the French colonies and across the North Atlantic, as well as European services. Aircraft included Douglas DC-3s and DC-4s, which were later joined by Lockheed Super Constellations and, during the mid-1950s, by turboprop Vickers Viscounts. Not surprisingly, the airline was the first to use the Sud Aviation Caravelle short haul jet airliner, which entered service in 1959.

Air France is currently owned by the French Government, with a 70 per cent interest, and private investors with a 30 per cent interest, while itself having a 25 per cent interest in the French domestic airline, Air Inter (q.v.), and interests in Air Algérie (15 per cent), Air Madagascar (30 per cent), Air Mauritius, Air Vietnam (25 per cent), Middle East Airlines (q.v.) (30 per cent), Royal Air Cambodge (40 per cent), Royal Air Maroc (21 per cent), Tunis Air (49 per cent), and several smaller African airlines. The route network includes Europe, North America, and Latin America, with the Caribbean, those parts of Africa and the Far East not served by U.T.A. (q.v.), Australia, and the domestic Postale de Nuit F-27 Friendship services for the French Post Office, using a fleet of Boeing 747s, 707s, 727s, Aerospatiale Caravelles, and Fokker F-27 Friendships, and with B.A.C.-Aerospatiale Concordes and Airbus Industrie A.300Bs on order.

Air-India: Air-India dates from 1932 and the formation of Tata Airlines to operate airmail services. Few passengers were carried during the early years, and it was not until 1938 that de Havilland Dragon Rapides were introduced for passenger services. World War II ended the company's normal services, but operations were continued on behalf of the Government and the armed forces, with the experience gained in operating larger aircraft proving valuable when normal peacetime operations were resumed. Tata Airlines became a public company after the war, and in 1946 became Air-India Ltd. Trans World Airlines provided assistance during the immediate post-war period, while Air-India operated Douglas DC-3 and DC-4 aircraft.

After India's independence, the Government took a 49 per cent shareholding, with an option on another 2 per cent, and the title Air-India International was adopted. Services to London with Lockheed Constellations were soon introduced, followed by a service to East Africa. All Indian airlines were nationalised in 1953, with Air-India International and six purely domestic airlines being formed into two state-owned corporations, the domestic Indian National Airlines Corporation and the Air-India International Corporation. The present title was adopted in 1962.

Currently, Air-India operates an international network of services to Europe and the United States, East Africa, the Far East, and Australia, with a fleet of Boeing 747s and 707s. A subsidiary, Air-India Charters, was formed in 1971 to promote tourist traffic, and aircraft are chartered from the parent company as required.

Air Inter–Lignes Aériennes Intérieures: the French domestic airline, Air Inter was formed in 1954 by Air France (q.v.), S.N.C.F. (French Railways), the Compagnie de Transports Aériens, and banks, private companies, and regional groups. Services did not start until 1958, using chartered aircraft, and then lasted for only a short period, with a genuine start not being made until 1960. Vickers Viscount turboprop airliners were introduced in 1962, with more Viscounts and Nord 262s following later. Aerospatiale Caravelles have also been operated, with Fokker F-27 Friendships supplementing the remaining Viscounts, which will be replaced upon the arrival of Dassault Mercure jet airliners. Currently, Air France and S.N.C.F. have a 25 per cent interest each, while U.T.A. (q.v.) holds 15 per cent.

airline: an operator of civil transport aircraft, whether freight or passenger, on a chartered or scheduled basis. Usually the term is confined to operators of aircraft of more than 12,500 lb, and below this the term is air taxi (q.v.) operator or third level airline. The various types of operation are also further indentified by the so-called 'Five Freedoms of Air Transport' (q.v.).

The first airline in the modern sense, i.e. operating aircraft as opposed to airships, was Aircraft Transport and Travel (q.v.),

a British company, but the airline with the longest existence as a single entity is K.L.M. Royal Dutch Airlines, which dates from 1919 when an Aircraft Transport and Travel aircraft inaugurated an Amsterdam–London service for K.L.M. The largest airline today is the Soviet operator, Aeroflot (q.v.), although the largest commercial operator is the American company, United Air Lines (q.v.). Major airlines are usually members of the International Air Transport Association (q.v.).

airliner: a civil transport aircraft used for carrying passengers, but excluding light aircraft and airships.

The first civil transport aircraft were converted World War I Airco D.H.4 bombers, used by Aircraft Transport and Travel and K.L.M., amongst others. Aircraft more readily recognisable as airliners appeared in the form of civil versions of the Vickers Vimy (q.v.) bomber, and the then new de Havilland D.H.66 (q.v.) Hercules. There was no shortage of airliner designs during the inter-war period, due in part to the increasing importance of the aeroplane, and also because the production and service life of an aircraft at that time was very much shorter than for an aircraft today – the main exception to this rule pre-war being the Douglas DC-3 (q.v.), although a number of smaller aircraft, including the original Lockheed Electra (q.v.) and the de-Havilland D.H.84 (q.v.) Dragon series also put up a good production performance for the time. Many of the landplanes of the period were slow and lumbering, but comfortable, and in this category can be placed the Handley Page H.P.42 (q.v.) and the Armstrong-Whitworth Argosy (q.v.) both of Imperial Airways. Much of the glory, and the more spectacular routes, went to the flying-boats (q.v.) of which the Short Rangoon and Calcutta (of the early 1930s) and the Short Empire 'C' class and Boeing 314 (of the late 1930s) can be counted amongst the most famous and successful.

The landplanes can perhaps be grouped into a number of phases, as various concepts and ideas became fashionable. A typical example was the American passion for speedplanes for fast transport of passengers and mail, although both were in small quantities and passenger capacity varied between six and eight seats. The aircraft were usually monoplanes, always single-engined and with the one pilot sitting in what was usually an open cockpit. Nevertheless the aircraft concerned were attractive on the whole. A typical speedplane was the Lockheed Vega. More useful work, although sometimes of an unspectacular nature, could be credited to the tri-motors, of which the more famous were the Ford Trimotor, the Fokker F.VIIB (q.v.), and the Junkers Ju. 52/3M (q.v.), although the Armstrong-Whitworth Argosy was another example. Such aircraft efficiently established regional air transport in Europe and the United States, and tackled some of the longer routes as well, before the advent of the Douglas DC-1 (q.v.), DC-2 (q.v.), and DC-3 (q.v.), and the smaller but faster Lockheed Electra (q.v.). A variety of aircraft in the de Havilland D.H.84 (q.v.) Dragon series ranged from three to about fifteen passengers in capacity.

The modern airliner came into existence before World War II, in the form of the Boeing 307 (q.v.) Stratoliner, with four engines and cabin pressurisation; while the other four-engined aircraft, such as the Focke-Wulf Fw. 200 (q.v.) Condor, the Junkers Ju. 90 (q.v.), and the attractive de Havilland Albatross (q.v.), of the immediate pre-war period, also gave an indication of things to come, although not quite so advanced as the Stratoliner. A twin-engined airliner for the shorter routes was the de Havilland D.H.95 Flamingo. All of these types suffered from the advent of World War II, which put an end to their full commercial exploitation and development, although the Douglas DC-3 and DC-4 (q.v.) and the Curtiss C-46 (q.v.) actually benefited from the conflict. Wartime civil transport aircraft included de Havilland D.H.98 (q.v.) Mosquito converted bombers for B.O.A.C., and civil conversions of the Consolidated Liberator and Boeing Fortress bombers.

After the war, apart from converted bombers such as the Lancastrian development of the Avro Lancaster, there were more new designs, including the Avro York, Vickers Viking (q.v.), Airspeed Ambassador (q.v.), Douglas DC-6 (q.v.), Lockheed Constellation (q.v.) and Super Constellation, Boeing 377 (q.v.) Stratocruiser, and the Handley Page Hermes. The early 1950s saw the de Havilland Comet I (q.v.), the world's first jet airliner, which proved to be a failure in service, and the highly successful Vickers Viscount (q.v.), which also had the distinction of being the world's first successful turboprop airliner. (There had hitherto been turboprop versions of the Avro Tudor, albeit an unsuccessful design.) A spate of turboprop aircraft followed, including the Lockheed Electra II (q.v.), Vickers Vanguard (q.v.), Bristol Britannia (q.v.), and Canadair CL-44 (q.v.) for the medium and longer routes, with twin-engined Avro (now Hawker Siddeley) H.S. 748s (q.v.), Fokker F-27 (q.v.) Friendships, Handley Page Heralds (q.v.), and N.A.M.C. YS-11s (q.v.) for the shorter routes. There were also a whole range of designs from the Soviet Antonov, Tupolev, and Ilyushin design bureaux. Against this trend, Douglas produced the DC-7 (q.v.) piston-engined airliner and Convair the 240/340/440 Metropolitan series. Meanwhile, new forms of air transport had appeared, including the vehicle ferries using Bristol 170 (q.v.) Freighters initially, and later Carvair conversions of the Douglas DC-4.

The late 1950s saw the re-birth of the jet age, with the Sud Caravelle (q.v.) short and medium-haul jet airliner, the de Havilland Comet 4 (q.v.), which was the first jet to enter service on the North Atlantic run, and its rival Boeing 707 (q.v.), Douglas DC-8 (q.v.), and Convair (q.v.) 880 airliners. The Soviet Union already had the Tupolev Tu-104 (q.v.) in service. The early 1960s saw the appearance of a whole range of airlines aimed to bring jet transport to the world's short and medium range routes, including the B.A.C. One-Eleven (q.v.), the Hawker Siddeley Trident (q.v.), the Douglas DC-9 (q.v.), and the Boeing 727 (q.v.) and 737 (q.v.), to be followed much later by the Fokker F-28 (q.v.) Fellowship, and during the early 1970s by the V.F.W.-Fokker (q.v.) 614. As the scope of air transport, and above all comfortable air transport, spread, the turboprop and piston engine put in a reappearance further down the scale with aircraft designed for feeder services, including the de Havilland Canada DHC-6 (q.v.) Twin Otter, the Jetstream (q.v.), the Short Skyvan (q.v.), the Beechcraft King Air (q.v.), and the Dassault Hirondelle (q.v.) with turboprop engines, and the piston-engined Britten-Norman BN-2 (q.v.) Islander and Dornier Do. 28D (q.v.) Skyservant.

The late 1960s saw the advent of the large 'jumbo' jet airliner, the Boeing 747 (q.v.), with accommodation for up to 530 passengers, although the problem of selling so many seats caused most airlines to limit the aircraft to about 350 seats. During the early 1970s, the airbus has followed the 747 in the form of the McDonnell Douglas DC-10 (q.v.) and Lockheed TriStar (q.v.), with the Airbus Industrie A.300B (q.v.) following in 1974. The mid-1970s should also see the B.A.C.-Aerospatiale Concorde (q.v.) supersonic airliner enter airline service.

So much progress has not been without its casualties. The Hughes Hercules (q.v.) flying-boat, with the largest wingspan of any aircraft ever built, the Saunders-Roe Princess (q.v.) flying-boat, and the Bristol Brabazon (q.v.) airliner were all aircraft which were too far ahead of their time, and too big, to enter production. All three were built, and flew. The Boeing 2707 (q.v.) supersonic airliner was cancelled because of technical and financial problems associated with building such a large and fast aircraft.

Future developments involve reducing aircraft noise, and the space required for the take-off and landing of an aircraft, since the size of airports and the number required are also causing pressure to be put on air transport by the community as a whole. This involves a whole range of concepts, such as S.T.O.L., V.T.O.L., and Q.T.O.L., respectively short, vertical, and quiet take-off and landing (q.v.). An interim concept is R.T.O.L. – reduced take-off and landing (q.v.). Aircraft will

also get bigger, for there are already plans for a civil version of the giant Lockheed C-5 (q.v.) Galaxy military transport, which is currently the largest aircraft in existence.

airmail: although balloons were used for communications purposes at the Siege of Paris in 1870 and 1871, and small balloons had been used by H.M.S. *Assistance* while searching in the Canadian Arctic for the ill-fated Sir John Franklin expedition in 1850, neither these nor a military air dispatch service inaugurated between Vienna and Kiev in 1918 using Brandenburg biplanes can be counted as airmail services in the modern sense. Nor can a 1911 Bleriot monoplane flight in the United States over the six miles between Garden City and Mineola, or a British flight that same year between Hendon and Windsor.

The first international scheduled airmail flight in the world was operated in November by the British airline, Aircraft Transport and Travel (q.v.), between London and Paris, although by this time a purely domestic service inaugurated in the United States by the U.S. Post Office between Washington and New York was some eighteen months old. In late 1920 a Hubbard Air Service Boeing Type C aircraft inaugurated the first United States' international airmail service by flying from Seattle to British Columbia.

The importance of airmail to airline development between the wars is generally and hugely underestimated. The United States airline industry received a massive boost during the mid-1930s when the U.S. Post Office transferred the airmails from the United States Army, which had suffered several accidents while flying the airmails, to airlines on a contract basis. The British, French, and Dutch airlines had already benefited from airmail traffic at this time because so much of their route mileage was used to link the parts of wide-flung colonial territories. The British 'Empire Air Mail' scheme, agreed in 1934 and inaugurated in 1937, enabled cheap-rate airmail letters to be sent to any address in the British Empire, and by 1938 Imperial Airways was undoubtedly the world's largest carrier of airmail. Imperial had been so encouraged by the airmail prospects that it had taken the then unprecedented step of ordering twenty-eight giant Short Empire 'C' Class flying-boats straight off the drawing-board, and in the event, further orders of an improved version of the aircraft were soon required. A through London to Australia airmail service had been started as early as 1934, however. A trans-United States airmail service between San Francisco and New York was started in 1924, three years after an experimental flight along the route. Britain's first regular domestic airmail service was started by Highland Airways in 1934, between Inverness and Kirkwall.

Since the war the relative importance of airmails to the airlines has declined with the massive growth in air freight and business and leisure travel which has taken place, although there can be no doubt that airmail is still a useful supplement to earnings. Originally de Havilland had intended to build a transatlantic jet mailplane after World War II, but decided that this would be uneconomic, and built the ill-fated Comet I instead. During the late 1940s, B.E.A. (q.v.) operated experimental helicopter airmail services within the United Kingdom.

airmiss: a near accident, deemed as such if two aircraft are in close proximity and would have collided had not one or both taken sharp avoiding action.

Air New Zealand: originally formed in 1940 as Tasman Empire Airways Ltd, or T.E.A.L., Air New Zealand was initially owned by the Governments of the United Kingdom, Australia, and New Zealand, and was intended to operate services between Australia and New Zealand. Ownership was left in the hands of the Australian and New Zealand Governments in 1954, when Britain withdrew, although co-operation with B.O.A.C. (q.v.) continued. Throughout these early years of its existence, T.E.A.L. was principally a flying-boat operator. Australia withdrew from a share in the airline's ownership in 1961, and in 1965 the present title was adopted. Air New Zealand currently operates a fleet of McDonnell Douglas DC-10 and DC-8 aircraft, having recently retired its flying-boat replacement Lockheed Electras, on an external route network to Australia, the Far East, the United States, and British Commonwealth territories in the Pacific.

air observation post: the first military use of airpower was that of a balloon as an observation post for the French Army at the Battle of Fleurus in 1794, and later at the Battles of Marberge and Mayence. A further use occurred during the American Civil War between 1860 and 1866 with the Unionist Army of the Potomac. The British Army used balloons frequently after 1878.

The use of balloons and aircraft for artillery-spotting and tactical reconnaissance – the air observation post concept – gives an army on the plain many of the advantages hitherto only conferred by command of high ground. During the early days of World War I, the idea was that aircraft should be used for A.O.P. and reconnaissance duties.

A.O.P. is probably less important than it was, although Cessna 0–1 (q.v.) Bird Dog light aircraft have been used in the Vietnam War. The greater mobility of modern armies and the vulnerability of light aircraft to ground fire, as the equipment available to the infantry becomes more potent, must account in part for this relative decline in importance. The remaining tasks are increasingly falling to the light helicopter, such as the Hughes 0H-6A Cayuse (a version of the Hughes 500 (q.v.)), the Bell 47 (q.v.), Sioux, and the Westland-Aerospatiale S.A.341 (q.v.) Gazelle.

airplane: U.S. term for aeroplane or aircraft (q.v.).

air pocket: a term generally given to turbulence which causes the aircraft to bounce. *See also* clear air turbulence.

airport: variously also termed a landing strip, airfield, air station, or aerodrome, from which aircraft may land and take off. A landing strip tends to be a single runway, usually rough-surfaced and in a remote area; airfield and air station tend to have military connotations, air station often being naval; and aerodrome is a seldom-used expression dating from the inter-war days of flying, usually referring to a place for private fliers. An airport is effectively the aerial equivalent of a seaport, handling freight and passengers, and possessing customs and immigration control facilities.

One of the first airports was that at Hounslow Heath, to the west of London, from which Aircraft Transport and Travel (q.v.) operated, before moving to Croydon, to the south of London, in March 1920. One of the first aerodromes for private fliers was built at Heston, also near London, in 1928, by Airwork (q.v.), a predecessor of British Caledonian Airways (q.v.). Southampton was one of the main points of departure for the Imperial Airways flying-boat services between the wars.

Today, the world's busiest airport is Chicago's O'Hare. London's main airport at Heathrow is also one of the busiest, and in fact in terms of the value of the cargo handled each year is Britain's second most important port, after the seaport of London and before Liverpool. Other major airports include Orly, near Paris, and Kennedy, one of the New York airports.

In the future, the concept of an airport with runways for take-offs and landings may become obsolete with the advent of vertical take-off civil aircraft which can rise straight up from a landing pad.

Air Rhodesia: Air Rhodesia was formed in 1967 to take over the Rhodesian services of the Central African Airways Corporation, operated since the end of World War II with B.O.A.C. assistance, which had been broken up with the ending of the Federation of Rhodesia and Nyasaland. Air Rhodesia is owned by the Government of Rhodesia, and operates domestically and to South Africa and the neighbouring Portuguese territories in Africa, using a fleet of Vickers Viscounts and Douglas DC-3s, with three Boeing 720s delivered early in 1973.

air safety: the year 1785 was not a particularly good one for air safety, for it was then that Pilâtre de Rozier and his companion Romain were killed while trying to cross the English Channel in a balloon using the dangerous combination of a hydrogen-filled envelope and a Montgolfier-style brazier. Greek mythology credits Icarus with flying too close to the sun and crashing to his death after the wax in his wings melted, but Pilâtre de Rozier and Romain have the unwanted distinction of being the first recorded fatalities in the air. The first man to be killed in an aeroplane accident was the American, Lieutenant Selfridge, in 1908, while being flown in a Wright biplane by Orville Wright. There have been no accidents during any of the American space missions which have resulted in injury or worse, but three astronauts were killed during ground testing of an Apollo capsule. It is believed, but unproven, that several Soviet astronauts may have died prior to Gagarin's (q.v.) successful flight.

The question of air safety is one which concerns many, and the average person feels sufficiently out of his element in the air to experience something rather more than just concern. The responsible authorities tackle air safety by a system of air traffic control (q.v.), airport licensing and inspection, and rigorous checks on operators and aircraft, including stringent testing prior to the award of a certificate of airworthiness. Accidents also receive thorough investigation, and if necessary an airline can be grounded, as can an aircraft type. Testing of modern aircraft types is so thorough, and the accumulated knowledge of aerodynamics and experience so great, that accidents to an aircraft in mid-air are rare indeed, although there is now an increasing hazard of mid-air collision as traffic increases, and a considerable number of airmiss (q.v.) incidents are reported each year.

Most modern accidents occur while the aircraft is on approach (q.v.), when incorrectly set altimeters (q.v.) or other errors may result in an accident, particularly if the approach is over high ground; landing and take-off are also periods of relative hazard. It should not be thought that an accident means the worst for the occupants of an aircraft, for in spite of the high speeds involved, the chances of survival in such an accident are fairly high. Jet aircraft using kerosene are less prone to fire than piston-engined aircraft. A few airlines use gasoline in jets on grounds of economy, but at present no such airlines operate into the United Kingdom, and some countries have discouraged the practice by taxation policies which leave kerosene as the cheaper fuel. The British and American authorities require that an airliner can be evacuated within ninety seconds, using only half of the available exits on the basis that the fire risk means that one cannot reasonably expect all exits to be available – although if they are, this is a welcome bonus.

Some of the more famous accidents have happened to airships (q.v.), including the British R-38 in 1921 and R-101 in 1931, and the German Hindenburg in 1937. The de Havilland Comet I (q.v.) airliner was withdrawn from service after a series of mid-air disasters. No air accident has rivalled the worst incidents at sea for loss of life yet, but fortunately no airliner can yet match an ocean liner's passenger capacity, for the most that even a Boeing 747 'jumbo' jet could carry would be 530 passengers, and the problem of filling so many seats means that many airlines are using only 350 or so seats, leaving the extra space for lounge and bar accommodation. There have, however, been a number of accidents with a death toll in excess of the 100 mark.

airscrew: the propeller (q.v.) of an airship or aircraft, either of the 'pusher' or 'tractor' type, depending on whether the propeller is mounted behind or in front of the engine.

air-sea rescue: *see* search and rescue.

airship: the first recorded attempt to build an airship, or cigar-shaped balloon, was in 1834 in France, but was unsuccessful. In 1835 a British airship, 'The Eagle', powered by paddles or flappers, with the crew pulling oar-like levers, was built but never flown. It was not until 1852 that Henri Giffard (q.v.) produced a successful airship which made an ascent near Paris and managed to travel at 4–5 mph using its single steam-powered engine. Although the Giffard could be classified as the first dirigible, or steerable, airship, the low power of its engine prevented all but very limited manoeuvres.

Some considerable progress had been made with airships by 1883, when the two brothers Albert and Gaston Tissandier (q.v.) flew an electrically-powered airship, although this still lacked sufficient power. The following year, Renard and Krebs built 'La France', with a more powerful electric motor, and this was actually able to fly away from and return to its point of departure. A Brazilian living in Paris, Alberto Santos-Dumont (q.v.), circled the Eiffel Tower in an airship in 1901, and in 1902 the Lebaudy (q.v.) brothers started to build the first really practical airships, using Daimler petrol engines. The first British airship was built by the British Army at Farnborough in 1906 by S. F. Cody (q.v.) and actually used the same engine as the first British aeroplane.

Count Ferdinand von Zeppelin (q.v.) founded his Luftschiffbau Zeppelin factory in Germany in 1898. During the years preceding World War I the giant Zeppelins were used on passenger services within Germany, carrying a total of 40,000 passengers and flying 170,000 miles without a single accident of note. World War I saw Zeppelins used on reconnaissance and bombing raids against British East Coast towns and cities, and against London, with some success until the Royal Flying Corps found that the hydrogen in the airships ignited easily when they were fired upon. The Royal Naval Air Service bombed the Zeppelin sheds.

After World War I, the airship was deprived of what could have been extensive commercial applications by the development of the aeroplane. One of the most famous of the post-war airships was the Graf Zeppelin, completed in 1928, and which operated successfully until 1937, when it was withdrawn from service. In 1930 the British R.100 made a double crossing of the Atlantic, but a British rival, built at public expense, the R.101, crashed at Beauvais in Northern France in 1930. The largest Zeppelin of all, the Hindenburg (q.v.), crashed while mooring at Lakehurst, New Jersey, in 1937, with considerable loss of life; the disaster was made worse by the use of hydrogen in the airship due to an embargo by the United States which prevented the sale of the much safer helium gas to Germany.

The Hindenburg disaster effectively ended the commercial history of the airship. Militarily, little operational use was left, and even though successful, World War I experiments of flying aircraft on to airships, and hooking on underneath, were not pursued further. During World War II, airships were either used as barrage balloons or for paratroop training. Since the war the American Goodyear concern, with extensive tyre and rubber interests, has built several airships in the United States and one in Britain; these act as platforms for television and film cameras, and also fulfil a publicity function.

Recently, interest in the airship as a containerised cargo carrier with nuclear-powered engines has been re-awakened, but it is too early to say definitely whether anything will come from the investigations being conducted into the potential of this development, although it is unlikely.

airspace: the air over a national territory.

Airspeed: a British aircraft manufacturer of the inter-war, World War II, and early post-war period, after which the company was acquired by de Havilland. Notable products included the Airspeed Envoy and Oxford communications and navigational training aircraft, and the Courier, the first British civil aircraft to have a fully retractable undercarriage. World War II production included the Horsa troop-carrying glider. A post-war product was the Ambassador (q.v.), which was B.E.A.'s last piston-engined aircraft type, excepting two special duty Herons.

airspeed: the speed of an aircraft through the air, as opposed to the speed over the ground, which is the real speed. Airspeed will be more than ground speed if there is a strong headwind, but less than ground speed if there is a tailwind. On

take-off and landing, the airspeed is the crucial factor.

air station: *see* airport.

airstation: U.S. term for aerostation (q.v.).

airstrip: the simplest and most primitive airfield, usually consisting only of a rough strip for landing and take-off – *see* airport.

air superiority: air superiority is the position of having full control over national airspace, and usually control over that of an opposing state, while it can also relate to air control of a battle area. It entails restricting the enemy's movements, particularly in the air, leaving the ground and air forces of the combatant nation possessing air superiority to operate with minimal interference from enemy aircraft. The restrictions imposed on the nation lacking air superiority stress the importance of achieving air superiority, and in June 1967 the Israeli forces made the attainment of air superiority their prime objective, which enabled them to gain victory in a mere six days.

An air superiority fighter is an aircraft designed for the purpose of giving its user early air superiority, although in this context the term can be meaningless should both sides have such aircraft available to them. The successful attainment of air superiority seldom depends on one aircraft, nor on aircraft alone, since it requires reliable intelligence about enemy deployment, plans, and capability, interceptor and fighter aircraft for defence purposes, and bomber and ground attack aircraft for pre-emptive strikes against airfields and other air defence installations. Good organisation, leadership, and training can sometimes be more important than aircraft quality or quantity in this respect.

air support: the two concepts of air superiority (q.v.) and air support are not unrelated, since the attainment of air superiority makes air support, usually of ground forces, that much easier without undue interference from opposing aircraft. Air support can take any or all of three forms – support by combat aircraft against enemy artillery or armoured forces, the dropping or other transport of essential supplies, sometimes in difficult terrain or in an area with effective enemy ground forces making surface transport uncertain, or fast air transport of ground forces, often using helicopters, for rapid reinforcement or redeployment of the available surface forces.

The Royal Air Force tends to use the term for transport operations, and most of the squadrons in the R.A.F. Air Support Command, prior to its amalgamation with Strike Command late in 1972 to form an Operational Command, had previously belonged to R.A.F. Transport Command.

air taxi: the term generally given to a light aircraft or helicopter, operated on charter with a pilot and weighing below 12,500 lb. A scheduled service operated with air taxi-sized aircraft is termed a 'third level' service, and the operator a 'third level' airline. There is nothing derogatory in this term, which embraces the fastest-growing sector of air transport, and such operations are encouraged in many countries.

air-to-air missile: an important item in modern aircraft armaments, and often the only armament of an interceptor (q.v.), the air-to-air missile is fired by one aircraft at another, and modern missiles of this type are usually guided by radar or an infra-red homing device, or sometimes a combination of the two, on to the target aircraft. The need for such weapons arose with the higher speeds of modern aircraft, as in such circumstances machine guns or cannon proved to be relatively ineffective. An in-between stage was the unguided air-to-air rocket, used on a fairly large scale by the Luftwaffe's Messerschmitt Me. 262 (q.v.) jet fighters against U.S.A.F. Boeing B-17 (q.v.) Flying Fortress and Consolidated B-24 (q.v.) Liberator bombers during the closing stages of World War II.

Early operational air-to-air guided missiles included the U.S.A.F.'s Hughes GAR-1 Falcon for the Northrop F-89 Scorpion fighter during the mid-1950s, and the de Havilland Firestreak (q.v.) for the R.A.F.'s Gloster Javelin and the Royal Navy's de Havilland Sea Vixen (q.v.) fighters.

Modern weapons of this type include Red Top (q.v.), Sidewinder (q.v.), 'Alkali' (q.v.), 'Anab' (q.v.), 'Ash' (q.v.), and 'Atoll' (q.v.), while new developments yet to enter service include the British SRAAM 75 (q.v.).

The weapon is not without its limitations. Early models could only operate in clear weather or in a straight line, and at low level over tropical rain forest or other sources of humid heat, efficiency is still reduced.

air-to-surface missile: the air-to-surface missile, or stand-off bomb, is increasingly used as a combination of conventional manned aircraft attack and the ballistic missile concept. A point about using stand-off bombs is that it enables a bomber to release its weapon before reaching the most heavily defended part of an enemy's territory, which is assumed to be that around potential targets. Advantages include a lower cost than that of a ballistic missile, a recall system instead of placing reliance on a destructive mechanism which may or may not work, and the use of aircraft which can be used on other types of bombing duties.

A rocket-powered and radar-guided air-to-surface missile was first put into service with the United States armed forces late in World War II; this was the Bat, which was used against Japanese warships with some success. Work started in 1946 on an improved missile, resulting in the Bell GAM-63 Rascal, which entered service with the U.S.A.F. in 1957, carried by Boeing B-47 Stratojet bombers. The R.A.F.'s Hawker Siddeley Dynamics Blue Streak (q.v.) entered service in 1962 with Britain's Vulcan (q.v.) and Victor (q.v.) nuclear bombers. Current missiles of this type include the Anglo-French Martel (q.v.), Hound Dog (q.v.), 'Kitchen' (q.v.), and 'Kipper' (q.v.), but the definition should also be extended to include anti-tank missiles such as the Nord SS-11 and SS-12 (q.v.) series, which are wire-guided to their target after being fired from Scout (q.v.), W.G.13 (q.v.), or Alouette (q.v.) helicopters.

Airtourer, Victa: the Airtourer was the winner of a 1953 Royal Aero Club competition for a two-seat light aircraft. First flight of a wooden prototype took place in Australia in 1959, since when the aircraft has been in production as an all-metal training machine using a single Continental 0-200-A or Lycoming 0-235 engine of 100–115 hp, giving a speed of 120 mph. Production has since been moved to New Zealand, and the Royal New Zealand Air Force is using the aircraft as a basic trainer.

air traffic agreement: international air transport is largely regulated by a series of inter-governmental air traffic agreements, most of which are on a bilateral basis. This is an effective means of regulating competition on international air services. A good example of a bilateral agreement is the London–Paris route, on which the daily total of flights is divided equally between Britain and France, with the result that when the British Government decided that the state-owned British European Airways should have a private enterprise competitor, the new airline's flights had to come from what had hitherto been B.E.A.'s share of the quota. On the North Atlantic, a less restrictive arrangement exists between the Governments of the United Kingdom and the United States, which allows for reasonable competition, but airlines, including those of the United States, have considerable difficulty in gaining authority to operate services into countries with more restrictive attitudes, notably Eire, Italy, and Australia.

The International Air Transport Association (q.v.) is largely able to decide upon fares and other conditions of travel, although sometimes new fare agreements are subjected to the approval of the governments of the countries concerned.

air traffic control: the main foundation for air traffic control systems as they exist today is the 1944 Conference on International Civil Aviation, which committed each state to provide an air traffic control service on grounds of safety and traffic throughput. The coverage by air traffic control may be divided into four groups, airways (q.v.), control areas (terminal or oceanic), control zones, and flight information regions. Control areas are generally to be found in the region of airports where several airways converge, or over vast stretches of water, over which aircraft have to maintain separation distances. A control zone is the actual take-off and landing area around an airport, while a flight information region is an area in which advice is available to aircraft passing through the area.

Air Vietnam: the airline of South Vietnam, Air Vietnam was founded in 1951 to take over the domestic and regional international services of Air France. Air France has a small interest in Air Vietnam of about 6 per cent, although originally it was some 25 per cent, while the South Vietnamese Government holds some 92·5 per cent, and local interests the remainder. Services are operated within the country and to Cambodia, Laos, Malaysia, Singapore, Thailand, Taiwan, Hong Kong, and Japan, using a fleet of Boeing 727s, Douglas DC-6s, DC-4s, and DC-3s, and some light aircraft.

airway: basically, an airway is an invisible corridor about ten miles wide, along which civil aircraft fly under air traffic control (q.v.) supervision. Beams of radio waves are usually provided to assist in keeping aircraft on the airways, which extend up to 40,000 feet from ground level, and run from radio beacon to radio beacon. Airways are usually identified by a colour code, such as Green 2 or Amber 1.

Airwork: a British company, Airwork was formed to provide a full range of technical services for the private flier, including an aerodrome at Heston, on the west side of London, which was opened in 1928 and soon became one of Europe's leading light aviation centres. Airwork assisted in the formation of Egypt's Misrair and Indian National Airlines before World War II, and with Sudan Airways after the war. During the war the company assisted with airfield management for the Allies, and after the war formed its own airline, which was a predecessor of the present British Caledonian Airways (q.v.).

The company today still undertakes a great deal of defence contract work, perhaps the most famous contract being assistance in getting the Royal Saudi Air Force's Lightning interceptors and air defence system to operational status.

airworthiness: the state of serviceability of an aircraft at any given time, and the general suitability of a design for its prime object, i.e. to fly.

Albatros: an early German aircraft manufacturer which came into prominence after production in 1914 of the successful Albatros D.I racing plane, which was used as a fighter during World War I. The D.I was soon followed by the Albatros D.II, D.III, D.IV, and D.V, although the last of these biplanes was unable to match the increasing superiority of the Allies' equipment, and the factory eventually had to start manufacture of the rival Fokker D.VII. Other Albatros aircraft included the C.III bomber, and the firm was one of many to produce the Rumpler Taube (Dove) for the German war effort.

Albatross, de Havilland D.H.91: the four-engined Albatross low-wing monoplane entered Imperial Airways service in 1938, on routes from London to Europe. A feature of the attractive twenty-two passenger aircraft was a bonded plywood fuselage. The four engines provided a total of 2,100 hp, giving a maximum speed of 210 mph and a range of 1,000 miles. The aircraft had a very advanced appearance, but further development was prevented by World War II. Imperial Airways only used five of these aircraft.

Albatross, Grumman HU-16: the Grumman Albatross, or HU-16A/B in U.S.A.F. classification since 1962, before which the classification was SR-16A/B, first flew in 1947 as a search and rescue amphibian. The basic type was the SA-16A, but later these were converted to SA-16B standard with a higher weight and improved performance. The U.S.N.'s standard aircraft entered service as the UF-1, but were later classified as HU-16C and HU-16D, while the U.S. Coastguard versions are HU-16Es. Anti-submarine versions entered service in 1961 with the Norwegian, Spanish, and Canadian armed forces, while other users of the aircraft include Argentina, Brazil, Chile, the Federal German Republic, Indonesia, Italy, Japan, the Philippines, Portugal, and Taiwan. Two 1,425 hp Wright R-1820A piston engines give the Albatross a maximum speed of 240 mph, a range of over 3,000 miles, and a payload of more than 10,000 lb.

Alcock, Captain Sir John (1892–1919): on 14–15 June 1919 Captain John Alcock piloted a Vickers Vimy (q.v.) twin-engined biplane bomber from St John's, Newfoundland, to Clifden, Co. Galway, in the first non-stop crossing of the Atlantic Ocean. The flight took 16 hours 12 minutes. Alcock was knighted for this achievement, but later in that same year was killed in a flying accident in Northern France while Chief Test Pilot for Vickers. Alcock's navigator on the North Atlantic crossing was Lieutenant Arthur Whitten Brown (q.v.).

Aldrin, Colonel Edwin, U.S.A.F.: Colonel Aldrin was a member of the Apollo XI (q.v.) space mission, and the second man to put foot on the moon on 21 July 1969. His companions were the flight commander, Neil Armstrong (q.v.), and Michael Collins (q.v.).

Alia Royal Jordanian Airlines: Alia Royal Jordanian Airlines commenced operations in late 1963 with a fleet of two Handley Page Heralds and a Douglas DC-7, replacing an earlier airline, Air Jordan, on both domestic and international services. Originally a mixed private and state enterprise, Alia is now completely state-owned. The name 'Alia' means 'high flying'.

Currently, Alia operates throughout the Middle East and to Europe, with a fleet of Boeing 707s and Aerospatiale Caravelles.

Alitalia–Alitalia Linee Aeree Italiane: Alitalia dates from 1946 when the airline was formed with assistance from British European Airways, while Linee Aeree Italiane was formed with T.W.A. assistance – in each case with the foreign airline holding a 30 per cent interest in the Italian airline, leaving the Institute for Industrial Reconstruction as the remaining shareholder. The two airlines operated separately until 1957, when they merged to form the present airline, partly to avoid the increasing tendency for the route networks to overlap, although L.A.I. tended to operate to North Africa and the United States, while Alitalia operated to South America, the Middle East, and Africa; but both airlines operated within Europe and domestically. The merged airline had an assorted fleet, including a selection of Douglas Commercial designs, Vickers Viscounts, and Convair Metropolitans.

Alitalia's first jets were introduced in 1960, including both the short-haul Sud Aviation Caravelle and the long-haul Douglas DC-8. These have since been followed by developments of the DC-8, as well as the DC-9 and DC-10, and the Boeing 747.

Currently Alitalia is owned by I.R.I. and a number of banking and industrial concerns, and has as subsidiaries the charter airline Società Aerea Mediterranea and Somali Airlines, as well as a number of travel and hotel companies. It operates a network of international services (domestic routes are now the duty of Aero Transporti Italiani) extending throughout Europe, to North and Latin America, Africa, the Middle and Far East, and to Australia, and uses Boeing 747, McDonnell Douglas DC-10, DC-9, and DC-8 (including the 'Super Sixty' series), and Aerospatiale Caravelle airliners.

Alize, Breguet Br. 1050: see Br. 1050 Alize, Breguet.

'Alkali': the standard Soviet Bloc air-to-air guided missile, used by interceptors and fighters and using homing radar to attack the target aircraft.

Allison: a subsidiary of Detroit Diesel, Allison is not in the big league of aero-engine manufacturers, which could be said to consist of only three companies in the United States and the United Kingdom. Allison powerplants for military use include the TF-41-A-1 of 14,500 lb thrust and the TF-41-A-2 of 15,000 lb thrust, being unreheated licence-built versions of the Rolls-Royce military Spey for the L.T.V. A-7 (q.v.) Corsair II, and the T63-A-5A 317 shp and T56-A-154 910 shp turbines for helicopters, of which the Bell

206A Kiowa and Hughes 500 Cayuse use the T63. Civil powerplants include the 909-B1 of 8,000 lb thrust, the turboprop 501-D22 of 4,050–4,500 shp for the Lockheed Electra and Hercules, and the Model 250-C18 of 317 shp and the 250-C20 of 400 shp. The company is also collaborating with Rolls-Royce on a joint development of the J99 lift jet for V.T.O.L. aircraft.

Alouette II, Aerospatiale SE. 3130: a five-seat single-engined helicopter, which first flew in March 1955, and of which almost one thousand have been sold in France and abroad for military and civil use. Powerplant is a single 360 shp Turbomeca Artouste IIC shaft turbine, giving a maximum speed of 100 mph and a range of 300 miles. Nord SS-10 and SS-11 wire-guided anti-tank missiles can be carried. The Alouette II has been built under licence by SAAB and Republic Aviation.

Alouette III, Aerospatiale SE. 3160: a seven-seat development of the Alouette II, and like the Alouette II also a Sud Aviation design. The Alouette III has been even more successful than its smaller relation, with more than 1,400 produced and sold. The single 870 shp Turbomeca Artouste IIIB shaft turbine provides for a maximum speed of 130 mph and a range of 450 miles. Nord SS-11 and SS-12 anti-tank missiles can be carried.

Alphajet: a collaborative Franco-German advanced jet trainer project designed to replace the Super Magisters of the Armée de l'Air and the Lockheed T-33A and Cessna T-37s of the Luftwaffe in the mid-1970s. The manufacturers involved are Dassault and Breguet of France, and Dornier of Germany, and the aircraft has a high swept wing and tandem seating for two in the cockpit. Two 2,300 lb thrust SNECMA-Turbomeca Larzac turbofans will provide a speed in excess of Mach 1·0, and more than 400 aircraft will be required by the two air forces.

altimeter: one of the basic instruments on any aircraft, since it gives the height of the aircraft above sea level. Altimeters work on barometric pressure, and therefore considerable care must be taken to adjust the altimeter to the pressure prevailing at any one time and in any one place in order that accurate readings may be obtained from the instrument. Incorrectly adjusted or read altimeters are a prime cause of accidents, usually while the aircraft is on approach.

altitude: the height of an aircraft above sea level.

Alvis: a British manufacturer of aero-engines, although in a comparatively limited way. The most notable Alvis product before the company ceased aero-engine production was the Leonides, used in the Hunting Pembroke and Sea Prince twin-engined light transports of the early 1950s, and the Hunting Provost single-engined basic trainer of the same period. The engine was also used in the Scottish Aviation Pioneer and Twin Pioneer light transports. Originally a manufacturer of quality motor-cars and military vehicles, only the latter activity remains now that the company is a part of the British Leyland Motor Corporation.

AM.3, Aerfer-Aermacchi: the AM.3 is a short take-off air observation post and liaison aircraft with a single Continental GTS10-520-C piston engine, giving a maximum speed of 180 mph, with a range of 500 miles. This three-seat aircraft is currently entering Italian Army service, for which 100 will be built.

Ambassador, Airspeed AS.57: the Ambassador was one of the immediate post-World War II generation of airliners, and first flew in July 1947, although the aircraft did not actually enter service with British European Airways until the early 1950s. The aircraft passed to new owners at the end of the decade, and one of their duties was acting as flying test-beds for the Napier Eland, Bristol Siddeley Proteus, and Rolls-Royce Dart and Tyne turboprops – although strangely it was decided not to develop a turboprop version of the aircraft. The Ambassador's own engines were two Bristol Centaurus piston engines of 2,625 hp which gave the high-winged aircraft a maximum speed of 250 mph, with a range of 1,350 miles.

American Airlines: the history of American Airlines, today one of the largest airlines in the western hemisphere, dates from the formation of the Aviation Corporation in 1929, following the acquisition of many small, sometimes one-route, airlines. The oldest of these, the Robertson Aircraft Corporation, dated from 1921, although most of them were formed during the late 1920s. Amongst the airlines were Canadian Colonial Airways, Colonial Air Transport, Embry-Riddle Aviation Corporation, Interstate Airlines, Gulf Air Lines, Continental Airlines, and Northern Air Lines. The following year, the Aviation Corporation established an operating subsidiary, American Airways, which started on the task of rationalising routes and equipment, while also acquiring other small airlines, including Century Air Lines, Century Pacific Lines, and Standard Airlines. After an Act of Congress in 1934 forced the Aviation Corporation to isolate its manufacturing and operating interests the present American Airlines Corporation was formed as a separate and independent entity.

American Airlines came into existence with a network stretching across the United States from coast to coast. A licence was awarded for the first international services, to Mexico City from El Paso and Dallas, in 1942, but services could not be started until after World War II ended. A further step towards creating an international network had already been taken by buying a 51 per cent interest in American Export Airlines, which dated from 1937, from American Export Lines; American Export Airlines had been formed to operate services to Europe, but by 1940 had only succeeded in gaining a temporary permit for a service to Lisbon. However, A.E.A. introduced the first New York to London commercial service in 1945, using Douglas DC-4s, and American's shareholding was increased to 62 per cent in 1948, when the title of American Overseas Airlines was taken for A.E.A. Pan American World Airways, A.O.A.'s main rival, acquired the airline in 1950.

After the war, the Convair 340 airliner was designed specifically to meet American's requirements, as the Douglas DC-3 had been some years earlier, and the Douglas DC-7 and Lockheed Electra were later to be. However, this period was one largely of rationalisation of services, under which many small towns were dropped from American's extensive network, to be handed over to the local service and feeder airlines, or even to third-level operators, who could serve such places more economically than a large transcontinental airline could ever hope to do. The trunk route network was strengthened at the same time by the addition of a Chicago–San Francisco route in 1955, and then by a non-stop New York–San Francisco service. During the early 1960s a merger with another major airline, Eastern Air Lines (q.v.), was mooted, but rejected by the Civil Aeronautics Board in 1963.

Currently, American Airlines operates from coast to coast across the United States, and to Toronto, Mexico City, and Acapulco, as well as to Hawaii, Fiji, Australia, and New Zealand, which services were awarded to the airline in 1970. In the Caribbean American has the network of Trans Caribbean Airways, which it acquired in 1971. American's fleet includes the McDonnell Douglas DC-10, which was first ordered by American in 1968, Boeing 747, 707, 720, and 727, and B.A.C. One-Eleven airliners.

amphibian: as with animals, the term amphibian denotes an aircraft equally at home on land or water. It does not refer to the beaching ability of many flying-boats, using either special trolleys or wheels fitted for the purpose.

The first recorded amphibian was an aircraft designed and built by the American, Glenn Curtiss (q.v.), in 1911 or 1912, which was basically a landplane with floats fitted and wheels retained. A feature of the early Curtiss amphibians, and indeed many American seaplanes and amphibians, was the preference for a single large float instead of the British and European idea of twin floats. The first true amphibian, in that it had retractable landing wheels and a properly-designed hull, was the Sopwith Bat Boat, built just before the outbreak of World War I.

A number of aircraft manufacturers were interested in amphibian design and construction between the wars, including Supermarine (q.v.) and Grumman (q.v.), while the Consolidated PBY-5 Catalina (q.v.) amphibian was probably the most successful aircraft of this type ever built.

Since the war the spread of aviation, justifying the cost of building proper airports, or at least landing strips, and the advent of the helicopter, needing no runways at all, have tended to make the amphibian something of a rarity. Only two amphibians are in production at present (although it is not so long since the Grumman Albatross (q.v.) was in production), and these are the Canadair CL-215 (q.v.) and the Thurston-Schweizer Teal (q.v.).

AN-: international civil aviation registration index mark for Nicaragua.

An-2 'Colt', Antonov: the An-2 first appeared in 1947, as a fourteen-seat utility transport using a single ASH-62-IR piston engine, giving the biplane a range of 550 miles and a maximum speed of 160 mph. The aircraft has been used by almost every Communist Bloc country, and in addition has been produced in Poland and the People's Republic of China. Total production has been variously estimated at between 5,000 and 10,000 aircraft. Originally an aircraft of simple construction and materials, the more recent models have had laminated plastic fuselages. The An-3 (q.v.) is a turboprop development.

An-3, Antonov: there is no NATO identification name for this aircraft at present. Basically the An-3 is a turboprop An-2 (q.v.), using the laminated plastic fuselage of the later An-2s, and retaining the biplane layout. However, a nosewheel replaces the An-2's tailwheel.

An-12 'Cub', Antonov: a development of the An-10, using four Ivchenko AI-20 turboprops and having a maximum payload of 22,000 lb, with a range of more than 2,000 miles and a maximum speed of 360 mph. Unusually, some versions of the An-10 and An-12 have a tail turret with two 20 mm cannon. The aircraft has a good rough field performance, and is in military and civil use in the Soviet Union, Eastern Europe, Algeria, Egypt, India, Indonesia, and Iraq. The first recorded appearance was in Aeroflot (q.v.) service in 1960, but the Soviet military probably had the aircraft for some time previously.

An-14 'Cold', Antonov: a light utility transport which first appeared in 1958, and has since been extensively developed, with an increase in passenger capacity from eight to fifteen seats, and the replacing of the original two Ivchenko AI-14RF piston engines with two turboprops in the An-14M version. A Turbomeca Astazou version, the An-14A, has been abandoned. The original version of the aircraft has been in Aeroflot service since 1965, generally on feeder and air taxi services. In common with all other Antonovs, except the An-2/3 biplanes, this is a high-winged aircraft. Performance varies from 150 mph and a 300 mile range in the original to about 200 mph and a range which is probably around 500 miles or so.

An-22 'Cock', Antonov: this was the largest aircraft in the world at the time of its unveiling in 1967, using four 15,000 shp Kusnetsov NK-12MA turboprops giving a maximum payload of 176,000 lb, a speed of 409 mph, and a range of up to 6,000 miles. In Soviet Air Force and Aeroflot service.

An-24 'Coke', Antonov: a light transport aimed at the same market as, and bearing a marked resemblance to, the Fokker F-27 Friendship and the Handley Page Herald. The An-24 first appeared in Aeroflot service in 1963. Two Ivchenko AI-24 2,550 shp turboprops give a maximum speed of 360 mph and a range of 1,800 miles. Passenger capacity is up to fifty persons. The aircraft is extensively used by the Soviet Bloc and allied states.

'Anab': a Soviet Bloc air-to-air guided missile, understood to use alternating radar and infra-red homing devices, and equipment for the Yak-28 (q.v.) 'Firebar' and Mikoyan MiG-21 'Fishbed' interceptors.

Anders, Major William A., U.S.A.F.: one of the three-man crew of Apollo VIII (q.v.), and therefore one of the first three

men to break free of earth's gravitational pull. The Apollo VIII mission lasted from 21 to 27 December 1968.

Andover, Hawker Siddeley: the Andover is a stretched military version of the civil HS 748 (q.v.), with a modified tail to incorporate a loading ramp and a unique kneeling undercarriage to enable the loading ramp to adjust to the height of vehicles used for loading or unloading the aircraft. The prototype first flew in late 1963, and the first of thirty-one production aircraft for the R.A.F. first flew in mid-1965. Two Rolls-Royce Dart Da.12 Mk.201 turboprops give this low-wing aircraft a speed of around 300 mph and a range of up to 1,200 miles, while forty-four fully-equipped troops can be carried.

Andree, Salomon: leader of an unsuccessful attempt to cross the North Pole by balloon (q.v.), Salomon Andree, a Swede, and his two companions left Spitzbergen in 1897, but disappeared without trace, until in 1930 their bodies, documents, and equipment were recovered from beneath the ice. Apparently the balloon had been forced down by the weight of frozen mist.

angled flight deck: a post-war British invention for aircraft-carrier operation, replacing the old straight-through, stern to stem, flight deck for landings and take-offs. The angle varies from ship to ship, being as little as 4° on the French carrier *Arromanches*; the Royal Navy's through-deck cruisers will also have a very low angle. Angling of the flight deck gives better lift to aircraft, eliminates the need for a net to prevent aircraft which are overshooting running into aircraft parked forward on the flight deck or on the forward catapults ready for take-off, and greatly improves the chances of an overshooting aircraft making a further attempt at landing. It is also a point that, should the worst happen, the aircraft lands in the water clear of the ship, enhancing the pilot's chances of survival. The larger aircraft carriers have waist catapults on the angled flight deck, as well as forward.

anhedral: a downward droop of the wings of an aircraft to give an inverted 'V' effect when viewed from in front or behind – usually a marked feature of Soviet-designed aircraft, although the giant American B-52 (q.v.) bombers and C-5 (q.v.) transports also have this feature.

Ansett Airlines of Australia: Australia's largest private enterprise airline, Ansett Airlines of Australia was formed in 1936 by the present chairman of Ansett Transport Industries, Sir Reginald Ansett, as Ansett Airways. Services started in 1936, although expansion was hindered somewhat by World War II. Two flying-boat operators, Barrier Reef Airways and Trans Oceanic Airways, were acquired in 1952 and 1963 respectively, but the major acquisition was that of A.N.A. – Australian National Airways – in 1957, after which Ansett's then position as Australia's third largest airline was secured, and the title Ansett-A.N.A. used until the present title was adopted in 1968. Other acquisitions have since included Butler Air Transport, now operated as a subsidiary, Ansett Airlines of New South Wales, and Butler's own subsidiaries, Queensland Airlines and Guinea Airways; while in 1960, Mandated Airlines and two other New Guinea operators, Gibbes Sepik Airways and Madang Air Services, were acquired and are now operated as Ansett Airlines of Papua New Guinea. The most recent acquisition was MacRobertson Miller Airlines in 1963.

Today, Ansett Airlines of Australia and its subsidiaries operate an extensive domestic route network throughout Australia in competition with the state-owned T.A.A. (q.v.), including scheduled helicopter services in Queensland and New South Wales, with a fleet of Boeing 727s, McDonnell Douglas DC-9s, Fokker F-27 Friendships and F-28 Fellowships, Lockheed Electras, Douglas DC-3s and DC-4s, Aviation Traders Carvairs, Short Sandringham flying-boats and Short Skyvans, de Havilland Canada DHC-6 Twin Otters, and Sikorsky S-61Ns, and a Bell Jet-Ranger. The subsidiaries include Ansett Airlines of South Australia, operating scheduled and domestic services within the State; Ansett Airlines of New South Wales, which operates within the State, including the remaining flying-boat operations; and Ansett Airlines of Papua New Guinea; as well as MacRobertson Miller Airlines, operating in Western Australia and the Northern Territory.

anti-aircraft measures: essentially a part of air defence systems. Although anti-aircraft guns, such as oerlikons and multiple pom-poms, are by no means useless, greater emphasis is placed today on various types of surface-to-air (q.v.) guided missiles, usually using wire or radar guidance. Typical of these weapons are Bloodhound (q.v.), Black Knight (q.v.), Tigercat (q.v.), Seacat (q.v.), Rapier (q.v.), 'Galosh' (q.v.), 'Ganef' (q.v.), and 'Guideline'.

anti-ballistic missile defences: a more sophisticated defence system than air defence in terms of preventing intrusion by enemy aircraft. Essential requirements include radar to detect and track incoming ballistic missiles (q.v.), with computer assistance for rapid identification of targets, and missiles capable of intercepting and destroying the attacking missiles. It is also necessary to be able to distinguish between decoys and the real thing, and to be able to cope with electronics countermeasures designed to protect the attacking missiles. A choice of defence exists, with large-scale defences entailing interception of every missile while outside of the atmosphere; area defence, with last resort interception in which heavy fallout over the target must be acceptable; or a combination of both of these methods, as with the American Safeguard (q.v.) system. Long-range, outside the atmosphere interception in the Safeguard system is left to Spartan (q.v.) missiles, with short-range, last-minute defence in the final phase of attack by Sprint (q.v.) missiles.

Antionette: an early manufacturer of both aircraft and aero-engines – until recently involvement in both these spheres of aviation activity was by no means uncommon. The company's title came from the name of a daughter of the owner (one Gastambide), but the main driving force behind the concern and the reason for its fame was the designer and engineer, Leon Levavasser, who was a firm believer in the monoplane. Although the early machines from about 1907 were not very successful, by 1909 the Antionette monoplanes were proving themselves to be worthwhile machines. Ailerons and wing warping were used on various models, but in many other respects the Antionettes did much to pave the way for the aircraft as we know it today, with tractor propeller and the mainplane in front of the tailplane. An Antionette was unsuccessful in an attempt on the English Channel crossing in 1909, and although Blériot sometimes used Antionette engines in his products, the successful cross-Channel aircraft was not one of these. However, the first successful aeroplane flight in Europe, in 1906 by Santos-Dumont, used an Antionette engine, and at the 1909 Reims international aviation meeting (q.v.) an Antionette set the world's first 100 km speed record.

anti-submarine warfare: two world wars have awakened the governments of the world to the devastating effects on shipping of even a small attacking force of submarines. The Soviet Union today has about ten times as many submarines as Germany at the outset of World War II, and in the Western Alliances anti-submarine warfare methods have been improved considerably in order to counter this very real threat. Although sonar (q.v.) equipment in fast surface vessels is still used extensively, and the problem of the nuclear-attack submarine is countered to a degree by hunter-killer submarines, also with nuclear power, the postwar period has also seen a considerable improvement in the part which aircraft can play; aircraft on maritime reconnaissance are no longer dependent on visual sightings of a submarine on the surface.

A problem with sonar in surface vessels was the distortion caused by the propellers of the search vessel, and aircraft dropped sonar buoys do eliminate this problem completely. Fixed-wing (q.v.) maritime reconnaissance aircraft can drop sonar buoys into the sea, to transmit any findings to the aircraft, which may be a Lockheed P-2 (q.v.) Neptune or P-3 (q.v.) Orion, or an S-3A (q.v.), Viking, a Grumman S-2 (q.v.) Tracker, a Breguet Br. 1050 (q.v.) Alize or Br. 1150 (q.v.) Atlantique, or a Hawker Siddeley Nimrod (q.v.) or Shackleton (q.v.). Large anti-submarine helicopters, such as the Westland-Sikorsky SH-3D (*see* S-61) Sea King, can 'dunk' their sonar equipment while the helicopter hovers; the operator listens, and then the sonar is retracted. Smaller helicopters, such as the Westland W.G.13 (q.v.), use surface radar to pick up a submarine on the surface or using a snorkel.

Nuclear-powered submarines seldom need to surface while on an operational cruise, and additional equipment which is fitted to modern maritime-reconnaissance (q.v.) aircraft includes magnetic anomaly detectors (q.v.) and autolycus (q.v.) equipment.

Attack, once a submarine is detected, can be by depth charge or missile from a fixed-wing aircraft, or by torpedo, usually from a helicopter, although most anti-submarine aircraft can also be so equipped.

Antonov: the Soviet Antonov design bureau first came into prominence in 1947 with the first flight of its single-engined biplane, the An-2 'Colt', which has since been produced in Poland and China and extensively developed. The next design of any importance was the An-10 and its successor, the An-12 (q.v.) 'Cub' four-engined, high-wing transport, while the An-14 (q.v.) 'Cold' was an attempt at an An-2 replacement. The An-24 (q.v.) 'Coke' is a light transport for forty to fifty passengers, while the An-22 (q.v.) 'Cock' is a very large transport aircraft, and at one time held the world record for aircraft size.

ANZUS: the Pacific Security Treaty, ANZUS, was designed as a substitute for NATO membership for Australia and New Zealand in 1951, with the United States as the third party and the driving force. The United Kingdom has never joined, although other treaties, such as ANZUK and SEATO (q.v.), effectively cover relations between the United Kingdom and the two British Commonwealth territories concerned. The signatories are committed to aid one another in the event of an attack on their territory or armed forces in the Pacific area, and the Australian and New Zealand forces in South Vietnam were a part of their ANZUS commitment. Military co-operation within ANZUS is no doubt one of the reasons for the prevalence of American equipment in the Australian and New Zealand armed forces, instead of that from their traditional supplier, the United Kingdom.

A.O.P.: *see* air observation post.

AP-: international civil registration index mark for Pakistan.

Apache, Piper PA.235: the Piper Apache executive aircraft, with six seats, has been out of production since 1959, when it was superseded by the Aztec (q.v.), with a larger engine and detailed improvements. Early Apaches used two Lycoming 0·320 engines of 150 hp, and a straight tail, while the later models used two Lycoming 0·540 engines of 235 hp and a swept tail, similar to that on the Aztec. Speed was in the region of 180 mph, with a range of up to 1,000 miles.

Apollo: the American Apollo programme was evolved with the object of putting a man on the moon by the end of the 1960s. In each launch a three-stage Saturn rocket was used, with lift-off from Cape Kennedy in Florida, and the missions ended with splash-down in the Pacific, and the recovery of the crew by United States Navy helicopters and frogmen from an aircraft carrier. Separate lunar landing modules and command capsules were used, with the command capsule orbiting the moon while the landing took place; after the return of the landing crew to the command capsule, the lunar landing module was fired back to the moon's surface.

Apollo VII (q.v.) was the first flight in the series, with the first American three-man orbit of the earth, while Apollo VIII (q.v.) made the first manned lunar orbit. Apollo IX was used to test the transfer facilities between the command capsule and the lunar landing module, and Apollo X had the lunar landing module descend to within nine miles of the moon's surface. The actual moon landing fell to

Apollo XI (q.v.) which landed in the Sea of Tranquillity on 20 July 1969, followed by Apollo XII on 19 November 1969, in the Sea of Storms. Apollo XIV, on 5 February 1970, landed near to the Apollo XII landing site, and Apollo XV, on 30 July 1971, landed in the Sea of Serenity, near Hadley Rille. On 25 April 1972 Apollo XVI landed in the Sea of Tranquillity. The Apollo XIII landing had to be cancelled due to a fault with the motor of the command capsule, requiring the use of the lunar landing module for all but the final descent on the return journey.

The programme itself has suffered serious cuts due to budgetary limitations, and also to the decreasing utility of each flight and the declining enthusiasm of public and politicians. Although the funds devoted to space exploration come in for strong criticism, it is a worthwhile point that the money used might never have been collected in the first place other than for space exploration, while there has been immense national prestige and technological fall-out, much of it in the form of unexpected bonuses. Mankind is probably never satisfied unless it is travelling towards the next horizon, anyway.

The Apollo XV, XVI and XVII (q.v.) missions used moon rover vehicles for the astronauts, with the last mission, Apollo XVII, including the first geologist to land on the moon, Dr Harrison H. Schmitt.

Apollo VII: the Apollo VII launch was the first United States three-man space flight, and as with all Apollo flights, a Saturn rocket was used. The flight took place between 11 and 22 October 1968, and paved the way for the lunar landing programme. Crew members were Captain Walter Schirra, U.S.N., Major Donald Eiscle, U.S.A.F., and a civilian scientist, Walter Cunningham, in command.

Apollo VIII: the Apollo VIII crew were the first men in history to break free from the Earth's gravitational pull when they commenced their flight on 21 December 1968, to orbit the moon. The flight ended on 27 December after a total of ten lunar orbits had been made. The crew were Colonel Frank Borman (q.v.), U.S.A.F., in command, Commander Arthur Lovell

Apollo XI

(q.v.), U.S.N., and Major William Anders (q.v.), U.S.A.F.

Apollo XI: after transfer trials with the lunar module in space on the Apollo IX flight, and Apollo X's descent to some nine miles from the surface of the moon in May 1969, the Apollo XI flight culminated in man's first landing on the moon. Lunar module 'Eagle' landed in the south-west corner of the Sea of Tranquillity at 02·56 hours, G.M.T., on 21 July 1969, and the flight commander, Neil Armstrong (q.v.), was the first man to set foot on another planet. He was followed by Colonel Edwin Aldrin (q.v.), U.S.A.F. Lt-Colonel Michael Collins (q.v.), U.S.A.F., piloted the command module 'Columbia' in orbit of the moon while the landing was in progress. The flight lasted from 16 July to 24 July 1969.

Apollo XVII: the last of America's Apollo manned spaceflights to the moon's surface, Apollo XVII was launched on 7 December 1972, landing on the moon on 11 December, and taking off on 14 December, with a return to Earth on 19 December. The lunar module was named 'Challenger', and the command capsule, 'America', conducted sounding tests of the lunar surface while in orbit. Crew comprised Captain Eugene A. Cernan, U.S. Navy, the mission commander and a member of the Apollo X mission crew, command module pilot Commander Ronald E. Evans, U.S.N., and a civil geologist, Dr Harrison H. Schmitt.

approach: the descent of an aircraft coming in to land.

apron: a part of an airport (q.v.) used for parking aircraft, usually of concrete or tarmac construction.

Ar. 234, Arado: the world's first jet reconnaissance aircraft to be designed as such, the Arado Ar. 234 saw service with the Luftwaffe in late 1944 and 1945, sometimes on bombing duties, although range and payload were such that the aircraft was unable to make any appreciable contribution to the German war effort. It is sometimes claimed that the Ar. 234 was the world's first true jet bomber, but the reconnaissance label is perhaps the more accurate.

Arado: a German aircraft manufacturer extant during the inter-war period and during World War II, Arado's most famous aircraft include the V-1 mailplane, built for a Berlin–Istanbul service in 1929, and the Ar. 234 (q.v.), the world's first jet reconnaissance-bomber, which first flew in 1944.

Arava, Israel Aircraft Industries: the Arava is a high-wing, twin-boom, utility transport with S.T.O.L. capability, using two United Aircraft 620 shp PT6A-27 turboprops in the civil, and two PT6A-29 turboprops in the military version. The aircraft made its first flight in 1970, three years after design work was initiated. The civil version can carry twenty passengers, while the military version can accommodate sixteen fully-equipped paratroops, or more than a ton of freight. The maximum speed is 220 mph, with a range of up to 800 miles.

area ruling: originally developed in the United States during the early 1950s, the concept of area ruling is now being advocated for transonic supercritical wing (q.v.) civil aircraft, particularly by Boeing, although the principle can be applied on its own. Basically, area ruling is another method of reducing drag, mainly from the fuselage, by streamlining in accordance with the likely airflow. Early trials resulted in a Convair YF-102, with a maximum level flight speed of Mach 0·92 before redesign, being able to exceed Mach 1·0 in a climb.

Argosy, Armstrong-Whitworth: the original Armstrong-Whitworth Argosy was the manufacturer's first airliner. The Argosy I entered Imperial Airways' service on the European routes in 1926, to be followed three years later by the Argosy II, which had twenty-eight seats against the twenty of the earlier version. The three-engined biplane used three Armstrong-Siddeley Jaguar III engines in the I, and Jaguar IVs in the II.

Argosy, Hawker Siddeley: the Argosy was evolved as the Armstrong-Whitworth

A.W.640 Argosy – a private venture civil project intended primarily for the freight market, and the first Series 100 aircraft flew for the first time in 1959. The first ten aircraft were shared between the American operator, Riddle Airlines, and B.E.A. in 1961, and during that year the first military Argosy for the Royal Air Force made its first flight. Ultimately a total of fifty-six aircraft were built for the R.A.F. An up-rated civil version, the Argosy Series 200, flew in 1964, and B.E.A. ordered five.

Although primarily a freight aircraft, the Argosy could carry up to ninety passengers if seats were fitted. A high-wing, twin-boom design, the civil version had opening doors at both ends of the fuselage, but the military version had only the tail doors. All versions used Rolls-Royce Dart turboprops, with the four R.Da.Mk.101 of the military version providing 2,680 shp each, and the four 532–1 of the civil Series 200 aircraft giving 2,230 shp each. The Series 200 could carry up to 31,080 lb, at which the range was about 500 miles, while the maximum speed was slightly in excess of 300 mph; the military version could carry 29,000 lb of freight, and had a range of almost 2,000 miles (when nearly empty), and a speed of about 270 mph. Most of the military aircraft have been replaced by the Lockheed C-130 (q.v.) Hercules, and in B.E.A. service the Series 200s have been replaced by Merchantman conversions of the Vickers Vanguard (q.v.).

Argus, Canadair CL-28/CP-107: *see* CP-107 Argus, Canadair.

Ariana – Ariana Afghan Airlines Co. Ltd: originally formed in 1955, Ariana was initially owned 49 per cent by an Indian charter airline, Indamer, which had previously operated pilgrim charters from Kabul to Jeddah. Then, as now, the Government of Afghanistan was the majority shareholder. Operations started in 1956, using Douglas DC-3s on domestic routes, services to the Persian Gulf and India starting later the same year.

Pan American acquired the Indamer shareholding in Ariana during 1957, this being followed by the introduction of Douglas DC-4 and DC-6 aircraft, while services were extended to Europe. The current fleet includes Boeing 727s, a Convair 440, and a Douglas DC-3, while services are operated to the Soviet Union, to major points in Western Europe and the Indian sub-continent, and to the Persian Gulf and Turkey.

armament: aircraft armaments originally consisted of a rifle or a revolver carried by the pilot or the observer of a reconnaissance aircraft in 1914, at the outset of World War I, although there had been some experiments with machine guns fitted to Wright biplanes in the United States in 1911. The first bombs were shells fitted with fins, often dropped over the side of the cockpit by an observer. Techniques quickly became more refined, and soon a definite fighter (q.v.) aircraft had come into existence, with bombers (q.v.) also taking their place as a distinctive type before the war was very far advanced. The German Zeppelin (q.v.) airships were already in use as bombers by this time.

By the outbreak of World War II, aircraft armaments had widened in scope to include rockets, for air-to-air use eventually, as well as for ground attack duties, mines, depth charges, and torpedoes. Heavier machine guns, or cannon, had become standard. The first simple air-to-surface (q.v.) missile had been used by the end of the war, while the first nuclear weapons had also been used. After the war air-to-air (q.v.) guided missiles, air-to-surface (q.v.) missiles, and stand-off bombs were extensively developed. Aircraft on ground-attack duties were fitted with rocket pods, allowing not only an increase over the number of projectiles which could be carried, but also quicker re-arming of an aircraft. The aeroplane became increasingly a feature of anti-submarine warfare (q.v.), usually equipped with anti-submarine torpedoes.

One post-war development for use on counter-insurgency operations has been the gunship (q.v.), usually a conversion of a transport aircraft with machine guns firing through the window openings, or a machine gun-equipped helicopter. In order to help evade surface-to-air (q.v.) missiles, many modern combat aircraft are being fitted with electronic countermeasures.

Armstrong, Neil: a civilian scientist, Neil Armstrong was the commander of the Apollo XI (q.v.) space mission, which resulted in the first manned lunar landing on 21 July 1969; Armstrong himself was the first man actually to put foot on another planet.

Armstrong-Whitworth: the British firm of Armstrong-Whitworth first came into prominence during the 1920s with its Argosy (q.v.) airliner for Imperial Airways. During the years which followed the company's Siskin fighter and Atlas army co-operation biplane for the R.A.F. enjoyed some fair degree of success, while links with Hawker (q.v.) resulted in its playing a part in Hart light bomber production during the early 1930s. The Atalanta of 1932 was the company's second airliner project, also for Imperial Airways, and this four-engined, high-wing monoplane marked a considerable step forward and away from the biplanes of the 1920s. A later aircraft, the Ensign of 1938, was also a four-engined, high-wing monoplane airliner for Imperial Airways, but in addition had the distinction of being the largest all-metal monocoque aircraft at the time.

On the outbreak of World War II in 1939, the Armstrong-Whitworth Whitley bomber formed the mainstay of the Royal Air Force's bomber strength. After the war an airliner project, the Apollo, made little progress, but another Argosy (q.v.), using turboprop engines and seeing airline and R.A.F. service, was an important part of the workload at the time of the company's merger into what became the Hawker Siddeley Group in 1960. Prior to this there had been more co-operation with Hawkers on production of the Sea Hawk (q.v.) jet fighter for the Royal Navy, Federal German Navy, and Indian Navy.

arrester hook: a hook fitted to the underside of an aircraft's fuselage, near to the tail, and usually retractable, to allow arrester wires (q.v.) to be engaged, particularly when landing on an aircraft carrier. The system is also used as a substitute for braking parachutes when landing some types of land-based fighter aircraft.

arrester wires: a series of wires stretched across the stern part of an aircraft carrier's flight deck to 'arrest' landing aircraft. The earliest recorded use of arrester wires for this purpose was with Eugene Ely's (q.v.) landing on the U.S.S. *Pennsylvania*, when the wires were stretched between anchoring 100 lb sandbags.

'Ash': the Soviet 'Ash' air-to-air guided missile equips the Tupolev Tu-28 'Fiddler' long-range fighter. Radar homing is used.

A.S.M.: *see* air-to-surface missile.

assault ship: a vessel used for landing troops; in U.S. Navy and Royal Navy service the stern is partially submersible to allow landing craft to be floated on and off. Extensive deck area for helicopters is also a feature.

astronautics: the field, or science, of space travel.

A.S.W.: *see* anti-submarine warfare.

Atlantic air services: air services across the North Atlantic on a regular basis took some time to evolve, even though the first non-stop crossing was made in 1919 by Alcock (q.v.) and Brown (q.v.), using a Vickers Vimy (q.v.) bomber. The German airline, Lufthansa, used seaplanes, catapulted from mail steamers as they approached the American or European coasts, as a means of speeding the mails during the late 1920s and the 1930s, and on the South American routes, mail was flown to Las Palmas by a Dornier Wal flying-boat of the Condor Syndicate to catch the steamer service which had left Europe some days previously. In 1939 a joint B.O.A.C.-Pan Am airmail service was started using a Short Empire 'C' Class and a Boeing B-314 flying-boat, the former type being refuelled in flight. Wartime transatlantic flights were operated largely by B.O.A.C.'s Consolidated Liberators and B-314 flying-boats, which carried important military personnel both ways, and returned aircraft ferry crews to North America.

Post-war transatlantic flights started in 1946, and during that year British South American Airways started operations to South America. B.O.A.C. introduced turboprop North Atlantic services in 1957 with Bristol Britannia (q.v.) 312 airliners, following this in 1958 with pure-jet services using de Havilland Comet 4s, which were very soon followed by Pan American's Boeing 707s (q.v.). Supersonic services are scheduled to start during Spring 1975 with the B.A.C.-Aerospatiale Concorde (q.v.).

Atlantic flights: the first transatlantic flight was made in May 1919 by a United States Navy Curtiss NC-4 flying-boat, using the southern route via the Azores and taking about 58 flying hours for the 4,000 miles. The first non-stop flight came the following month, when Captain John Alcock (q.v.) and Lieutenant Arthur Whitten Brown (q.v.) flew in a Vickers Vimy (q.v.) bomber from St John's, Newfoundland to Clifden, Co. Galway, Ireland, taking 16 hours. That same month the British airship, the R.34, crossed the Atlantic from Scotland to New York between 2 and 6 July, and from New York to Norfolk between 9 and 13 July. A non-stop solo crossing was made in May 1927 in a Ryan monoplane, 'Spirit of St Louis', by Colonel Charles Lindbergh.

The first non-stop crossing of the South Atlantic was also made in 1927, in October, with Dieudonné Costes and Lieutenant Joseph La Brix flying a Breguet from St Louis, Senegal (in West Africa) to Port Natal, Brazil – a distance of 2,150 miles in 19 hours. Condor Syndicate Dornier X twelve-engined flying-boats made flights to South America in 1932 and after, while Zeppelins were also used on the route.

Commercial proving flights on the North Atlantic were started in 1937 by Imperial Airways and Pan American, using Short Empire and Sikorsky S-42 flying-boats respectively. In July 1938, Imperial Airways used a Short Mayo (q.v.) composite aircraft (q.v.) on the route. North Atlantic airmail services started in 1939, but were only operated briefly before the outbreak of World War II, although ultimately the effect of the war was to regularise North Atlantic air travel. *See also* Atlantic air services.

Atlantique, Breguet Br. 1150: *see* Br. 1150 Atlantique, Breguet.

ATLAS: the ATLAS consortium was formed during the late 1960s by Air France, Lufthansa, Alitalia, and Sabena to co-ordinate maintenance of their Boeing 747 (q.v.) 'jumbo' jet fleets. The Spanish airline, Iberia, joined in 1972 after ordering the European airbus, the A.300B (q.v.). A similar consortium is K.S.S.U. (q.v.). It is possible that during the longer term these consortia could lead to international mergers by the airlines concerned, particularly those within the European Economic Community.

Atlas: a United States space satellite launcher rocket, replacing the Redstone rocket, which had put America's first man into space, but not into orbit. An Atlas rocket put the MA-6 Friendship satellite with Colonel John Glenn (q.v.) on board into the first American orbit of the Earth in 1962; a duplicate flight followed later in the year. There were further flights in 1963, with twenty-two orbits from the last satellite in the programme, after which the Atlas rocket was replaced by the Titan, which put the Gemini (q.v.) series of two-man spacecraft into orbit from early 1965 onwards.

Atlas, Armstrong-Whitworth: the Armstrong-Whitworth Atlas, a twin-seat, single-engined biplane, was used during the late 1920s and the 1930s by the Royal Air Force and Royal Canadian Air Force for army co-operation and ground attack duties, until eventually replaced by the Hawker Audax (q.v.).

'Atoll': one of the first generation of Soviet air-to-air guided missiles, using an infra-red homing device and having a poor performance in bad visibility. 'Atoll' missiles equipped MiG-17 (q.v.) fighters.

Atom bomb: the first use of the atom bomb was by the U.S.A.A.F. against the Japanese city of Hiroshima on 6 August 1945; a second bomb followed on 9 August, on another Japanese city, Nagasaki.

Attacker, Supermarine

Boeing B-29 (q.v.) Superfortress bombers were used in both cases.

Basically, an atom bomb depends on the rapid splitting of very heavy atoms, as opposed to fusion in the hydrogen bomb (q.v.) and therefore the atom bomb is sometimes known as a nuclear fission device. The material is usually a rare form of uranium-235 or plutonium-239. Even a small atom bomb or warhead can produce an effect equal to several thousand tons of T.N.T., and it is usual to measure the effect of these weapons in terms of kilotons, equal to thousands of tons of T.N.T. The first British A-bomb was detonated in Australia in 1956.

Attacker, Supermarine: the Attacker was the Royal Navy's first jet aircraft designed specifically for operation from aircraft carriers. A single-seat, single-engined aircraft with straight leading edges on wings and tail, it was a fairly undistinguished design, and by the mid-1950s had been replaced by Hawker Sea Hawks (q.v.) and de Havilland Sea Venoms (q.v.).

Audax, Hawker: one of the series of closely related and very attractive single-engined biplanes produced by Hawkers during the early 1930s, in which the lines of the famous Hurricane (q.v.) fighter could already be distinguished. The Audax was the army co-operation aircraft of the family, meeting a 1931 requirement to supplement, and eventually replace, the Armstrong-Whitworth Atlas (q.v.). The aircraft was also acquired by Iraq, Persia, and South Africa, and the Hartbees close support aircraft was evolved from the Audax for service in South and East Africa. In common with the rest of the series, a Rolls-Royce piston engine was used.

Auster: a British light aircraft manufacturer in existence during the 1940s and 1950s, producing a series of high-wing monoplanes for private users and for military use, including A.O.P. duties with the British Army. The company was merged into the ill-fated Beagle Aircraft (q.v.) in 1960.

Austrian Airlines – Österreichische Luftverkehrs: the present Austrian Airlines, or

A.U.A., dates only from 1957, after Austria had been without an airline since Olag Österreichische Luftverkehrs, an airline dating from 1923 and operating between Vienna and Central European capital cities, ceased operations on the German occupation of the country in 1938. A.U.A.'s first international route was to London, using Vickers Viscount turboprop airliners. Currently the airline, which is almost 100 per cent state-owned, operates a fleet of McDonnell Douglas DC-9 jet airliners, which replaced the earlier Viscounts and Caravelles, on a network of European routes.

autoflare: a form of automatic landing equipment, largely developed with B.O.A.C. for the Vickers VC.10 (q.v.), which takes the aircraft down to the flare (or position where the nose is pulled up before touchdown). The system is duplexed, which means that there are two systems and in the event of the failure of one, control is handed back to the pilot, and it is this which prevents the aircraft from making complete landings using the system, since it is important not to put the aircraft in a position from which the captain cannot initiate overshoot procedure in the event of an equipment failure. The autoland (q.v.) system used on B.E.A. Tridents (q.v.) is a complete landing system.

autogiro: the autogiro was the brainchild of a Spaniard, Juan de la Cierva (q.v.), and the first flights by an autogiro, or gyroplane (q.v.), were in Spain in 1923. Although the term autogiro is the more commonly used, it originated as a marketing name for gyroplanes marketed by Cierva. Apart from his design and production activities, Cierva was the first passenger to ride in an autogiro when, in 1926, he was flown in one of his own products, a C.6D. Much of Cierva's work after 1925 took place in the United Kingdom.

The thinking behind the autogiro was the elimination of stalling on take-off and landing, and in an autogiro, as with a helicopter, the lift comes from the rotary wings, although the autogiro differs from the helicopter in that the wings rotate in the slipstream from a conventionally-mounted engine and propeller, and take-off is reduced as opposed to vertical. The conventional aircraft's control surfaces – rudder, elevators, and ailerons – were retained on the Cierva autogiros. The need for the autogiro was largely removed by improvements in conventional aircraft design and performance, and by the advent of the helicopter (q.v.), but recently the gyroplane has enjoyed something of a renaissance, with single- and twin-seat Cierva Rotocraft and Wallis products on the market, although largely as 'fun' aircraft for private owners.

autoland: as the name suggests, a system of automatic landing. There are several such systems on the market at present, of which the most notable are those by the British Smiths Instruments and the American Bendix Corporation; it was the former which paved the way with early experiments, and later perfection, on B.E.A. Hawker Siddeley Trident (q.v.) airliners. The system is triplexed, meaning that there are three separate systems and if any one is inoperative or in error, it is automatically overridden by the other two. Unlike autoflare, autoland actually puts the aircraft on to the runway. The systems for the Trident are no longer experimental, and have been cleared for normal operations into several of Europe's leading airports, giving the user an edge over his competitors during periods of poor visibility.

autolycus: an important item in modern anti-submarine warfare (q.v.) techniques, the autolycus is a short-range instrument for detecting diesel fumes from a submarine either on the surface or recently submerged, or using snorkel equipment. American jargon for the equipment is, aptly enough, 'sniffer'. Although the equipment is useless against nuclear-powered submarines, the majority of submarines are still conventionally-powered, and this makes the autolycus worth while.

automatic landing: see autoland and autoflare.

automatic pilot: see autopilot.

autopilot: the automatic pilot is respon-sible for movement of controls and for maintaining the course of an aircraft in flight, taking much of the strain out of longer-distance flights. Automatic pilots were increasingly used after World War I, and were fitted to British civil aircraft from as early as 1920, when an Aveline Stabiliser was fitted in a Handley Page aircraft. The first automatic pilot was designed and tested by Lawrence Sperry, an American, on a Curtiss floatplane in 1913.

auxiliary air force: usually the title given to a reserve or volunteer force, although in today's terminology the expression would usually be 'air force reserve'. The Royal Auxiliary Air Force operated as a reserve for the Royal Air Force for many years, and auxiliary air forces existed in certain British colonies prior to World War II. The current Hong Kong air arm is still organised in this way.

auxiliary power unit: equipment for maintaining power supplies in an aircraft, for starting and ventilation, while the main engines are stopped and ground equipment is not available. Modern airliners are increasingly being operated into airfields in the less developed parts of the world, making this equipment desirable as an aid to the intensified air transport coverage. Usually the equipment takes the form of a small low-thrust turbojet.

AV-8A Harrier, Hawker Siddeley: see Harrier, Hawker Siddeley.

Avco Lycoming: one of the leading American aero-engine manufacturers, particularly in the design and production of piston engines for light aircraft, Avco Lycoming also produces a number of turbine engines, including the ALF-502 turbofan of 6,000 lb to 7,200 lb thrust and the ALF-301 turbofan of 2,700 lb to 3,000 lb thrust, as well as the LTC-4 series of turboprops in the 3,600 shp to 10,000 shp range. A major user of Avco Lycoming light aircraft engines is Piper (q.v.), one of the major light aircraft producers in the world, while some Beechcraft (q.v.) designs also use the company's engines.

Avenger, Grumman: a World War II product for the United States Navy, later adopted by the Royal Navy, the Grumman TBF Avenger had the reputation of being the best World War II torpedo-bomber. The mid-wing aircraft, with internally-stowed torpedo, used a Wright Cyclone engine of 1,700 hp, and carried a crew of three, including a rear-gunner, in a long cockpit. Bombs, rockets, and mines could also be carried, and the aircraft was ultimately used on a variety of missions, including close support of ground forces.

Avia: a Czechoslovak company which existed between the wars, Avia produced a number of the major aircraft types used by the Czechoslovak Army Air Force after 1920 and before the German invasion of 1938. The more notable products included the B.H.21 fighters, the B.H.10 and B.H.11 trainers of the 1920s, and the B.H.33 and B.H.34 fighters of the 1930s. Shortly before the final German invasion, the Avia A-135 fighter and A-300 bomber entered C.A.A.F. service. A small civil airliner project was the B.H.16 of 1924.

Aviaco – Aviacion y Comercio: Aviaco was formed in 1948 by a group of Bilbao businessmen as an all-cargo charter airline using Bristol 170s, not starting passenger operations until 1950, with services to Madrid and Barcelona, which were followed by other domestic services to the Canary and Balearic Islands, and an international service to Marseille. Further domestic expansion followed with services based on Madrid, and the first jet aircraft, Sud Caravelles, were leased from Sabena during the early 1960s. Vehicle ferry services to Palma were operated during the middle and late 1960s using Aviation Traders Carvairs.

Currently Aviaco, which is now owned 75 per cent by Iberia, operates a fleet of Convair 440 Metropolitans and Fokker F-27 Friendships on a mainly domestic route network, with some services to North Africa, and also operates inclusive tour charter flights.

Avianca–Aerovias Nacionales de Colombia: although Avianca itself was formed in 1940, the airline's history through one of its two predecessors, S.C.A.D.T.A., dates from 1919, making it the first airline in the Americas and the second airline in the world to start operations. Although the initial equipment consisted only of a single Junkers seaplane, rapid expansion followed, due to the difficult terrain of Colombia, and in spite of the problems of operating into the capital, Bogota, some 9,000 feet above sea level – which S.C.A.D.T.A. reached using a Fokker. A flying-boat service to Florida was introduced in 1925.

Apart from Servicio Aereo Colombiano, the other predecessor airline in 1940, other airlines acquired by Avianca during its history have included S.A.E.T.A. in 1954 and, recently, Sociedade Aerea del Tolima, and Sociedad Aeronautica de Medellin. The latter was acquired by Avianca's subsidiary Aerotaxi, which operates air taxi and third-level services, while another subsidiary, Helicol, which is owned jointly with the Keystone Helicopter Corporation, operates helicopter charters. Avianca itself is owned 38 per cent by Pan American World Airways (q.v.).

Currently, Avianca operates an extensive domestic and Latin-American network, with services to North America and Europe, using a fleet of Boeing 707, 720, and 727 jet airliners, Hawker Siddeley 748 turboprop airliners, and Douglas DC-4s and DC-3s; while Aerotaxi uses Beechcraft, Cessna, and de Havilland Canada aircraft types; and Helicol has Bell helicopters.

aviation: the field of heavier-than-air powered aircraft.

aviation meetings: a feature of the early years of aviation were the aviation meetings, of which the first was held at Reims in 1909, with the support of the champagne industry. It would be an over-simplification to consider these meetings as an early form of the modern flying display, since the generous prizes offered were a tremendous incentive for the pioneers of flight, many of whom were battling against public indifference with limited resources, and any orders received for the aircraft which showed their paces at the meetings were something of a bonus. Other meetings followed that at Reims, including those at Nice and at Milan in 1910, at which records for height, speed, and distance covered were made.

aviator: one who flies as part of the crew of a heavier-than-air flying machine.

Avicar, C.A.S.A. C.212: *see* C.212 Avicar, C.A.S.A.

Avion III: the second of Clement Ader's (q.v.) unsuccessful designs, the steam-powered Avion III failed to leave the ground in spite of two attempts to do so in 1897, although in 1906 Ader claimed that he had flown 300 metres (just under 1,000 feet) in the aircraft. The Avion III retained the bat-wing design of the 'Éole', and was twin-engined with each engine driving a propeller.

avionics: that branch of electronics which relates to aviation, and includes navigational and control equipment, such as Loran and Doppler (q.v.) head-up displays, and identification equipment such as transponders (q.v.).

Avions Fairey: the Belgian subsidiary of the famous British aircraft manufacturer Fairey Aviation (q.v.), Avions Fairey dates from 1931, following an order for the Aviation Militaire for Fairey Fox fighters, which were built in Belgium. Before World War II broke out, Avions Fairey also built Battle light bombers for the Aviation Militaire. Since the war the company has taken part in European licence construction programmes of American-designed equipment for NATO air forces, including the Lockheed F-104 (q.v.) Starfighter. The company is also a subcontractor on the Breguet Br. 1150 Atlantique maritime-reconnaissance aircraft.

One of Avions Fairey's managing directors, Ernest Tips, designed the Tipsy Nipper and Tipsy Trainer light aircraft, some of which were built by the company.

Avions Marcel Bloch: *see* Avions Marcel Dassault.

Avions Marcel Dassault: formed after the end of World War II by Marcel Dassault (q.v.) (formerly Marcel Bloch) Avions Marcel Dassault proceeded to produce a range of military aircraft, including the Ouragan fighter-bomber, the Mystère fighter and its Super Mystère development, and the Étendard carrier-borne fighter. A light transport, communications and navigational training aircraft produced by the company during the 1950s was the Flamant. Although these aircraft enjoyed some modest success, the company did not reach its present important position until the late 1950s, with the Mirage (q.v.) fighter, which remains in production and of which there have been fighter-bomber, ground attack, conversion trainer, interceptor, reconnaissance, and nuclear bomber versions, with variable-geometry and V.T.O.L. developments. A broadening of the company's activities into the civil field has been marked by the success of the Mystère 20 (q.v.) business jet, the Hirondelle (q.v.) light transport, and the new Mercure (q.v.) short and medium range airliner. The company's first collaborative project is with Germany on the Alphajet (q.v.) advanced trainer. Guided weapons work is also undertaken by A.M.D.

Avions Marcel Dassault merged with the Breguet (q.v.) concern in early 1972.

Avions Max Holste: *see* Reims Aviation.

Avions Pierre Robin: originally the Centre Est light aircraft concern under the Sud Aviation wing, Avions Pierre Robin became a separate entity in 1969, and manufactures a number of light aircraft designs, including the old Centre Est products such as the Rallye and the Horizon, as well as the new Royale four/five-seat, single-engined light aircraft.

Avro: after producing a number of designs, including a triplane and a cabin biplane in which Lieutenant Wilfred Parkes was able to make the first controlled recovery from a spin, Avro, named after its founder, the Englishman A. V. Roe, came into prominence during World War I with its Avro 504 (q.v.) biplane trainer. After the war the company produced the 563, a twelve-seat airliner;

the Avenger, at its time the fastest single-seat fighter in the world; and the Avian, a single-seat sport and racing monoplane used on several pioneer flights. During the mid-1930s there followed the Avro 642 (q.v.) and the Avro 19 (q.v.), which led to the Anson light transport for the Royal Air Force, while the R.A.F.'s Avro 504s were replaced by Avro 620 biplanes, although these in turn were to be replaced by de Havilland Tiger Moths.

During World War II, Avro aircraft included the Lancaster heavy bomber, which was the Royal Air Force's best bombing aircraft during the war, and was followed at the close of the war by the Lincoln heavy bomber and York transport. An extensively-developed version of the Lancaster bomber became the peacetime Avro Shackleton (q.v.) maritime-reconnaissance aircraft for the Royal Air Force and the South African Air Force. Less successful was the Avro Tudor airliner, of which there were jet and turboprop developments. This was originally designed for British South American Airways, but suffered from structural failure.

Avro's entry into the jet age started with its experimental 707A/B delta-wing fighter, from which the Vulcan (q.v.) nuclear bomber was evolved for R.A.F. service. A highly-successful light turboprop transport, the 748, appeared during the late 1950s, and with the Vulcan was Avro's contribution to the product range of the Hawker Siddeley Group, into which Avro was merged in 1960.

Avro 19: the Avro 19 was the civil version of the Anson light bomber and light transport aircraft of the late 1930s and World War II. Two Armstrong Siddeley Cheetah piston engines gave a maximum speed of 170 mph, and a range of 660 miles, to this low-wing monoplane, which could carry eight passengers. A few sur-vived for more than twenty years after the end of World War II.

Avro 504: although the Avro 504 enjoyed a brief spell as a fighter and reconnaissance aircraft in 1914, at the outset of World War I, it became more widely known as the Royal Flying Corps', and later the Royal Air Force's, standard trainer for the war and the immediate post-war period. The most common variant of this single-engined, twin-seat biplane was the 504K, with a 100 hp Gnome rotary engine. A highlight of its offensive career, before settling down to training duties, was making one of the first raids on Zeppelin sheds for the Royal Naval Air Service.

Avro 642: the Avro 642 was a high-wing monoplane airliner which entered service on Imperial Airways' European routes in 1934. Sixteen passengers could be carried, and power was supplied by two Armstrong Siddeley Jaguar engines.

Avro 748: *see* HS 748 and Andover.

AWACS: the U.S.A.F.'s AWACS (airborne warning and command system) project is based on a planned modification of the Boeing 707 (q.v.), and will also replace the present airborne-early-warning Lockheed EC-121 (q.v.) Constellations. In-service date should be during the mid-1970s.

Azor, C.A.S.A. C.207: *see* C.207 Azor, C.A.S.A.

Aztec, Piper PA-23: the Piper PA-23 Aztec appeared during 1959 as a successor to the Apache (q.v.), retaining much of the Apache's appearance, but with a longer nose and more powerful Lycoming 250 hp IO-540 engines. The aircraft is a six-seat executive and air taxi type, with a maximum speed slightly in excess of 200 mph, and a range of 1,200 miles.

B

B-: international civil registration index mark for Nationalist China (Taiwan).

B-: U.S.A.F. designation for bomber aircraft, although to some extent preceded by A- (q.v.) designation during World War II.

B-1, North American Rockwell: long-range, supersonic (Mach 2·0 plus), variable-geometry bomber under development for the U.S.A.F. The first flight of this four-engined aircraft is scheduled for 1975, with deliveries during the late 1970s. Little detailed information is available.

B-10, Martin: developed from the experimental Martin XB-10, winner of the Collier Trophy for 1932, the B-10, with its all-metal, mid-wing monoplane construction, was the first of the new generation of bombers. The aircraft entered United States Army Air Corps service in 1935. Two Wright Cyclone 750 hp radial engines produced a maximum speed of 215 mph, which was superior to most contemporary fighters. Amongst the several fairly new features could be counted a rotating gun turret in the nose.

B-17 Flying Fortress, Boeing: the B-17 was developed from Boeing's 299, the manufacturer's first four-engined bomber, which appeared in 1935 with Pratt and Whitney 750 hp Hornet radial engines. The B-17 itself was to form the backbone of the U.S.A.A.F.'s bomber force in Europe, and had a very distinctive heavy armament, with up to twenty machine guns to counter German fighter attack while on daylight bombing raids outside of the range of Allied fighter escorts. Four 1,250 hp Wright radial engines powered the production models, which were also used by the R.A.F. After the war a number of aircraft were converted for civil transport duties, while others were modified for United States Coastguard search and rescue duties (Fortress PB-1G) and for maritime-reconnaissance patrols with the Royal Air Force. A development of the B-17 was the B-29 (q.v.) Superfortress.

B-18, Douglas: the Douglas B-18 bomber was a military development of the highly successful DC-2/3 series of airliners (although with a somewhat more bulbous appearance), which first entered service with the U.S.A.A.C. in 1936. Two Wright radial engines gave a maximum speed of more than 200 mph, and up to a ton of bombs could be carried over a range of 1,000 miles. The aircraft also served with the Royal Canadian Air Force during World War II.

B-24 Liberator, Consolidated: the B-24 Liberator was the most widely-used American heavy bomber of World War II – a high-wing aircraft with heavy protective armament and four Pratt and Whitney radial engines. A total of some 18,000 B-24s saw service with the United States Army Air Force, the Royal Air Force, and the United States Navy (PB4Y-1). It was possible for the aircraft to carry a ten-ton bomb load, and the maximum useful range was 2,000 miles. British Overseas Airways Corporation 'civilianised' Liberators operated wartime Atlantic air services (q.v.), carrying diplomatic and top military personnel both ways across the Atlantic, as well as returning ferry pilots to North America.

B-25 Mitchell, North American: the B-25 Mitchell medium bomber was produced in a wide range of versions for a number of air forces, including both the U.S.A.A.F. and the R.A.F., during World War II, and a few survive in Central and South America. A twin-engined, high-wing bomber, its 1,700 hp Wright R-2600-29 radials provided a top speed of 275 mph, with a range of 1,350 miles and a warload of 4,000 lb.

B-26 Marauder, Martin: one of the more potent medium bombers of World War II, the B-26 Marauder was also used on ground-attack duties. The high-wing aircraft, with two 2,000 hp Pratt and Whitney radial engines, could fly at more than 300 mph, carry up to 4,000 lb of bombs, and had a defensive armament of anything up to eleven machine guns.

B-29/50 Superfortress, Boeing: the largest bomber of any nation during World War II, the Boeing B-29 Superfortress entered U.S.A.A.F. service in late 1943 and saw little service in Europe, but the aircraft was in any case intended to strike back at the industrial heartland of Japan. A streamlined fuselage and four 2,500 hp Wright radial engines contributed to a maximum speed well in excess of 300 mph, and the aircraft could carry up to 20,000 lb of bombs over relatively long ranges. Amongst the features of the aircraft were remote control gun turrets. On 6 and 9 August 1945, Boeing B-29 Superfortress bombers dropped the first atom bombs (q.v.) on Japan, and ended World War II.

A development of the B-29 was the B-50, with four 3,800 hp Wright R-4360 radial engines, which, in addition to its combat role, was used as a launch aircraft for post-war Bell supersonic research aircraft, and later as a tanker aircraft to boost the operational range of the Boeing B-47 (q.v.) Stratojet bombers, in which duty several of the aircraft remained active well into the 1960s. The KB-50 tanker versions were equipped with two General Electric J47-GE-23 turbojets to boost their performance to a maximum speed of 445 mph, and a range of about 3,000 miles.

B-30 Baltimore, Martin: *see* A-30 Baltimore, Martin.

B-36, Convair: the B-36 was something of an anachronism on appearance since it had been designed to bomb Germany in the event of Britain falling during World War II, and did not fly until 1946! Entry into U.S.A.F. service was during the late 1940s and the start of the 1950s. Six Pratt and Whitney 3,800 hp R-4360-35 radial engines with pusher propellers were later augmented by four General Electric J47-GE-19 turbojets, giving the aircraft a maximum speed of 435 mph, a range of 5,000 miles, and the ability to carry a 10,000 lb bombload. Trials were held as a composite aircraft (q.v.) equipped with a Republic F-84 (q.v.) Thunderjet fighter for protection when operating away from normal fighter cover. Although the B-36, like the B-50, was a founder-member of the U.S.A.F.'s Strategic Air Command (S.A.C. (q.v.)), the aircraft was dated by the advent of jet bombers such as the B-47 (q.v.) and B-52 (q.v.).

B-47 Stratojet, Boeing: the Boeing B-47 Stratojet first flew at the end of 1947, and was the first swept-wing jet bomber. Deliveries to the U.S.A.F. started in 1950. The aircraft played an important part in the Korean War, and Douglas and Lockheed production helped towards the eventual total of more than 2,000 aircraft for the U.S.A.F. Six General Electric 6,000 lb thrust J47-GE-25 turbojets gave a maximum speed of 600 mph and a range of 4,000 miles, while up to 20,000 lb of bombs could be carried. A remote-control tail turret was equipped with 20 mm cannon. The aircraft was progressively withdrawn from service during the early 1960s after spending some years as the mainstay of Strategic Air Command. Special versions included the RB-47B and RB-47E for photographic reconnaissance, the RB-47H for electronics reconnaissance, the WB-47 for weather reconnaissance, and the missile transport DB-47.

B-52 Stratofortress, Boeing: the B-52 first flew in 1952, and since then it has retained the distinction of being the world's largest bomber aircraft. Eight 17,000 lb thrust Pratt and Whitney TF33-P-3 turbojets give a maximum speed of 660 mph and a range of up to 12,500 miles. The aircraft can carry a heavy conventional bombload, nuclear bombs, or Hound Dog (q.v.) or Skybolt air-to-surface missiles. Some 750 aircraft were built for the U.S.A.F.'s Strategic Air Command during the late 1950s and early 1960s. Although some have since been withdrawn, the remaining aircraft have been responsible for much of the very intensive bombing during the Vietnam War. A distinctive feature of the high-wing aircraft is very severe anhedral. A remote-control tail turret uses 20 mm cannon. Many B-52s carry highly sophisticated and effective electronics countermeasures, making surface-to-air guided missile attack difficult, although large salvos of unguided missiles have been used with success.

B-57 Canberra, Martin: the B-57 is a licence-built version of the English Electric Canberra (q.v.) jet bomber, which was the world's first medium jet bomber and was built in the United States by Martin and General Dynamics for bombing versions include B-57A, -57B, -57C, and -57E, and reconnaissance duties RB-57A, RB-57D, and RB-57F, the last-mentioned having a considerably extended wingspan for high altitude work.

B-58 Hustler, General Dynamics: the B-58 Hustler began life in 1952 as a Convair (q.v.) design to meet a U.S.A.F. requirement for a supersonic strategic bomber, and was the first aircraft in the world to fit into this category. First flight was in late 1958, and production covered just over 100 B-58 aircraft, with a few TB-58A conversion trainers. Four reheated 15,600 lb thrust General Electric J79-GE-5B turbojets give a maximum speed of 1,385 mph (Mach 2·1), with a range of 2,000 miles and capacity for a nuclear bomb, a stand-off bomb, or air-to-surface missiles. The aircraft is now in reserve.

B-66 Destroyer, Douglas: the Douglas B-66 Destroyer was a land-based development of the United States Navy's A-3D Skywarrior – a reverse of the normal practice in aircraft development – and used two 10,000 lb thrust Allison J71-A-13 turbojets for a maximum speed of 620 mph, a range of 1,500 miles, and a warload of 15,000 lb. First flight was in June 1954, and 250 aircraft were built during the mid-1950s. There were three operational versions, the RB-66B and RB-66C for reconnaissance, and the B-66C for general bombing duties; the RB-66A was a development version only. During the late 1960s the aircraft were replaced in active service by McDonnell Douglas F-4/RF-4 (q.v.) Phantom IIs.

B.A.C.: *see* British Aircraft Corporation.

B.A.C. 145/167 Jet Provost: the Jet Provost was evolved as a jet-engined version of the Hunting Percival Provost trainer which appeared in service during the 1950s with the R.A.F. and several other air forces, including those of Burma, Iraq, and Rhodesia, and the Irish Army. Trials were conducted with Mk.1 and Mk.2 versions of the Jet Provost before the Mk.3 was developed and standardised on as the R.A.F.'s basic training aircraft in 1959, although basic instruction on piston-engined aircraft was later reverted to. Some five hundred Mk.3 and Mk.4 Jet Provosts, and the counter-insurgency T.Mk.51 and T.Mk.52 variants, were built for the R.A.F. and other air forces. In 1960 Hunting Percival became a part of the British Aircraft Corporation (q.v.). An advanced development with pressurised cockpit, the 145, first flew in February 1967, and deliveries started in 1969 to the Royal Air Force as the T.Mk.5, while an armed version, the T.Mk.55, was sold to the Sudanese Air Force. A version of the 145 designed specifically for counter-insurgency duties is the 167 Strikemaster, which first flew in October 1967, and has been supplied to Kuwait, Muscat and Oman, Saudi Arabia, Singapore, and South Yemen.

The Jet Provost is a low-wing, twin-seat aircraft with a side-by-side cockpit, using a single Rolls-Royce Viper turbojet. The original Jet Provost used a 1,750 lb thrust Viper to give a maximum speed of 430 mph; the 145 uses a 2,500 lb thrust Viper 202 for a maximum speed of 440 mph and a range of 700 miles; and the Strikemaster uses a 3,140 lb thrust Viper 535 for a maximum speed of 475 mph and a maximum range of 1,380 miles.

Bader, Group Captain Douglas, D.S.O., D.F.C., R.A.F.: in spite of losing both legs in a flying accident in a Bristol Bulldog in 1931, Douglas Bader became one of the leading fighter aces of World War II, using artificial legs. His personal score amounted to some 22½ German aircraft, and he gained the D.S.O. and bar, and D.F.C. and bar. He was shot down over France in 1941 while commanding a Spitfire wing, and remained a prisoner of the Germans until 1945, in spite of two escape attempts, one of which nearly succeeded. After the war he commanded a group, including a squadron of Meteor jet fighters, before leaving the R.A.F. and subsequently becoming manager of an oil company's civil aviation operations and a journalist.

'Badger', Tupolev Tu-16: *see* Tu-16 'Badger', Tupolev.

balance of power: a basic precept of British foreign and defence policy in the nineteenth century and the first half of the twentieth century was to counter the ambitions of the strongest state in continental Europe, and these policies usually became most apparent in military terms, with each side struggling to gain superiority, or at least maintain equality. Often the balance could only be achieved by taking on an ally. Usually the opposition was between France and Germany. Today, the main competition and the struggle for a balance of power is between the United States and the Soviet Union, each with their respective allies in NATO (q.v.), SEATO (q.v.), and CENTO (q.v.) on the one hand, and the Warsaw Pact (q.v.) on the other. In fact, Russia's military superiority and the need of the West to place reliance on nuclear weapons to counter even a Soviet conventional attack has resulted in more of a balance of terror (q.v.) than a balance of power in the true and traditional sense.

It is United States policy, with which the United Kingdom largely concurs, to maintain a balance of power in the Middle East between Israel and the Arab States: this includes not only arms supplies to Israel, but also to Jordan and Saudi Arabia, although the latter has been largely unaffected by the Arab–Israeli Wars. In South-East Asia, American policy is to prevent Communist-backed take-overs of neutral or pro-Western nations in Indo-China, often with Australian and New Zealand assistance under ANZUS (q.v.). It could be fairly said, therefore, that while the balance of power between the major powers has been largely replaced by the balance of terror, a balance of power is maintained by the major powers between the second- and third-rate powers of conflicting ideologies.

balance of terror: originally coined as a criticism by pacifist elements, the term 'balance of terror' is now generally accepted as describing the situation under which neither opposing alliance will attack because of the certainty of a counter-attack which would result in unacceptable damage. The term is entirely one for the nuclear age since by implication nuclear weapons have to be used, and therefore the balance of terror is closely allied in meaning to nuclear deterrent. Equal power, as in a balance of power, is not essential, but since the deterrent effect is dependent for its continued existence on a minimum nuclear capability being maintained, and the opposition not being allowed to reach an overwhelming degree of strength or capability which might tempt an attack, a certain degree of continual development and preparation is necessary.

Undoubtedly a World War III in Europe would have started by now if it were not for the deterrent effect of nuclear weapons and efficient and quick response delivery systems. The situation is one of British and American forces against those of the Soviet Union, with those of France and Communist China having a more limited potential, being outside of the mainstream of development. A danger lies in the West's inability to counter a conventional Soviet attack by other than nuclear means, which could lead to a rapid escalation of any conflict involving the major powers directly and in opposition. On the other hand, there can be little doubt that the use of nuclear weapons in 1945 by the United States against Japan brought an abrupt end to World War II and spared many British, American, and Commonwealth lives which would have been lost in continued conventional war against a still-powerful enemy.

Balkan Bulgarian Air Transport: formed in 1947 as a joint Soviet and Bulgarian undertaking, B.V.S., with services starting in 1948, Balkan Bulgarian Air Transport was known as T.A.B.S.O. from 1949 until 1968. Initially the airline was mainly concerned with domestic operations, but increased emphasis on international services followed the handing over of complete control to Bulgaria in 1954. During the initial stages, Lisunov Li-2 and Ilyushin Il-14 airliners were operated, and the first turboprop equipment, Ilyushin Il-18s, was not introduced until 1962.

Currently, Balkan Bulgarian operates a fleet of Tupolev Tu-134s, Ilyushin Il-18s and Il-14s, Antonov An-24s, An-10s, and An-2s, and Mil helicopters on a network which includes domestic services and services throughout Europe, North Africa, and the Middle East.

ballistic missile: a ballistic missile is an unmanned rocket-propelled projectile which flies to its target on a high parabolic trajectory, which usually involves a substantial part of the flight being outside of the atmosphere, particularly in the case of intercontinental ballistic missiles (q.v.). The delivery system is usually a three-stage rocket, with a first stage booster and a third stage containing the warhead and penetration aids designed to assist delivery by foiling defensive measures. The missile is usually guided for the first part of its journey, after which course and target become easily predictable, particularly for anti-ballistic missile defences (q.v.). Penetration aids usually include either chaff or radar-jamming devices, using multiple warheads in an attempt to flood the defences, and often these are independently targeted, shielding the warhead from any type of neutron kill device which might immobilise it, or detonating the warhead as soon as attack appears to be imminent in an attempt to get at least a partial kill.

The first operational ballistic missile was the German V-2 rocket of World War II, which was used with considerable effect against London although, of course, only conventional warheads were used. Current ballistic missiles include the submarine-launched Polaris (q.v.) and Poseidon (q.v.), the latter having a multiple independently-targeted warhead, Minuteman (q.v.), the French submarine-launched Mer-Sol Balistique Strategique (q.v.), the Soviet 'Serb' (q.v.) and 'Sawfly' (q.v.) submarine-launched missiles, and the 'Sasin' (q.v.) and 'Savage' (q.v.) land-based missiles.

ballistic missile early warning: the American B.M.E.W.S. is a combination of long-range radar and a telecommunications network to provide rapid identification of I.C.B.M.'s and their targets. *See also* anti-ballistic missile defences.

ballistic missile defences: *see* anti-ballistic missile defences.

balloon: two Frenchmen, the brothers Joseph and Étienne Montgolfier (q.v.) of Lyons, invented the hot air balloon, or Montgolfière as it became known, after observing how pieces of paper rose with the smoke and heat of a fire. After experiments with paper bags, they produced a paper and linen sphere which, with a wool and straw fire smouldering beneath it, made an ascent on 5 June 1783. The two brothers sent up an even larger balloon working on the same principle at Paris, which carried three animals, a cock, a sheep, and a duck, for a two-mile flight before landing them safely. Further successful ascents were made with tethered balloons, including some in which men were carried up to a height of around 100 feet, before it was decided to proceed with an untethered manned flight, for which two convicts were to be used until two aristocrats, François Pilâtre de Rozier (q.v.) and the Marquis d'Arlandes, volunteered to make the journey, which proved to be a successful flight over Paris on 21 November 1783, lasting for 5 miles.

Only a matter of days separated the first and second aerial voyages in history. J. A. C. Charles (q.v.), a professor at the University of Paris, designed a small hydrogen balloon from a rubberised fabric invented by his partners, the Roberts brothers, and this made a successful ascent from Paris on 27 August 1783. Charles was in keen competition with the Montgolfier brothers, and believed that they were using an unknown gas as a lifting agent, though he knew that it was not as light as the then new hydrogen. Charles managed to overcome the very real difficulty of producing hydrogen in sufficient quantity for a man-carrying balloon, and on 1 December 1783 such a balloon made a $31\frac{1}{2}$-mile flight, the first 27 miles being with

Charles and Ainé Robert, and the remaining $4\frac{1}{2}$ miles being only with Charles, who had decided to fly on alone when only enough lift remained for one man.

The Charlière type of balloon, as it became known, was the type most widely used, because hydrogen is very much lighter than air and therefore more satisfactory than depending on hot air for lift. However, hydrogen is also dangerous because it becomes explosive once in contact with the atmosphere – by means of a leak in the balloon's fabric, for example. During the twentieth century, the much safer although less efficient helium was to replace hydrogen in airships (q.v.) and balloons. Features of the Charlière copied by subsequent balloons included a venting valve in the crown, a barometer altimeter, a net slung over the balloon, from which hung the car, and ballast, which could be used to lighten the balloon as the ascent or flight progressed. The Montgolfière was not doomed to extinction, however, because of its cheapness; indeed the first British aeronaut, James Tyler, used a Montgolfière when making an ascent from Edinburgh in 1784. Frequently throughout the nineteenth century the inefficient coal gas had to be used in balloons because of the cost and difficulty of producing hydrogen.

An Italian, Vincenzo Lunardi (q.v.), made the first hydrogen balloon ascents in the British Isles in 1784. That same year the first women made ascents in tethered balloons in Paris on 20 May, and on 4 June a Madam Thible made an ascent with a Monsieur Fleurant at Lyons in a Montgolfière. Pilâtre de Rozier and a companion, Romain, suffered the unsought distinction of the first balloon fatality on 15 June 1785, while attempting a crossing of the English Channel in a balloon using the unbelievably dangerous combination of hydrogen and a brazier slung under the balloon to heat the gas. However a Frenchman, Jean-Pierre Blanchard, and an American, John Jeffries, had made a successful crossing of the English Channel from Dover to France on 7 January, using a Charlière.

The first recorded military use of balloons came in 1784, as air observation posts (q.v.) for the French Army at the Battle of Fleurus (but these were tethered balloons, and a French army captain, Joseph Coutelle, was made 'Chef de Bataillon de Aérostiers de la République'. During the American Civil War, between 1860 and 1866, the Unionist Army used balloons for observation duties with the Army of the Potomac.

Attempts to use balloons for communications purposes included that by H.M.S. *Assistance* in 1850, which while searching for the ill-fated Sir John Franklin expedition in the Canadian Arctic, sent up small unmanned balloons carrying messages giving the ship's position. During the Siege of Paris in 1870 and 1871 sixty-eight balloons were built in a factory set up in the Gare du Nord, and were launched (all but one of them manned) to take correspondence and passengers out of the city and over enemy-held territory; thirty-six landed successfully in France, while the rest landed in enemy-held territory in other countries, as far away as Norway in one case.

The British Army started its experiments with balloons in 1878 at Woolwich Arsenal, giving the United Kingdom the longest continuous history of military aviation of any country, and by the following year the Royal Engineers had a fleet of five balloons. The balloons were included in a British Army expedition to Bechuanaland in 1884, and one in 1885 to the Sudan. The United States set up a Balloon Section in the United States Army in 1892, while the Netherlands followed in 1896, as did Spain, while Imperial Russia waited until the turn of the century.

The balloon has been entirely eclipsed for military and commercial purposes first by the airship and then by the aeroplane. Salomon Andree's (q.v.) attempted balloon crossing of the North Pole in 1870 was ill-fated, because of frozen mist forcing the balloon down. Balloons have only really survived as a cheap means of launching weather instruments for meteorological purposes, but latterly have enjoyed a minor comeback for pleasure purposes, usually on the lines of the Montgolfière.

Baltimore, Martin A-30: see A-30 Baltimore, Martin.

Bandeirante, C.T.A. C-95: see C-95 Bandeirante, C.T.A.

Baron, Beechcraft: a popular light twin-engined business, executive, and private aircraft, the Baron is also used for military instruments training and communications duties, particularly with the U.S.A.F. The Baron can carry up to six persons and has a maximum speed of 240 mph and a maximum range of 1,200 miles. The B55 version, which is also the U.S.A.F.'s T-42A instrument trainer, uses two 260 hp Continental I0-470-L piston engines, while the C55 uses 285 hp Continental I0-520-Cs. The Travel Air is a 180 hp Lycoming I0-360-powered version with a straight fin, as opposed to the Baron's swept fin.

Barracuda, Fairey: a carrier-borne torpedo bomber with a single piston engine and a high wing, the Barracuda replaced the Royal Navy's Fairey Albacores, starting in 1943. Amongst the main events during the aircraft's career, which lasted until 1953, can be counted its participation in the Allied landings at Salerno in September 1943, operating from H.M.S. *Illustrious*; and the successful attack on the German battleship *Tirpitz* in the Kaafiord, Norway, in 1944, by two waves of twenty-one Barracudas each from H.M.S. *Emperor*, *Fencer*, *Furious*, *Pursuer*, *Searcher* and *Victorious*, with eighty carrier-borne fighters for protection.

barrage balloon: an anti-aircraft measure introduced during World War II by both sides, the barrage balloon (having a strong resemblance to the airship or Caquot balloon rather than a balloon of Montgolfière or Charlière pattern) kept attacking aircraft at a height which hampered accurate bomb-aiming and also presented anti-aircraft gunners with good targets. Any aircraft attempting to fly under the balloons, and the anti-aircraft fire, ran the risk of having its wings ripped off by the strong anchoring cables of the balloons.

base: a military term, usually U.S., for an airfield, particularly of the more sophisticated and well-equipped variety.

Basset, Beagle: the R.A.F. communications version of the Beagle 206 (q.v.).

Battle, Fairey: the Battle was a low-wing, single-engined, light bomber monoplane designed to fulfil a 1933 Air Ministry requirement. It was obsolescent at the time of entry into R.A.F. service in 1937, in spite of the then popular view of the aircraft as the bomber with fighter performance – which owed more to appearances than to actual performance. Battles formed a prominent part of the Advanced Air Striking Force which accompanied the British Expeditionary Force in France prior to the retreat on Dunkirk. After Dunkirk the type was relegated to target-towing and training duties. A carrier-borne fighter development was the Fairey Fulmar.

B.E.2, Royal Aircraft Factory: the B.E.2 had the distinction of being one of the British Army's first reconnaissance aircraft, and as such it played a prominent part in World War I, the first Royal Flying Corps unit to land in France being equipped with the aircraft. A 90 hp Royal Aircraft Factory engine powered the twin-seat biplane, which was later armed with a machine gun for the observer, who nevertheless sat in the forward seat behind the engine! Unlike many reconnaissance aircraft, the B.E.2 did not prove to be a particularly good fighter type.

Be-6 'Madge', Beriev: a post-war Soviet maritime-reconnaissance flying-boat, the gull-winged Be-6 (NATO code name 'Madge') was powered by two 2,000 hp Shvetsov ASh-73 radial engines which provided a maximum speed of just over 250 mph and a maximum range of 3,000 miles. A retractable radome was usually fitted for surface scanning, and some models were also fitted with M.A.D., while bombs, mines, torpedoes, or depth charges could be carried under the wings. The aircraft was phased out of service during the 1960s, but a turboprop development, with many improvements, exists in the Be-12 (q.v.) 'Mail'.

Be-10 'Mallow', Beriev

Be-10 'Mallow', Beriev: the Be-10 (NATO code name 'Mallow') first appeared in 1961, and has since formed a small part of the Soviet maritime-reconnaissance force. A twin-jet flying-boat with a very modern hull design, two 14,330 lb thrust AL-7PB turbojets give a maximum speed of up to 566 mph and a range of some 4,000 miles, and the aircraft's anti-shipping missiles can be fitted under the wings. A number of world records for flying-boats have been gained by the aircraft, including those for speed and altitude, with and without payload, which can amount to 15 tons.

Be-12 Tchaika 'Mail', Beriev: a modern Soviet maritime-reconnaissance amphibian, the Be-12 (NATO code name 'Mail') uses two 4,150 shp Ivchenko AI-20 turbo-props, giving a maximum speed of some 340 mph, and a range of around 3,000 miles. Originally based on the Be-6 (q.v.), the Be-12 has an advanced hull form and other detailed improvements.

B.E.A.–British European Airways: originally formed in 1946 to operate the European network of B.O.A.C. (q.v.), British European Airways also took over some R.A.F.-operated services and, in 1947, the route networks of the Associated Airways Joint Committee Companies, which included Railway Air Services, Scottish Airways, Channel Islands Airways, Isle of Man Air Services, Great Western Air Lines, and Southern Air Lines. The mixed fleet which resulted included Douglas DC-3s, Avro 19s, de Havilland Rapides, and ex-German Junkers Ju. 52/3Ms, which were soon supplemented by Vickers Viking airliners and, in 1952, B.E.A.'s last piston-engined mainline aircraft, the Airspeed Ambassadors. Meanwhile, B.E.A. had operated trial flights with the prototype Vickers Viscount turboprop airliner during 1951, and during the early 1950s became the first airline in the world to introduce a turboprop airliner, the Viscount 700 series, to regular airline service. Later deliveries included the Viscount 800 series, the larger Vickers Vanguard, and the all-freight Armstrong-Whitworth Argosy turboprop aircraft.

Throughout its history, B.E.A. has devoted a considerable amount of effort to helicopter operations, starting with the B.E.A. Helicopter Experimental Unit in 1947, which initially operated airmail services with Westland Dragonfly machines, later developing some passenger services, mainly as airport feeders, and using Westland Whirlwind helicopters before finally withdrawing such services in 1956. A subsidiary, B.E.A. Helicopters, was formed in 1964, mainly to operate Sikorsky S-61N machines to the Isles of Scilly; but more recently a considerable volume of business has developed ferrying supplies to North Sea oil rigs, and also on air taxi work.

The airline introduced its first jet aircraft, de Havilland Comet 4Bs, in 1960, and has since followed these with Hawker Siddeley Tridents, including the IC, 2E, and 3B versions, and B.A.C. One-Eleven 500s.

Today, British European Airways, which has been state-owned throughout its history, is being operated as a part of the British Airways Board (q.v.), and is organised into several operating divisions, the main ones being: B.E.A. Mainline; British Air Services, which is owned two-thirds by B.E.A. and one-third by private investors, and is the holding company for Cambrian Airways and Northeast Airlines; B.E.A. Super One-Eleven Division, which is mainly concerned with operating the airline's German domestic network to and from West Berlin; Scottish Airways Division; Channel Islands Airways Division; B.E.A. Airtours, the all-charter subsidiary formed in 1969; B.E.A. Cargo Division; B.E.A. Helicopters; Sovereign Group Hotels; and B.E.A. Travel Sales Division. The airline is the largest fleet-operator outside of the Soviet Union and the United States, but is being progressively merged with B.O.A.C. (q.v.).

B.E.A. has an extensive domestic and European network, plus German domestic services to and from West Berlin, and services to North Africa and the Middle East. The fleet includes the Hawker Siddeley Trident 1C, 2E and 3B, the B.A.C. One-Eleven, the Vickers Vanguard and Merchantman, the Vickers Viscount, the Boeing 707 (B.E.A. Airtours), the Sikorsky S-61N and Bell JetRanger (B.E.A. Helicopters), and Short Skyliner aircraft. Lockheed L-1011 TriStar airbuses are on order for 1974 delivery.

Beagle 206: the first flight of the Beagle 206 took place in August 1961, and production of the 206C started in 1965. Prior to this, the Royal Air Force ordered twenty of a military communications version, the Basset, which first flew in 1964. Altogether, some eighty aircraft were built before the company collapsed. A low-wing, twin-engined monoplane, the 206 had accommodation for between five and eight passengers, and power from two 310 hp Continental GIO-470-A piston engines, giving a maximum speed of 220 mph and a range of up to 1,600 miles.

Beagle Aircraft: Beagle Aircraft was formed in 1962 from an amalgamation of Auster, F. Miles, and British Executive and General Aviation, to produce the Auster range of light aircraft – all of which were high-wing, single-engined monoplanes with three or four seats – and the Beagle 206 (q.v.) light twin-engined aircraft. It was also hoped to develop the lighter 246 twin, and its single-engined counterpart the 123, but only the 206 entered production. Nevertheless, a light single-engined private-owner and training aircraft called the Pup appeared and enjoyed some success, while a military trainer version, the Bulldog, seemed also to have the makings of a success. For much of its existence Beagle was owned by Pressed Steel, and after this concern was acquired by the British Motor Corporation, Beagle was sold to the British Government, going bankrupt due to a shortage of working capital in 1970. The only design to survive the company's collapse was the Bulldog, which was taken over by Scottish Aviation (q.v.).

'Bear', Tupolev Tu-20: see Tu-20 'Bear', Tupolev.

Bearcat, Grumman: the Grumman F8F Bearcat single-seat fighter was the fastest piston-engined aircraft to enter U.S. service when the first deliveries were made to the United States Navy towards the end of World War II. A single 2,500 hp Pratt and Whitney radial engine gave a maximum speed in excess of 400 mph.

Beaufighter/Beaufort, Bristol: the Bristol Beaufighter and Beaufort were both derived from the Bristol Type 156, and entered R.A.F. service in 1940. These two aircraft were part of a small family of twin-engined mid-wing monoplanes developed by Bristol, the others being the Blenheim (q.v.) and its Brigand development. The Beaufort was employed by R.A.F. Coastal Command throughout World War II on maritime-reconnaissance duties and as a torpedo bomber, while the Beaufighter served as a long-range escort fighter, a night fighter, and a ground attack aircraft, frequently operating with warloads of anti-tank rockets from 1942 onwards. The Beaufighter was employed in Europe and North Africa mainly, and was eventually superseded by the de Havilland D.H.98 (q.v.) Mosquito on night fighter duties after 1943. Both the Beaufighter and the Beaufort used two 1,065 hp Bristol Taurus radials.

Beaver, de Havilland Canada DHC-2: see DHC-2 Beaver, de Havilland Canada.

Beech 18: the Beech 18 first flew in 1937, and remained in production for more than twenty years, over 7,000 eventually being built. Of these more than 5,000 were war-time-produced C-45 Expeditor light transport and communications aircraft or T-7 and T-11 Kansan navigational trainers. Accommodation for between five and eight passengers is available, plus a crew of two, while two 450 hp Pratt and Whitney R-985-AN-3 piston-engines give a maximum speed of some 230 mph and a range of up to 1,500 miles. Many Beech 18s remain in service.

Beechcraft: the origins of the Beech Aircraft Corporation date from 1932, when Walter Beech, who had previously been president of Travel Air, formed the company which carried his name. During the years which followed, a range of light aircraft, including the single-engined D17 five-seat cabin biplane, was established, but the company's first real success was the Beech 18 (q.v.) of the late 1930s, which remained in production for more than twenty years and remains in civil and military service to the present time.

Post-war successes included the Beech 35 Bonanza, which first flew in 1945 and introduced the now famous butterfly or 'V' tailplane which is frequently associated with Beechcraft designs, although most of the company's output uses conventional tailplanes. A twin-engined version, the Beech 50 Twin Bonanza, soon appeared, followed by the related Travel Air, with a conventional tail. The 1950s saw the T-34 Mentor single-engined tandem-seat military trainer and the Beech 55 Baron twin-engined business aircraft enter production.

Although one of the 'big three' light aircraft manufacturers in the world, Beechcraft products tend to be at the heavier end of the light aircraft market, with much emphasis on the business and executive user of aircraft rather than the private owner or club flier; and such aircraft include the successful Duke, Debonair, Queen Air, and King Air. The only really light aircraft in the range are in the Musketeer (q.v.) series, the earliest of which first flew in 1961, and their success has not been enough to change the traditional Beechcraft image. Even the Bonanza and Twin Bonanza have grown from their original four seats to being six-seat aircraft. Other Beechcraft activities include sub-contract work on major aircraft programmes, such as the Lockheed C-141 (q.v.) Starlifter and the McDonnell Douglas F-4 (q.v.) Phantom II, and on Bell helicopters. A light training missile, the Beechcraft AQM-37A, simulates attack by enemy air-to-air and surface-to-air missiles.

Beech-Hawker: the marketing name in the United States for Hawker Siddeley's range of business and executive jets (*see* BH-200, BH-600 and HS 125), which are sold as an extension of the Beechcraft range of light and business aircraft. A similar agreement exists between the McDonnell Douglas concern and Piaggio of Italy covering the PD-807 (q.v.) executive jet. In each case the aircraft is built by the European concern, but supplied and finished to meet customers' requirements by the American firm, which is spared the cost of developing an executive jet of its own.

Belfast, Short: the Short Belfast heavy transport aircraft was originally based on the Bristol Britannia design, but after extensive alterations while still on the drawing-board, only part of the wing structure and the tailplane of the Britannia remain. It was also intended that the Belfast should be a transport for the Blue Streak (q.v.) surface-to-surface missile, with nose and tail doors fitted to the aircraft, but cancellation of the missile order resulted in the aircraft's being built with tail doors only. First flight of the Belfast was in January 1964, with deliveries to the R.A.F. of the only ten aircraft produced starting in January 1966. The Belfast is now supposed to carry the heavy equipment for troops flown in by the R.A.F.'s VC.10s, and both these types of aircraft are operated by one composite squadron. However, as a troop-carrying aircraft in its own right the Belfast can accommodate up to 250 troops on two decks, the upper deck being removable. Four 5,735 shp Rolls-Royce Tyne R.TY.12 turboprops give a maximum cruising speed of 360 mph and a range of up to 5,300 miles, while the maximum load which can be carried is 78,000 lb. Plans for jet-engined and airbus versions of the aircraft were abandoned.

Bell 47 Sioux: the Bell 47 Sioux was the first commercial helicopter to be awarded a Type Approval Certificate – by the Federal Aviation Authority in 1946 – but most of the 47s produced since deliveries started in 1947 have gone to military users throughout the Western world, with American production supplemented by licence production by Westland (q.v.), Agusta (q.v.) and Kawasaki (q.v.). First flight of the helicopter was in 1945. A large number of versions have been built, with the U.S. Army's OH-13G and the U.S.N.'s TH-13M trainer using the same 200 hp Franklin 6V4-200-C32 engine as the civil 47G, while the 47G-2 with a 240 hp Lycoming TV0-435-BIA engine has a military counterpart in the 0H-13H. The 47J-3 is a civil development by Agusta with a five-seat cabin replacing the three-seat bubble of the standard machine, and with a covered tail boom.

More recent versions include the OH-13S, with a 250 hp Lycoming TVO-435-25 engine, and the TH-13T training version of the OH-13S or 47G-3. Performance varies, but on more recent versions there has been a 105 mph maximum speed and a 300 mile range.

Bell 204/205 Iroquois: the first flight of a Bell 204 helicopter took place in October 1956, after Bell had won a U.S. Army design competition the previous year. Deliveries of the UH-1A Iroquois to the U.S. Army started in June 1959, and a succession of different versions have followed with detailed improvements, one of the latest versions being the UH-1M. The HH-1K is a U.S. Navy rescue version, and the TH 1L a U.S. Navy trainer. The AH-1G (q.v.) Huey Cobra is a combat or attack helicopter development retaining the powerplant and rotor blades of the Iroquois, but having a narrow fuselage with a tandem seating cockpit, and stub wings for stability and as weapons attachment points. A single 1,100 shp Lycoming T53-L-11 shaft turbine gives the Iroquois a maximum speed of 140 mph with a maximum range of just over 300 miles, while the fuselage can accommodate up to eight passengers, or three stretchers, in addition to a crew of two.

A civil version is the Two-Twelve, with a United Aircraft PT6T-3 turbo-twin, and accommodation for up to fourteen passengers.

The Iroquois is in widespread service, the only notable exceptions to its use being the British and French armed forces in the West. There is licence production in Germany by Dornier (q.v.), in Italy by Agusta (q.v.) and in Japan by Fuji (q.v.). Agusta also undertakes Two-Twelve production.

Bell 206 JetRanger/OH-58 Kiowa: first flown during 1966, the Bell 206 JetRanger is one of the most attractive helicopters and one of the few such designs to be introduced to the civil market first, with a military application following, rather than vice versa. Production by Bell is supplemented by licence production by Agusta in Italy. The U.S. Army introduced its first Bell OH-58 Kiowa observation

helicopters during the early 1970s. A single 270 shp Allison 250-C18 turbine gives the six-seat helicopter a maximum speed of 150 mph and a range of up to 500 miles.

Bell 209 Huey Cobra: *see* AH-1G Huey Cobra, Bell.

Bell Aerospace Corporation: the Bell Aerospace Corporation has been a subsidiary of Textron Inc. since 1960, and is itself the holding company for three subsidiaries, Hydraulic Research and Manufacturing, Bell Aerosystems, and Bell Helicopter Company; the last named is the largest. The original Bell Aviation Corporation was formed in 1935 by Lawrence Bell. During World War II the Corporation produced the Bell P-59 Airacobra fighter, which was one of the first piston-engined fighters to have a tricycle undercarriage, and developed the Bell Airacomet, which was the United States' first jet aircraft, and was flown for the first time in 1944, although never entering U.S.A.F. service. After the war a number of research aircraft, including the X-1 and X-2, and the V.T.O.L. X-22, were produced. A Bell XS-1, air-launched from a Boeing B-29 Superfortress, was the first supersonic rocket-propelled aircraft. Another Bell project was the X-5, a variable-geometry and variable-camber research aircraft.

Helicopter production first started in 1941. The Bell 47 (q.v.) Sioux, which first flew in 1945, became the first helicopter to be awarded a Federal Aviation Authority Type Approval Certificate in 1946, and since then the 47 has remained in production for more than twenty years, with licence production in Europe by Westland and Agusta, and in Japan by Kawasaki, mainly for military users. Other helicopters, which have shared what appears to be an almost automatic success, include the Bell 204/205 (q.v.) Iroquois and its civil relation the Two-Twelve, the Bell 206 (q.v.) Kiowa or JetRanger, and the AH-1G (q.v.) Huey Cobra, which was the first of the attack helicopters developed as a result of American experience during the Vietnam War. Most of Bell's helicopters have been near the lower end of the size scale.

Bell Aerosystems undertakes air cushion vehicle manufacture, amongst other activities. This work started during the mid-1960s, with a licence from the British Hovercraft Corporation (q.v.) to produce the SR-N5 (q.v.) as the Bell SK-5 for the United States Army. More recently, Bell has been producing experimental designs for evaluation by the United States armed forces, including the SES-100B surface effect ship and the AALC JEFF(B) assault hovercraft. A Canadian subsidiary has produced the Bell 7380 Voyageur, a twin Pratt and Whitney ST-6 1,300 hp turboprop-powered A.C.V. capable of carrying a 25 ton payload, with a maximum speed of 50 mph. A further development, the 7480 Voyageur, uses United Aircraft of Canada ST-6 turboprops.

Bellanca: one of the smaller American light aircraft manufacturers, Bellanca started production in 1956 with the Cruisemaster, following this with the 260 and the 260A, before introducing the Viking low-wing four-seat monoplane, which is the company's most successful design so far and is in full production.

Beriev: a Soviet flying-boat design bureau whose products known to the West have consisted entirely of maritime-reconnaissance aircraft, including the B-6 (q.v.) 'Madge' and its Be-8 development, the Be-10 (q.v.) 'Mallow' jet-powered flying-boat, and Be-12 (q.v.) 'Mail' turboprop flying-boat. There do not appear to have been any very recent designs, and these aircraft seem only to have appeared in limited numbers.

Bf. 109, Messerschmitt: the Messerschmitt Bf. 109 (or Me. 109) first entered Luftwaffe service in 1937, replacing Heinkel He. 51s of the Legion Condor which were opposing Russian fighters in the Spanish Civil War. The aircraft also played its part in the massive displays of air power which preceded the German occupation of Austria in 1938 and Czechoslovakia in 1939, and provided fighter cover for the invasions of Norway, Denmark, the Low Countries, and France in 1940. It was the Supermarine Spitfire's opponent in the Battle of Britain, and although slightly faster than the British aircraft, and possessing the valuable feature of a cannon firing through the propeller boss, giving very accurate fire, it proved to be the inferior aircraft in combat, due to its being slightly less manoeuvrable, lacking any armour protection around the cockpit, and having a slight tendency to weakness in the tail section. Designed by Professor Wilhelm Messerschmitt (q.v.), the Bf. 109 remained in production throughout World War II, and was used as the manned part of the Mistletoe composite aircraft of 1944 and 1945, in which an explosive-filled Junkers Ju. 88 bomber was flown with a Bf. 109 on top of its fuselage, the two aircraft separating after the Ju. 88 was aimed at its target.

A 1,050 or 1,100 hp Daimler Benz piston engine gave the single-seat fighter a maximum speed in excess of 375 mph. Even better performance came from post-war versions produced in Spain by C.A.S.A., which used a 1,400 hp Rolls-Royce Merlin engine, of the type designed for the Spitfire!

BH-7 Wellington, British Hovercraft Corporation: A fully-amphibious air cushion vehicle, the BH-7 was the first wholly British Hovercraft Corporation (q.v.) design to be developed. It entered production for the British Inter-Service Hovercraft Trials Unit and the Imperial Iranian Navy in 1970 and 1971, the former operator using the craft purely for evaluation and research, while the Iranians employ armed versions on anti-piracy and anti-smuggling duties in the Persian Gulf. The craft is capable of carrying a 14 ton payload and has a maximum speed of 60 knots, while the range is some 400 miles at a 35 knot cruising speed. Powerplant is a single Rolls-Royce Marine Proteus 15M 541 of 4,250 shp, driving an air propeller. Civil versions are available should a customer appear, and these could carry about 100 passengers, or freight.

BH-8, British Hovercraft Corporation: a planned development intended to bridge the gap between the BH-7 (q.v.) Wellington and the SR-N4 (q.v.) Mountbatten air cushion vehicles, but not yet in production or under active development; its future will depend on a customer becoming available. Powerplants will probably be two Rolls-Royce Marine Proteus driving air propellers, and the payload would be either about 30 tons of freight or 250 passengers.

BH-125, Beech-Hawker: more usually known as the de Havilland D.H. 125 or the Hawker Siddeley HS 125 (q.v.).

BH-200, Beech-Hawker: a smaller development of the HS 125-400 (*see* HS 125) business jet, which also uses two Rolls-Royce Viper turbojets and has accommodation for six passengers. Development and production have not been actively pursued pending an up-turn in the market for this type of aircraft, which is largely, although not entirely, centred on the United States. The aircraft will probably emerge during the middle part of the 1970s.

BH-600, Beech-Hawker: a larger development of the HS 125-400 (*see* HS 125) business jet, using Rolls-Royce Viper turbojets mounted in the tail, with accommodation for up to twelve passengers, and a maximum cruising speed of about 520 mph. Production started in 1972, following the extensive testing in the United States which took place during 1971. The aircraft is fitted out to customers' requirements in the United States by Beechcraft (q.v.), who are Hawker Siddeley's agents and market the aircraft as an extension of their own range of light aircraft. Production is by Hawker Siddeley Aviation (q.v.) in the United Kingdom.

B.H.C.: *see* British Hovercraft Corporation.

Bienvenu: a French professor who, with his colleague Launoy, produced in 1784 a twin-rotor helicopter model which formed the basis of a developed model by Sir George Cayley (q.v.) in 1796; and from this came the basis of the modern helicopter.

bilateral air traffic agreement: *see* air traffic agreement.

biplane: the term biplane describes an air-craft with two wings, i.e. an upper and a lower wing, usually below and above the fuselage, although on some aircraft, such as the Armstrong-Whitworth Argosy (q.v.), the Handley Page H.P.42 (q.v.) and the Supermarine Walrus (q.v.), both wings were above the fuselage. The theory was that two wings were better than one! A biplane can provide high lift, but the drag at speed is excessive.

The only biplanes in production today are the Antonov An-2 (q.v.) and its An-3 (q.v.) development, although during the early years of flying the biplane was the conventional aircraft, and monoplanes, such as those from Antionette and Blériot, the oddities. Strangely, most of the designs pursued by hopeful aviators, or would-be aviators, during the nineteenth century were monoplanes, and one of the first biplane designs was that by Sir Hiram Maxim (q.v.), which although successful on trials never proceeded beyond the model stage. Other supporters of the biplane, for gliding as well as powered flight, included Chanute (q.v.), Ferber (q.v.), Pilcher (q.v.), and the Wright (q.v.) brothers. All of the Wright Flyers (q.v.) were biplanes.

It was not until the 1920s that the monoplane began to edge the biplane to the side, and most of the American speed-planes (*see* airliner) of the late 1920s were monoplanes, as were the Ford Trimotor and Fokker F.VII. The Royal Air Force's last biplane fighters, the Gloster Gladiators, however, saw action against the Axis Powers in Greece and Malta during World War II. The de Havilland Dragon series of light transport aircraft entered production during the mid-1930s, and some types remained in production until after World War II. British Tiger Moth and Belgian Stampe biplanes are still in service today with aerobatics enthusiasts.

Bird Dog, Cessna 0-1: *see* 0-1 Bird Dog, Cessna.

'Bison', Myasischev Mya-4: *see* Mya-4 'Bison', Myasischev.

Blackburn and General Aircraft: the Blackburn concern produced a number of early aircraft designs prior to World

Black Knight

War I, including some Antionette-pattern monoplanes and a biplane, before proceeding to produce the Blackburn Baby float-biplane for the Royal Naval Air Service during the war, and later following this with the Dart, a 450 hp Napier Lion-powered biplane torpedo bomber. These two aircraft marked the start of a long connection with the Royal Navy, with the accent being on carrier-borne aircraft, although there were a few non-naval aircraft built over the years.

Between the wars, Blackburn produced a number of widely differing designs, including the Iris, a three-engined biplane flying-boat which saw civil and Royal Air Force use, and the Lincock, a 215 hp radial-engined biplane fighter of the late 1920s. Blackbirn Ripon biplane torpedo bombers entered Fleet Air Arm service during the late 1920s, remaining until 1934. Another Fleet Air Arm plane, a monoplane fighter and dive-bomber, the Skua, first flew in 1937, when two hundred were ordered for the Fleet Air Arm. Although out-performed by more modern designs after the outbreak of World War II in 1939, nevertheless it had the distinction of being the first British aircraft to down a German aircraft; and on another occasion, in 1940, Skuas successfully dive-bombed the cruiser *Königsberg*. A development of, and successor to, the Skua was the Roc, which although offering little better performance, had a rear-facing power-operated gun turret behind the pilot.

The Firebrand, a carrier-borne torpedo-bomber with a good performance and impressive appearance, entered Fleet Air Arm service in 1945, too late for active service in World War II; but it remained until finally replaced by Westland Wyverns in 1953. The Beverley, which was in production for the R.A.F. during the mid-1950s, was something of an exception for the company, being a high-wing, four 2,850 hp Bristol Centaurus piston-engined heavy transport aircraft. The Buccaneer jet bomber was put into production for the Royal Navy during the early 1960s, by which time the company had become a part of Hawker Siddeley Aviation (q.v.). Originally a carrier-borne aircraft, the Buccaneer is now also in R.A.F. and South African Air Force service.

Black Knight: a British rocket developed and manufactured by Westland Aircraft, which has never been used for military purposes and is now most often used in small-scale space programmes and upper atmosphere research.

'Blechesel', Junkers J.1: *see* J.1. 'Blechesel' Junkers.

Blenheim, Bristol: the Bristol Blenheim series of light bombers was derived from the Bristol 142 of the mid-1930s, and the 142M military development. These were the first really light monoplane bombers designed and developed by a British company, and used two 935 hp radial engines in the Mk.I version to give a maximum cruising speed of 240 mph and a range of 1,000 miles. Later versions used 1,065 hp Bristol Taurus radials, which gave a maximum speed of nearly 300 mph. R.A.F. Bomber Command had six squadrons of Blenheims at the outbreak of World War II in September 1939, but the aircraft remained in service and production for most of the war, with later variants, such as the Mk.IV, having a heavier defensive armament and often undertaking fighter-bomber and long-range escort fighter duties. A successor aircraft which saw relative little action was the Bristol Brigand, developed from the Blenheim, while the Armstrong-Whitworth Albermarle glider-tug was also based on a planned Bristol development of the Blenheim.

Blériot, Louis (1872-1936): although best known as the man who made the first successful aeroplane crossing of the English Channel, from France to England on 25 July 1909, Louis Blériot was an aircraft designer and manufacturer in his own right, and amongst his pioneering efforts can also be counted the second cross-country flight in the world, on 31 October 1908, from Toury to Artenay and back. In each of his pioneering flights Blériot used one of his own distinctive monoplane designs, an XI with a 25 hp Anzani engine for the cross-Channel flight and XIII for the cross-country flight. Blériot's first successful aeroplane was the V, which made some hops, as opposed to flights, on 5 April 1907; by 25 July his VI was able to fly 150 metres, and in September, 184 metres. As with most of his designs, these two aircraft used Antionette (q.v.) engines. The 50 hp Antionette-powered VII is generally considered to have been the first modern monoplane.

Blériot also founded flying schools at Paris, in France, and Hendon, in England. Pilots from the Blériot and Graham-White flying schools made the first airmail flights in England from 9 September 1911, with the first flight using a Blériot machine, although it must be said that these were novelty flights rather than regular or proving public service operations. The first non-stop flight between London and Paris, on 12 April 1911, used a Blériot monoplane.

Blériot was a man of considerable integrity – his products were the first to be grounded when a weakness was detected in the wing of a type in what was then extensive French military service – after he himself had drawn attention to the fault and reported on the shortcoming. Matters were soon rectified.

'Blinder', Tupolev Tu-22: *see* Tu-22 'Blinder', Tupolev.

Blitzkrieg: Blitzkrieg, or Blitz, was the German expression for the concept of heavy night bombing of a target over a period of time, and it was first put into effect during the first part of World War II. Although it is sometimes considered that the Blitz started with the German attack on Norway in April 1940, the generally understood term relates to the intensive bombing of British cities between September 1940 and May 1941, where some 1,300 bombers of Luftflotten 2, 3, and 5 were used. The results of the Blitz on British cities were a serious disappointment to the Germans, and a source of considerable relief to the British, who had been told to expect crippling effects on industry and communications, coupled with a disastrous loss of life. It would be wrong to underestimate the suffering which resulted from the action, but it fell far short of the worst predictions, and British

industry was able to recover. It is possible that failings in the Germans' plan were the lack of a heavy bomber – the Junkers Ju. 88s and Heinkel He. 111s were basically light bombers – and the inability to concentrate the bulk of their forces over any one target at once. In any case only some 600 to 800 aircraft were airborne on any one night. The success of the later R.A.F. offensive against German cities can be credited to the heavy bomber and to massive bomber concentrations.

Bloch, Marcel: today better known as Marcel Dassault (q.v.), the name which he adopted after World War II, Marcel Bloch headed the French aircraft manufacturing firm which bore his name and was an important force in French aviation at that time, although not as important internationally as the present-day Avions Marcel Dassault. Bloch's first military design was the M.B.200, a twin-engined, high-wing bomber monoplane which saw service with the Armée de l'Air from 1934 to 1939, and the M.B.210 development for the French Navy. One of the only two modern French fighters available at the outset of World War II was the M.B.151, a low-wing monoplane with a single 1,080 hp Gnome Rhône radial engine; it was also used by the Greek and Rumanian air forces after production started in 1938. Other products of the firm included small general-purpose aircraft.

Blohm und Voss: the rise to power of Adolf Hitler and the rapid expansion of Germany's armed forces during the second half of the 1930s was accompanied by financial incentives for non-aviation firms to enter the industry, and the heavy-engineering firm of Blohm und Voss was one of these, specialising in large flying-boats. Two of the most famous designs produced by the company were the Bv. 138, a twin-boom, triple-engined, anti-shipping flying-boat used extensively in the Baltic from 1940 onwards, and the giant Bv. 222 transport flying-boat, which used six engines and had a range of almost 4,000 miles. Since the war the company has merged its identity into Messerschmitt-Bölkow-Blohm (q.v.).

Bloodhound, B.A.C.: originating as a Bristol Aeroplane (q.v.) design, pre-merger and now out of production, the Bloodhound Mk.2 surface-to-air guided missile is still operational with the Royal Air Force, and in Singapore, Sweden, and Switzerland. Developed from the Mk.1, which was also used by the R.A.F., but used pulse radar instead of the Mk.2's continuous-wave radar, the Mk.2 retains the ram-jet propulsion and four solid fuel booster rockets of the Mk.1. The Mk.2 Bloodhound has a range of up to 43 miles, and has destroyed targets flying from below 1,000 feet to above 40,000 feet. A proximity fuse is fitted.

Blowpipe: a lightweight surface-to-air missile, the Short Blowpipe is man-portable and can be used with the launch tube mounted on the operator's shoulder. A solid-propellant single-stage rocket, the missile has a diameter of only three inches and is fifty-five inches in length. Wire guidance is used, and the weapon is intended for localised defence over short ranges, although details of the performance have not been released.

Blue Steel: a few Hawker Siddeley Dynamics Blue Steel stand-off bombs or air-to-surface guided missiles remain in service, but only with a small number of Hawker Siddeley Vulcan 'V' bomber squadrons of the R.A.F. A liquid-fuel rocket with a range of up to 200 miles when released at high altitude, the Blue Steel was converted to low-level operations along with the R.A.F.'s bombers when this change in tactics was implemented, but at some cost in the effective range. A nuclear warhead can be fitted.

Blue Streak: originally intended as a British I.C.B.M., Blue Streak was cancelled for military use during the late 1950s when it became apparent that land-based missile sites were vulnerable to enemy attack. Since that time the rocket has earned itself a reputation for reliability as a first-stage launcher for smaller Earth satellites in the E.L.D.O. (European Launcher Development Organisation), but unfortunately this performance has not been matched by the European second-and third-stage rockets which have marred E.L.D.O. performance. Britain's withdrawal from E.L.D.O. did not seriously effect Blue Streak, due to the rocket's excellent reputation.

B.M.E.W.S.: the United States ballistic missile early warning system, which includes long-range radar for identification of missiles, computers for identification of their intended targets, and a communications system for contact with the launching sites.

B.M.W.: a name more usually associated with the production of motor-cycles and high performance motor-cars, B.M.W.'s connection with aero-engines nevertheless dates from World War I, when the Fokker D.VII used the company's engines, as did the Focke-Wulf Fw.190 of World War II, amongst other aircraft. The company can also claim to have been the world's first turbojet engine manufacturer. Since the war the company's aero-engine interests have been merged with those of M.A.N. (q.v.).

BN-2 Islander, Britten-Norman: design of this low-cost, twin-engined, high-wing utility transport began in 1963, with the first flight of the prototype taking place in June 1965, using two 210 hp Continental IO-360-B engines. Deliveries of the production aircraft, using the more powerful Lycoming 0-540-E4C5 piston engines, began in August 1968, since when some 400 of the ten-seat aircraft have been ordered for delivery to most parts of the world. Production was largely subcontracted to the British Hovercraft Corporation (q.v.) and to Rumania. A maximum speed of 180 mph and a range of 350 miles when fully laden are available. Sales of the aircraft were unaffected by Britten-Norman Aircraft's (q.v.) financial collapse in October 1971. A military development is the Defender, with provision for under-wing stores, while the Trislander is a three-engined stretched version with seats for eighteen; the third engine being mounted in the tail.

B.O.A.C.–British Overseas Airways Corporation: the history of B.O.A.C. dates from 1919 and the formation of the first airline, Aircraft Transport and Travel, although the start of B.O.A.C.'s ancestry is more generally taken from the founding of Imperial Airways in 1924 from an amalgamation of Handley Page Transport, Instone Air Line, Daimler Hire, and British Marine Air Navigation. Another of B.O.A.C.'s direct ancestors was British Airways, formed in 1935 from the merger of Hillman's Airways, Spartan Air Lines, and United Air Lines. Both airlines were nationalised as a result of the Cadman Report of 1938, which recommended that the two airlines should not compete on the same routes – the competition was mainly on the European routes – and should maintain a close working liaison. B.O.A.C. was created in 1939 from the two airlines, and commenced operations in 1940.

Imperial Airways had developed services to the Empire: between Cairo and Basra in 1927, to Karachi in 1929, to Calcutta, Rangoon, and Singapore in 1933, and to Brisbane in 1935. The Singapore to Brisbane section was operated by the then new Australian airline, Qantas Airways (q.v.). Meanwhile, another main route was developed, reaching Mwanza in 1931, and Nairobi and Capetown in 1932. A further route from Khartoum to Kano was also developed. These services had been developed as an extension of the 1,760 miles of mainly cross-Channel services with which Imperial Airways had been started in 1924. During the late 1930s, Imperial introduced Short Empire 'C' class flying-boats, 28 of which it had taken the then completely unprecedented step of ordering straight from the drawing-board for the Empire Air Mail service. In fact it had to re-order additional aircraft before the outbreak of World War II. The Empire flying-boat also joined in B.O.A.C. and Pan American transatlantic commercial flight trials.

During World War II, B.O.A.C. continued to operate the African routes, using flying-boats, while providing a connection at Khartoum with a new horseshoe-shaped route from Cape Town to Australia; and after Japan invaded Malaya, B.O.A.C. operated the service non-stop over the 3,512 miles between Ceylon and Perth. Other war services included a de Havilland Mosquito-operated Scotland–Sweden run, carrying personnel in both directions and returning from Sweden with ball bearings – this route involved flying over enemy-held territory. B.O.A.C. used Liberators on the North Atlantic to return ferry pilots to North America, while Boeing 314 flying-boats operated a North Atlantic V.I.P. service, and Douglas DC-3s operated to Lisbon.

After the war B.E.A. (q.v.) took over the European routes, while British South American Airways operated to South America until B.O.A.C. took over B.S.A.A. and its routes after the airline encountered difficulties with its Avro Tudor aircraft. Handley Page Hermes aircraft were introduced in 1950 to replace B.O.A.C.'s Short Solent flying-boats and Avro Yorks, while Boeing Stratocruisers were also introduced. In 1952 the airline became the first in the world to introduce jet aircraft, using de Havilland Comet Is; but after several disasters, the Certificate of Airworthiness for this aircraft was withdrawn in 1954, leaving B.O.A.C. in severe difficulties, having to cancel its South American services and to search for replacement aircraft.

The airline introduced the first North Atlantic turboprop service in 1957 with Bristol Britannias, following this with the first Atlantic jet service in 1958 with de Havilland Comet 4s. South American services were restarted in 1960, but withdrawn as an economy measure in 1964, when the then British United Airways (see British Caledonian Airways) introduced a service to Brazil, Uruguay, Argentina, and Chile. A North Pole service to Japan was introduced in 1969, and in 1970 a trans-Siberian service to Japan was also introduced. The airline was the first, in 1972, to confirm its order for a supersonic airliner, the B.A.C.-Aerospatiale Concorde.

Currently B.O.A.C., which is a subsidiary of the British Airways Board (q.v.) and is being merged with B.E.A. to form an integrated British Airways, maintains an interest in Air Mauritius, Air Pacific, Cathay Pacific, Gulf Aviation, Gulf Helicopters, Malaysian Airlines and

Boeing 80

Mercury Singapore Airlines, New Hebrides Airways, and T.H.Y. Turkish Airlines through a subsidiary, B.O.A.C. Associated Companies. Another (all-charter) subsidiary is British Overseas Air Charter, which leases aircraft and crews from B.O.A.C. when required. The current network, which includes services to Latin America, North America and the Caribbean, the Middle and Far East, Africa, and Australia, is operated by a fleet of Boeing 747 and 707, and Vickers VC.10 and Super VC.10 aircraft.

Boeing 80: a tri-motor biplane using three 525 hp Pratt and Whitney Hornet radial engines for a maximum speed of 125 mph and a load of up to fifteen passengers, the Boeing 80A first appeared on the Chicago–San Francisco route of United Air Lines in 1928. The aircraft was available with either an open or a closed cockpit, with the latter not always finding favour with aircrew.

Boeing 247: the Boeing 247, a low-wing, all-metal airliner with two 570 hp Pratt and Whitney Wasp radials, first appeared in 1933. Performance included a maximum speed of nearly 200 mph, and accommodation for ten passengers was provided. The aircraft first entered service with United Air Lines. It was one of the first civil aircraft to employ flaps to reduce landing speed without stalling – an innovation made necessary by the continual increase in wing loadings (q.v.). It was also Boeing's first retractable undercarriage airliner.

Boeing 307 Stratoliner: the Boeing 307 Stratoliner was the world's first fully-pressurised airliner, entering service with Pan Am and with Transcontinental and Western Air in 1940, before transfer to the U.S.A.A.F. for war service, which was followed by a spell with the original owners before resale. These aircraft had the long life of more than twenty years apiece. The pressurised cabin had an interior pressure of 8,000 feet while the aircraft was cruising at 20,000 feet.

Boeing 314: the Boeing 314 flying-boat was built for Pan American's North Atlantic services, and in 1939 Pan Am, and

Imperial Airways (using Short Empire (q.v.) flying-boats) operated a trial scheduled service. Although the advent of World War II prevented operations being placed on a permanent footing by the two airlines, B.O.A.C. eventually received three 314s for a North Atlantic and West African service operated during World War II for high-ranking military and diplomatic personnel.

Boeing 377 Stratocruiser: Boeing's first post-World War II commerical airliner, the 377 Stratocruiser, entered service in 1947. A 'double-bubble' double-deck fuselage provided accommodation for between fifty and a hundred passengers, depending on whether this landplane was fitted with a comfortable 'flying-boat' style of seating, or a more densely-seated interior. Four 3,500 hp Pratt and Whitney radial engines gave the aircraft a maximum speed of around 350 mph and a range of up to 4,000 miles. The Stratocruiser usually appeared on services with a high proportion of first-class travellers. Related aircraft included the C-97 Stratofreighter for the U.S.A.F. and its KC-97 tanker version. Although largely retired from civil and military service, a few C-97s and KC-97s remain in U.S. National Air Guard (reserve) service, and with the Israeli Defence Force/Air Force, while Aero Spacelines has modified a small number into a range of 'Guppy' freighters, which can carry loads too large for other civil aircraft.

Boeing 707/720: the Boeing 707 was evolved from the Boeing 367-80, a private-venture jet transport which flew for the first time in July 1954, and was the predecessor aircraft for the U.S.A.F.'s C-135 jet transport and KC-135 jet tanker aircraft. First flight of the production civil aircraft, the Boeing 707-120, took place in December 1957, using four 13,500 lb thrust Pratt and Whitney JT3C-6 jet engines, and although the aircraft was intended purely for American domestic operators, it was put into service on Pan Am's North Atlantic routes in late 1958. A shortened-fuselage version was built for Qantas, although these have since been replaced by standard 707-320 aircraft, the long range version which first flew in January 1959, using four 18,000 lb thrust Pratt and Whitney JT3D-3B turbofans. The 707-420 used four Rolls-Royce Conway 508 turbofans, and this aircraft, which was originally intended for B.O.A.C., also entered service with Air India, El Al, and Lufthansa. The 707-220 was a special version of the -120 for Braniff.

The first flight of the 720 took place towards the end of 1959. This aircraft was conceived as a smaller version of the 707–320, using Pratt and Whitney JT3C-7 turbofans. Both the 707 and 720 series have been produced in all-cargo and quick-change convertible cargo-passenger versions. Maximum cruising speeds are in the region of 550 mph and maximum speeds are about 600 mph. Ranges extend up to 4,000 miles normally, although far longer flights have been made in favourable circumstances, and there are some regular scheduled flights with non-stop lengths of more than 5,000 miles. An all-cargo 707-320C can carry up to 96,000 lb of cargo, while the passenger versions can accommodate between 110 and 220 passengers, depending on version, seat pitch, and class mix.

Boeing 727: deliveries of the Boeing 727 started in late 1963, with the series 100 aircraft using three 14,000 lb thrust Pratt and Whitney JT8D-1 turbofans mounted in the tailplane, giving a maximum speed of 630 mph, a normal range of around 2,000 miles, and a maximum payload of 32,000 lb in the cargo versions. Maximum use of 707 components was made, including much of the fuselage and the nose. A development, the series 200, has since entered service with up-rated engines and accommodation for up to 190 passengers, instead of the 130 of the series 100, in a stretched fuselage.

Boeing 737: the Boeing 737 is the smallest of Boeing's family of jets and uses two of the same type of Pratt and Whitney turbofans as the 727 (q.v), while as many of the other components as possible are common with the 707 or 727, although the engines are mounted under the wings. A series 200 version with stretched fuselage and 14,500 lb thrust Pratt and Whitney JT8D-9

engines is also in production and service. Accommodation varies between 80 and 113 passengers, while the maximum speed is 600 mph and the range is up to 2,000 miles. Cargo and quick-change convertible versions of this aircraft are available, as with all current Boeing production types.

Boeing 747: the Boeing 747, or 'jumbo' jet is the world's largest civil aircraft, and the first of the modern range of wide-bodied jets, as well as being the first production aircraft to use one of the quieter advanced technology engines – although these are not as quiet as the more sophisticated engines which are used on the airbuses. The first flight took place in February 1969, without there being a prototype aircraft. Four Pratt and Whitney JT-9 turbofans give the 747 a range of up to 5,000 miles, while accommodation can be provided for up to 530 passengers, although most airlines operate the aircraft as a 350-seat mixed-class airliner. Plans exist to stretch the short upper deck towards the tail, and also for a shortened airbus, with two engines, but retaining as many 747 components as possible. Maximum speed is 650 mph, and all-cargo and quick-change versions are available and in service. The latest version is a short-haul aircraft, the 747SR, with a strengthened fuselage for short stage-lengths.

Boeing 767: a design proposal for a three-engined airliner using the N.A.S.A. super-critical wing concept for flight at the speed of sound. The supercritical wing concept was developed by N.A.S.A. as the optimum design for high speed subsonic and low speed supersonic transport. No production decisions have been taken.

Boeing 2707: an ill-fated Mach 2·7 airliner project, the Boeing 2707 was selected by the United States Federal Aviation Agency in preference to a Lockheed design, as the result of a design competition. It was intended that the 2707 should carry up to 315 passengers in a fairly wide fuselage, with a range in excess of 4,000 miles, and use four General Electric GE4/75 engines of 60,000 lb thrust each. Variable-geometry wings were included in the design, which was cancelled by

Boeing Aircraft

Congress after work had started, because of heavily increased costs and design delays, owing in part to the fact that such a large aircraft was making a huge step forward. The B.A.C.-Aerospatiale Concorde (q.v.) uses a narrow fuselage and has a relatively small passenger capacity, while the cruising speed is Mach 2·0 to avoid the problems inherent with higher speeds.

Boeing Aircraft: today the world's leading manufacturer of civil airliners, Boeing's history dates back to the early 1920s, when the company was known as Boeing and Westervelt and engaged in airline as well as aircraft manufacturing activity. Over the years Boeing has produced every kind of aircraft with the exception of those for the private flier, although racing versions of some of the company's fighter designs were available.

The company's first aircraft was a float-biplane which used a 200 hp Hall-Scott engine, but this was soon followed by the Boeing 40, a passenger and mailplane design using a 420 hp Pratt and Whitney Wasp radial engine. This aircraft operated the first coast-to-coast night flights in the United States. During the late 1920s, Boeing joined the fashion of building tri-motor airliners with the Boeing 80 (q.v.), which used three 525 hp Pratt and Whitney Hornet radials to carry fifteen passengers at speeds of up to 125 mph. Another Boeing mailplane of the period was the 95, which used one Hornet radial. It was during this period, the late 1920s and early 1930s, that Boeing produced a series of fighter designs for the United States Army Aviation Corps and the United States Navy (*see* Boeing fighters).

Appropriately enough, in the light of subsequent history, Boeing was the manufacturer which made many of the first steps in the evolvement of the modern airliner, starting with the Boeing 247 (q.v.), a ten-passenger, twin 570 hp Pratt and Whitney radial-powered monoplane with a maximum speed of 180 mph, which included a retractable undercarriage and flaps amongst its features. By the end of the 1930s, the four-engined Boeing 307 (q.v.) Stratoliner was entering airline

service; this was the world's first pressurised airliner, although its further development and full exploitation was prevented by World War II.

Another significant Boeing activity before the war had been the design and production of bomber aircraft, starting with the B-9, which used two 575 hp Pratt and Whitney Hornet radials for a top speed of 180 mph, as well as having the distinction of being one of the first all-metal monoplane bombers. A further development was the Y1B-9. By the late 1930s, Boeing was actively testing the 299, which was the direct predecessor of the B-17 (q.v.) Flying Fortress of World War II, and through this of the later B-29 (q.v.) and B-50 Superfortress. Another aircraft with a significant wartime role was the Boeing 314 (q.v.) flying-boat, which had been used before the war by Pan Am in co-operation with Imperial Airways (which used Short Empire (q.v.) flying-boats) in pioneering transatlantic services, and during World War II was used by B.O.A.C. on transatlantic V.I.P. flights.

After World War II, the firm produced the Boeing 377 (q.v.) Stratocruiser (originally intended as the C-95, a military transport), which saw service with many of the world's major airlines, while also developing the B-47 (q.v.) Stratojet jet bomber for the U.S.A.F. This was followed by the huge B-52 (q.v.) Stratofortress heavy bomber, which remains in U.S.A.F. service. The jet bomber experience was invaluable in developing a long-range military jet transport, the C-135 (q.v.), and the KC-135 tanker, which led to the highly successful Boeing 707 (q.v.) airliner. The Boeing 707 was the second jet airliner to enter service on the North Atlantic in 1958, and has since become the world's leading long-haul airliner. Using as many 707 components as possible, both to cut development and production costs, and spares-holding costs for the airlines, Boeing developed a whole family of jet aircraft from the 707, including the shorter-range Boeing 720 (q.v.), the medium range Boeing 727 (q.v.) trijet, and the twin-jet Boeing 737 (q.v.) for short-haul routes. The 707's successor and the world's largest civil airliner is the Boeing 747 (q.v.) 'jumbo' jet, but the company was unlucky with plans for a supersonic transport, the Boeing 2707 (q.v.), which had to be cancelled due to the very heavy development costs involved. A supercritical wing project is the Boeing 767 (q.v.).

Other Boeing activities include guided missile manufacture, notably of the Bomarc surface-to-air missile, and work for N.A.S.A., which together prompted Boeing to change its name from Boeing Aircraft to the Boeing Company some years back in order to avoid a too-restrictive definition of activity. Since 1960 Boeing has been concerned with helicopter manufacture through Boeing-Vertol (q.v.), and is currently examining the possibilities of short take-off aircraft in partnership with de Havilland Canada and under NASA sponsorship, using a modified DHC-5 (q.v.) Buffalo.

Boeing Fighters: Boeing involvement in fighter production for the United States Army Air Corps and United States Navy started in 1922, with the manufacture of the MB-3A for the U.S.A.A.C., followed in 1923 by Boeing's own design, the 440 hp Curtiss 0-12 powered PW-9 biplane. The PW-9 was extensively modified and appeared in a number of versions throughout the 1920s, including the FB- series of fighters for the U.S.N. and U.S.M.C. A successor to the PW-9, the U.S.A.A.C.'s P-12, used a Pratt and Whitney Wasp radial engine of 500 hp, and remained in production from 1929 to 1932, with the U.S.N.'s version (designated the F4B) frequently being used as a fighter-bomber.

Experiments with high-wing monoplane fighters, the XP-9 and XP-15, followed and resulted in the U.S.A.A.C.'s lowwing, all-metal P-26 of 1933, and the lesssuccessful P-29 of 1935. Throughout the early 1930s the U.S.N. received a number of variants on the FB4 theme, and some experimental monoplanes, although Boeing monoplanes never entered U.S.N. operational service.

Boeing-Vertol: Boeing-Vertol originated as the Piasecki Helicopter Corporation in 1946, and a range of twin-rotor helicopters have been developed and put into production since that date. The 'Vertol' title was adopted in 1956, and the company became Boeing's Vertol Division in 1960 after its acquisition by Boeing in that year. A number of successful designs have included the CH-46D Sea Knight (Bv.107 (q.v.)) for the U.S.M.C. and the CH-47C Chinook (Bv.114 (q.v.)) for the United States Army.

Boelcke, Captain Oswald: a German Military Aviation Service squadron commander during World War I, Oswald Boelcke realised the value of squadron team work, and in 1916 formed the first of the famous German fighter circuses. Boelcke also believed in training his pilots in fighter techniques before taking them into combat, and undoubtedly his most illustrious pupil was Baron Manfred von Richthofen (q.v.). After gaining a total score of forty Allied planes, Boelcke was killed in a collision with one of his pupils on 28 October 1916.

Bölkow: see Messerschmitt-Bölkow-Blohm.

Bolschoi, Sikorsky: the Sikorsky Bolschoi first flew in 1913, and has the distinction of being the first large multi-engined aircraft. A biplane, the Bolschoi used four engines, mounted back to back in two pairs on the lower wing, and even at that early date had a cabin. The aircraft was a major step for Sikorsky on his way to designing his Ilya Mouromets (q.v.).

Bomarc, Boeing: development of the Boeing Bomarc surface-to-air guided missile originally started in 1949, with tests taking place in 1952, but there still exists in service with the U.S.A.F. and Canadian Armed Forces an improved version. Propulsive units are twin ramjets, with solid fuel booster rockets for takeoff. The rocket is fired vertically and is radio-guided on to its target with homing radar for the last stages of interception. Nuclear warheads can be used, but high explosive is more usual. Range is long, at up to 440 miles, while the Bomarc can operate effectively against aircraft flying at up to 100,000 feet.

bomb: the bomb first evolved before World War I, with first hand grenades and then artillery shells fitted with fins, for stability and rudimentary guidance, dropped from German Zeppelin airships over London and British East Coast cities, before the bomber (q.v.) and specially designed bomb appeared. Development was rapid, with 250 lb bombs available by the end of World War I; but with peacetime neglect, development between the two wars was minimal, so that the R.A.F.'s stock of bombs in 1939 contained much that was left over from 1918.

During World War II several major developments took place. Not only did bombs get larger, but their use and means of delivery became more specialised also. Large bombs included the British 12,000 lb Tall Boy and 22,000 lb Grand Slam, dropped from Avro Lancaster (q.v.) bombers, which exploded below ground level to create maximum effect. A bouncing bomb was developed by Sir Barnes Wallis (q.v.) for the British attack on the Ruhr Dams, so that the bomb could strike the dams with the greatest effect. Intensive use was also made of incendiary, or fire bombs, during the war. The United States had the Bat, a stand-off bomb or air-to-surface (q.v.) guided missile, which used radar-guidance and could carry 1,000 lb of explosives; it was first used in 1945 against Japanese warships. The use of stand-off bombs has since become important since they allow an attacking aircraft to release its warload before reaching the heavily-defended area around the target. The United States dropped the first atom bombs (q.v.) on Hiroshima and Nagasaki in August 1945, ending World War II.

Perhaps the most popular size of bomb during the war and since has been the 1,000 lb bomb. However, not everything dropped from an aircraft can be counted as a bomb in the accepted sense. Many aircraft dropped torpedoes or mines, fired rockets, or dropped napalm tanks, the latter being the modern low-cost, but none the less effective, equivalent of the incendiary bomb used so effectively during World War II.

The bomb dropped from the manned aircraft is still a very important part of modern warfare, in spite of the developments in surface-to-surface (q.v.) missiles which originated with Dr Wernher von Braun's V-1 and V-2 rockets of World War II. Although originally developed for high-flying bomber aircraft, air-to-surface missiles such as Blue Steel have been modified for modern low-level bombing missions, while a new generation of such weapons, including Martel (q.v.) and Bullpup (q.v.) are designed for this role, even though the demands on range are considerable.

bomber: although German Zeppelin airships were used early in World War I to drop bombs (q.v.) on British cities, until these craft proved vulnerable to machine gun fire; and hand grenades and artillery shells were dropped overboard by observers in reconnaissance aircraft, the bomber as a distinctive aircraft type took longer to evolve than did the fighter, which was closely related to many reconnaissance aeroplanes. No doubt this was due in part to the rarity of large multi-engined aircraft, of which the only examples were Sikorsky's designs, such as his Ilya Mourometz of 1916; but there were some single-engined bombers in due course, including the Airco D.H.4 and D.H.9, and a bomber version of the Royal Aircraft Factory's F.E.2a fighter appeared with the designation F.E.2b.

Important bomber aircraft of World War I included the series of designs produced by Handley Page, of which the most famous and successful was the 0/400, while the French produced the Breguet 14 and 19, and the Italian Caproni firm a number of designs including the Ca.5, a trimotor biplane with a twin-boom fuselage. The Germans struck back with such aircraft as the Gotha Type 4.

After World War I a number of aircraft appeared which were too late for active service, including the Vickers Vimy (q.v.), while many of the first American warplanes also came into this category, including the Martin MB and the Douglas T2D-1. Peace, and the international financial situation, meant that the rate of progress between the wars, and certainly during the 1920s, was slower than during the preceding decade. Nevertheless the Westland Wapitis replaced the R.A.F.'s D.H.9s, while Hawker Hart and Fairey Fox light bombers, Boulton Paul Sidestrand medium bombers, and Vickers Virginia and Handley Page Hinaidi heavy bombers entered service. In the Soviet Union, Andrei Tupolev designed the giant TB-3 monoplane.

The start of the bomber in its World War II form can probably be credited to to such aircraft of the 1930s as the Boeing B-9 and Martin B-10 (q.v.), both of which were twin-engined monoplanes. As monoplanes began to replace the biplane, the French introduced the Bloch M.B.200 and M.B.210, while the Italians built the Fiat B.R.20 Cicogna and the Savoia-Marchetti S.M.79-11 Sparviero, and the R.A.F. received Armstrong-Whitworth Whitleys (q.v.) and Fairey Battles (q.v.), with Handley Page Hampdens and Vickers Wellingtons (q.v.) starting to arrive in 1938. German bombers were developed as airliners to circumvent the conditions of the Armistice, and such aircraft included the Dornier Do. 17 and Heinkel He. 111. A real glimpse of the future of the heavy bomber came in 1935, with the start of test flying on the Boeing 299, the direct ancestor of the B-17 (q.v.) Flying Fortress.

Not all bombers were landplanes, for there were such distinguished seaplanes as the Heinkel He. 115, and the Italian Z. 506B Airone, although the tendency was for seaplanes and flying-boats to concentrate on anti-shipping and maritime-reconnaissance (q.v.) duties. Nor were all bombers of the same basic type, leaving aside differences of size and range, for there were many fighter-bomber designs, combining speed with a small warload. The Germans were great believers in the dive-bomber (q.v.), such as the Junkers Ju. 87 (q.v.) Stuka; there were many British and American dive-bombers as well, including the Blackburn Skuas of the Fleet Air Arm.

World War II was very much the war of the heavy bomber, such as the British Avro Lancaster (q.v.), Handley Page Halifax (q.v.) and Short Stirling (q.v.), or the American Boeing B-17 (q.v.) Flying Fortress and B-29 (q.v.) Superfortress, or

Consolidated B-24 (q.v.) Liberator. The lack of any real number of heavy bombers probably went heavily against the Axis, whose only heavy bombers were the Piaggio P. 108 four-engined bomber and the Heinkel He. 177, which had four engines mounted in pairs and driving two propellers, but which was far from being successful. Russia had the Petlyakov Pe-2 (q.v.) light bomber and Pe-8 (q.v.) four-engined heavy bomber. There were, of course, many good light bombers, including the British Bristol Blenheim (q.v.) and Beaufort (q.v.) and the de Havilland D.H. 98 (q.v.) Mosquito, the American Douglas A-20 (q.v.) Havoc and North American B-25 (q.v.) Mitchell, and the French Potez 63, while the Germans also had the Heinkel He. 111 and Junkers Ju. 88.

The Messerschmitt Me. 262 (q.v.) jet fighter first entered service in 1944 as a bomber, while the Arado Ar. 234 jet reconnaissance aircraft was also used primarily on bombing duties. Piston-engined bombers were still entering service at the end of the war, and for some time afterwards, however, including the Avro Lincoln and the Convair B-36 (q.v.). Nevertheless, the writing was on the wall for these, with the advent of such aircraft as the English Electric Canberra (q.v.) and the Boeing B-47 (q.v.) Stratojet during the early 1950s. Heavier jet bombers followed during the 1950s, with the Boeing B-52 (q.v.) Stratofortress, the Avro Vulcan (q.v.), the Handley Page Victor (q.v.) and the Vickers Valiant (q.v.), while the French had the Sud Vatour, and Russia had a succession of designs by Tupolev. The 1960s saw supersonic bombers such as the North American B-58 (q.v.) Hustler and the Dassault Mirage IV (q.v.) enter service, while many existing bombers were converted to low-level bombing missions in order to avoid detection by defending radar. The world's first purpose-built low-level strike bomber was the carrier-borne Hawker Siddeley Buccaneer (q.v.), which first entered production during the early 1960s.

That the bomber still has an important role to play is best illustrated by the fact that new aircraft types are currently under development, including the Panavia 200 (q.v.) Panther and the North American Rockwell B-1, both of which are variable-geometry designs, as is a new Tupolev design, NATO code-named 'Backfire'. After many years of stowing the whole warload inside a bomb bay, particularly during World War II, although the 22,000 lb Grand Slam used to protrude considerably, the heavy warloads of modern aircraft made complete internal stowage difficult, and so for much of an aircraft's warload there has been a return to the World War I concept of under-wing racks and wing strongpoints, but with the bombs themselves sometimes having an improved aerodynamic shape.

Bonanza, Beechcraft: the Beechcraft 35 Bonanza first flew in December 1945, and introduced the famous butterfly 'V' tailplane which has distinguished some Beechcraft designs. In common with all other post-war Beechcraft designs, it is a low-wing monoplane. Originating as a four-seat, single-engined aircraft, it has since become a four- to six-seater, and is available with one or two engines (the Twin Bonanza). A 285 hp Continental IO-520-B piston engine is used, and there is also a turbocharged version available. Performance includes a maximum speed of around 200 mph and a range of up to 1,100 miles. The Debonair was a lower-powered variant with a conventional tailplane, and a military trainer development is the T-34 (q.v.) Mentor.

Borman, Colonel Frank, U.S.A.F.: the Commander of the Apollo VIII mission, which was the first manned spacecraft in history to break free of the earth's gravitational pull, Colonel Frank Borman, U.S.A.F., was making his second space trip. His crew on the Apollo VIII flight, which orbited the moon and lasted from 21 to 27 December 1968, included Commander James A. Lovell, U.S.N., who had also accompanied Borman on the Gemini 7 flight which had made 206 orbits of the Earth between 4 and 18 December 1965, making a rendezvous in orbit with Gemini 6.

Boulton-Paul Aircraft: a British firm which was primarily engaged in component manufacture and became a part of the Dowty Group in 1961, Boulton-Paul Aircraft also produced a number of complete designs of its own, and although production runs were small, the designs were generally successful from the technical viewpoint. The Sidestrand twin-engined bomber biplane for the Royal Air Force, for which eighteen were built, first appeared in 1928 and remained in service until replaced by the Overstrand (q.v.) development in 1934. Meanwhile a mail biplane design which was greatly influenced by the Sidestrand, the P.64, first flew in March 1933, and although it crashed during trials, orders were received for the P.71A development. After World War II, delta-wing research aircraft with supersonic performance were built and flown.

Boxkite, Bristol: an early British biplane, based on the Henry Farman biplane, the Boxkite was fairly successful and was built for private and military users.

Br. 1050 Alize, Breguet: the Breguet Br. 1050 Alize carrier-borne, anti-submarine aircraft was developed from the abortive Br. 960 Vultur turboprop strike fighter design which first flew in 1951, work on Alize development commencing in 1954. A modified Vultur flew in 1955, but the first flight of an Alize prototype was in October 1956, with deliveries of the production aircraft to the Aéronavale starting in 1959. A number of aircraft have also been delivered to the Indian Navy. A single Rolls-Royce Dart R.Da.21 turboprop of 2,100 shp powers the aircraft, which has a crew of three, a range of up to 900 miles, and a maximum speed of just below 300 mph.

Br. 1150 Atlantique, Breguet: the Breguet Br. 1150 Atlantique long-range maritime-reconnaissance aircraft was developed starting in 1959 as a result of Breguet's winning a design competition for a successor to NATO's Lockheed Neptune. Although Breguet designed and assembled it, the production effort has been subcontracted amongst the other customer nations, which are Belgium, West Germany, and the Netherlands, as well as

France. The prototype first flew in October 1961, and deliveries to the Aéronavale started in December 1965. Two licence-built (by Hispano Suiza) Rolls-Royce Tyne R.Ty 20 Mk.21 turboprops of 6,105 shp each power the aircraft, giving a maximum speed of 380 mph and a range of 4,200 miles. A crew of twelve is carried.

Brabazon, Bristol: work started on the Bristol Brabazon airliner in 1945, with a first flight in September 1949, after which four years of extensive testing followed before the aircraft was scrapped. The decision to terminate the project was due to the aircraft's being too large for the available market – it would have been capable of carrying upwards of 200 passengers had it entered airline service. Eight 2,500 hp Bristol Centaurus engines mounted in the wing in pairs, side by side, and driving four contra-rotating propellers, powered the aircraft.

Brabazon of Tara, Lord: the first Englishman to fly an aeroplane, and the holder of the then Aero Club of Great Britain's Aviator Certificate No. 1 as J. T. C. Moore-Brabazon. Lord Brabazon of Tara was one of Britain's most esteemed aviation pioneers until his death in 1969. His first flying experiences were in a Voisin biplane in May 1909, but later in that same year he won a £1,000 *Daily Mail* prize for the first Briton to fly a British-built plane (a Short-Wright biplane) over a distance of one mile. Served as chairman of transport aircraft requirements committee, and Bristol Brabazon (q.v.) named after him.

Braniff International–Braniff Airways: dating from 1928, when it was formed by Tom and Paul Braniff as Braniff Air Lines, the original Braniff was soon acquired by the Universal Aviation Corporation, which shed many of its subsidiaries to create American Airways. The present airline dates from 1930, when it was re-formed completely independent of Universal, and commenced operations with a Tulsa-Wichita service. Expansion came during the mid-1930s with more services and, most important, a share in the airmail contracts awarded by the U.S. Post Office.

The present fleet name, Braniff International, was first used in 1948 to mark the airline's being awarded the Houston–Havana–Lima service, although a service to Mexico City and Buenos Aires had already been operated briefly during 1946. Throughout the 1950s further international points in Latin America were added to the network. Before this, expansion in the United States had been assisted by the acquisition in 1952 of Mid-Continent Airlines, a company dating back to 1936. The most significant acquisition, however, was that in 1967, of Pan American Grace Airways, owned jointly by Pan American Airways (q.v.) and the W. R. Grace Company, and dating from 1929. This gave Braniff a much-expanded Latin American network.

Currently, Braniff operates an extensive domestic and Latin American network, with services to Hawaii and military contract flights worldwide. The fleet consists of Boeing 747, 707, 720, and 727 aircraft, and McDonnell Douglas DC-8-62s; these having replaced an earlier fleet of B.A.C. One-Elevens which were built with the requirements of Braniff and British United in mind.

Brantly Helicopter Corporation: the Brantly Helicopter Corporation was formed in the United States during the early 1950s, initially building the B-2 two- or three-seat light helicopter, and following this with the stretched 305, with accommodation for up to five. In 1966 Brantly was acquired by the Lear Jet Corporation (q.v.), but production of Brantly designs has since reverted to a new firm, Brantly Operators Inc.

Braun, Dr Wernher von: one of today's leading space scientists, Dr Wernher von Braun was directly connected with the development of the German A-4 (V-2 (q.v.)) surface-to-surface missile before and during World War II. After the war, von Braun went to the United States to live and work – a better fate than that of some of his wartime colleagues who were pressed into Soviet service – and ultimately to play a leading part in the National Aeronautics and Space Administration's space programme, culminating in the Apollo moon landings.

break-even figure: the production figure at which an aircraft's development and tooling costs can be written off and profits earned, this is largely a figment of the imagination of politicians, stockbrokers, and financial journalists, and the term is avoided by aircraft manufacturers. A problem is that with most aircraft, by the time any break-even aircraft is arrived at, the product will have been developed somewhat and the costs of this work will have pushed the break-even point further away. Also, an aircraft reaching the end of its production life offers less scope for price increases to match inflation if any hope of further production is to be maintained against competition from newer models.

In reality, an aircraft manufacturer hopes to earn some profit from all but the earliest production models, and he is also very interested in the profit on spares, which are generally estimated to cost as much again as the original aircraft during the first ten years of operational life, and from refurbished aircraft when aircraft change owners. The financial performance of an aircraft for its manufacturer cannot therefore be properly assessed until some time after production has ended, and it is possible for the profit on spares and refurbishing to exceed by far profits on manufacture of the aircraft itself.

Breguet: the French firm of Breguet Aviation was active before the outbreak of World War I, with a number of biplane designs, including one of predominantly steel tube manufacture, by Louis Breguet, the company's founder. The Breguet 14 (q.v.) biplane of 1916 was one of the more successful French bombing and reconnaissance aircraft of World War I, as well as being one of the first aircraft to have flaps. A successor to the 14 was the 16, a night bomber which appeared too late for service during the war. The next major Breguet project was the 19 (q.v.), another bomber biplane which entered service with the Armée de l'Air in 1925; in September 1930 a much-modified version made the first non-stop Paris to New York flight.

Not all of Breguet's designs were con-

ventional, and the company was one of many to experiment with helicopters, producing a co-axial contra-rotating rotor design in 1931 which solved many of the problems associated with helicopter control. However there was no further development of this design. Nor was Breguet purely a landplane manufacturer: it produced the Breguet 530 Saigon flying-boat which saw service with Air France throughout the 1930s, and the Short Calcutta was built under licence as the Breguet Br. 521 Bizerte for the French Navy. A low-wing, all-metal monoplane bomber was the twin-engined Breguet Br. 460, although this aircraft, a landplane, was not one of the more notable designs of the period.

After the war Breguet designed the Br. 761 Provence, the Br. 763 Deux Ponts or Universal, and the Br. 765 Sahara series of twin-deck, mid-wing, four-engined transport aircraft for Air France and the Armée de l'Air. The Vultur turboprop strike-fighter was cancelled, but the Br. 1050 (q.v.) Alize was developed in its place, and retained much of the Vultur design. The company's Br. 1150 (q.v.) Atlantique design won a NATO design competition in 1959 for a Lockheed P-2 (q.v.) Neptune replacement, and has since been developed and put into production as a European collaborative effort. Breguet is B.A.C.'s partner in production of the Jaguar (q.v.) strike fighter, and with Dassault and Dornier is developing the Alphajet (q.v.) advanced trainer for the Luftwaffe and Armée de l'Air in the mid-1970s.

Breguet was merged with Avions Marcel Dassault (q.v.) in early 1972.

Breguet 14: the Breguet 14 day bomber first entered French service in 1916. A two-seat biplane with one engine, several versions were ultimately developed for reconnaissance duties as well as for the aircraft's original role. A total of thirty-two 16 lb bombs could be carried.

Breguet 19: a single-engined biplane bomber with two seats, the Breguet 19 entered Armée de l'Air service in 1925 and remained operational into the 1930s. In September 1930 an extensively modified 19 became the first aircraft to fly non-stop across the Atlantic from east to west, flying from Paris to New York.

Breguet 941S: the Breguet 941S short take-off transport aircraft first flew in June 1961, and since then has undergone an extensive test programme and numerous experimental flights, including some with America's Eastern Airlines and American Airlines, but there has been no indication of production being undertaken in the near future. A high-wing monoplane with an unpressurised fuselage, it is powered by four 1,200 shp Turbomeca Turmo IIID turboprops and uses deflected engine slipstream for the shortened take-off run. Up to forty passengers or 22,000 lb of freight can be carried, while the performance extends to a maximum speed of 300 mph and a range of up to 2,000 miles.

'Brewer', Yakovlev Yak-28: *see under* Yak-28 'Firebar', Yakovlev.

'Brisfit', Bristol Fighter: *see* Fighter, Bristol F.2B 'Brisfit'.

Bristol 170 Freighter/Wayfarer: originally intended to be a military freighter, the Bristol 170 high-wing monoplane started its development life during the closing years of World War II, surviving cancellation after the war as a private venture to make its first flight in December 1945. A number of developments of the basic aircraft took place, of which the most common was the stretched-nose Mk.32. All-freight and car-carrying versions, with clam-shell nose doors, were named the Freighter, while all-passenger versions were named the Wayfarer. Strangely, most production models went to military users. Only a few aircraft survive, using two 2,000 hp Bristol Hercules 734 piston engines for a maximum speed of 190 mph and a range of up to 1,730 miles, while a payload of up to 12,500 lb or, in the car-carrying versions, up to twenty passengers and three cars, can be carried. Unusual features include a non-retractable undercarriage and the position of the flight deck on top of the fuselage.

Bristol Aeroplane: one of the oldest British aircraft manufacturers, the Bristol Aeroplane Company started life producing aircraft based upon the Henri Farman (q.v.) biplanes, of which the first was the Boxkite (q.v.). This was followed by a series of biplane and monoplane designs, eventually leading to the Bristol Scout (q.v.) biplane of 1914, and the M.1. monoplane and F.2B Fighter (q.v.), or 'Brisfit', biplane which appeared towards the end of World War I. The 'Brisfit' remained in R.A.F. service until well into the 1920s.

After the war the company continued to be mainly concerned with military aircraft, although there were a number of civil projects which did not, however, get very far. The mid-1920s saw the appearance of the Bristol Bulldog (q.v.) biplane fighter, for the R.A.F. and foreign users and this was followed by another biplane, the Bullpup. A return to monoplane designs came with the Bristol 142, a low-wing, twin-engined design of the mid-1930s which, with the 142M, served as the prototype for the Blenheim (q.v.) light bomber which entered R.A.F. service in 1937. The Blenheim served throughout World War II, and was the first of a small series of related aircraft, including the Beaufort (q.v.) maritime-reconnaissance and torpedo-bomber, and the Beaufighter (q.v.) long-range escort and night fighter. A development of the Blenheim was the Brigand, although this aircraft arrived too late to play a major part in World War II.

An unspectacular but none the less useful and successful design during the immediate post-World War II period was the Bristol 170 (q.v.) Freighter, a twin-engined, high-wing monoplane which was famed for its service as a car ferry across the English Channel, the Mediterranean, and in New Zealand, but which was actually most used on military transport duties. Less well starred was the Brabazon (q.v.), an eight-engined airliner which was eventually abandoned after extensive flight trials because it was too big for the then air transport market. A later airliner, the turboprop Britannia (q.v.), was extremely popular with passengers, and was nicknamed the 'Whispering Giant' by the Americans. Military and

Bristol Siddeley Engines

civil transport developments of the Britannia, and a maritime-reconnaissance version, were produced under licence by Canadair (q.v.).

The company was not purely an airframe manufacturer. In common with many of the early manufacturers it had an engine division, which was later merged with Armstrong Siddeley to make Bristol Siddeley Engines (q.v.). A motor car division produced a low volume production quality car, but this was eventually separated from the aeroplane-manufacturing interests to lead a separate existence. Guided missile development and manufacture resulted in the Bloodhound (q.v.) surface-to-air guided missile which, with the Bristol 188 Mach 3·0 research aircraft, passed to the British Aircraft Corporation in 1960 when Bristol was one of the companies merged to form B.A.C.

Bristol Siddeley Engines: Bristol Siddeley Engines was formed by an amalgamation of Bristol Engines and Armstrong Siddeley Engines in 1959, and later absorbed Blackburn Engines, which had been a subsidiary of Blackburn and General Aircraft, in 1962. Another addition was the aero-engine subsidiary of de Havilland (q.v.) Aircraft. In 1966 the company was acquired by Rolls-Royce (q.v.) at which time its product range included the Olympus, for the Concorde (q.v.) supersonic airliner; the Pegasus, for the Harrier (q.v.) V.T.O.L. fighter; the Viper for jet trainers and executive aircraft; the M45, for the VFW-Fokker 614 (q.v.), and the Gnome, for turbine helicopters and hovercraft. A licence was held for production of the General Electric CF-6 airbus engine, but this was later allowed to lapse. Co-operation existed with the French firm of SNECMA on the Olympus, the M45H, and a range of engines for the Anglo-French helicopter programme's S.A. 330 (q.v.) and S.A. 341 (q.v.) aircraft. The company also led in vectored-thrust development for V.T.O.L., and in marine applications of aero-engines, such as the Proteus.

Britannia, Bristol: the first flight of the Bristol 175 Britannia took place in August 1952, following which B.O.A.C. received fifteen of the Britannia 102 medium-range airliners. The first flight of the long-range and stretched-fuselage 300 series occurred in July 1956, with deliveries to B.O.A.C. starting in 1957, and to other airlines in 1958. The 200 series, with the same length fuselage and range as the 300s, was a military transport for the R.A.F., of which most were built by Short Brothers (q.v.). Four Bristol Siddeley Proteus turboprops powered the aircraft: Proteus 705s of 3,900 shp in the 102s, Proteus 755s of 4,310 shp in the 200s, and Proteus 765s of 4,445 shp in the 300s. Maximum seating varied depending on the version of aircraft, and the class mix and pitch specified, but was in the region of 80 to 140. The aircraft had a maximum speed of 400 mph, and could carry 57,400 lb of freight up to 4,300 miles before refuelling.

Licence-built developments of the Britannia, which was nicknamed the 'Whispering Giant' by the Americans because of its low noise level, were by Canadair, with the CL-44 (q.v.), CP-107 (q.v.) Argus, and CC-106 (q.v.) Yukon, and Short Brothers, with the Belfast (q.v.). Most of these developments used Rolls-Royce Tyne turboprops.

British Aircraft Corporation: the British Aircraft Corporation was the result of the government-inspired mergers which took place in the British aircraft industry in 1960; the companies which were amalgamated to form the British Aircraft Corporation included Bristol Aeroplane (q.v.), English Electric (q.v.), Hunting Percival (see Percival Aircraft), and Vickers (q.v.), the latter including its subsidiary Supermarine (q.v.). Ownership of B.A.C. (which is divided into a number of manufacturing divisions for airframes and guided weapons) was, until 1972, with Britain's General Electric (not to be confused with the American concern) and Vickers, with 40 per cent each, and Rolls-Royce, with 20 per cent. General Electric's involvement was due to the acquisition of English Electric, while Rolls-Royce acquired the Bristol Siddeley interest when that company was acquired. Currently, ownership is divided equally between General Electric and Vickers.

Major production work includes the Concorde (q.v.) supersonic airliner, in conjunction with Aerospatiale (q.v.); the Panavia 200 (q.v.) Panther, with Messerschmitt-Bölkow-Blohm (q.v.) and Fiat (see Aeritalia); the Jaguar (q.v.) strikefighter with Breguet (q.v.); the One-Eleven (q.v.) series of airliners; the B.A.C. 145/167 (q.v.) jet trainer and tactical strike aircraft; and product support for the VC.10 (q.v.), Viscount (q.v.), and Britannia (q.v.) airliners, the Lightning (q.v.) interceptor, and the Canberra (q.v.) bomber. B.A.C. is a partner with M.B.B. and Fiat in Panavia (q.v.), and with SAAB and M.B.B. in the Europlane (q.v.).

Guided weapons work is a very important part of B.A.C.'s activities and, apart from Bloodhound (q.v.) product support, includes Rapier (q.v.), Sea Dart (q.v.), Swingfire, and Thunderbird (q.v.) guided missiles.

British Airways Board: the British Airways Board was formed in 1972 to operate as a state-owned holding company for B.E.A. (q.v.) and B.O.A.C. (q.v.) and their subsidiaries. An outright B.E.A.-B.O.A.C. merger under the name of British Airways is now well advanced.

British Caledonian Airways: Britain's third largest airline, and the largest independent airline in Europe, British Caledonian Airways was formed in 1970 after the takeover of British United Airways by Caledonian Airways, the resulting airline being known as Caledonian/B.U.A. until 1971, when the present title was taken. The object of the new airline is to continue the scheduled service and charter operations of British United and the charter operations of Caledonian Airways, while also competing with the state airlines on the main domestic and international trunk routes.

British United Airways dated from the merger of Airwork and Hunting Clan in 1960. Airwork dated from 1928, when the company was formed to provide the full range of services required by private fliers, and later assisted in the formation of Egypt's Misrair airline and Indian National Airlines and, after the war, Sudan Airways. During World War II the company acted as a subcontractor to the armed forces. Airwork's own airline operations started after the end of World War II, initially using Vickers Vikings and Handley Page Hermes aircraft on military and civil contract work, but also developing coach-class services to West Africa after 1952. During this period, Airwork co-operated with another airline in the same field of activity, Hunting Clan, which also operated an all-freight service to Africa under the name of Africargo. Before the merger of the two airlines, Airwork acquired Transair and Air Charter in 1958, and Morton Air Services, dating from 1933, in 1959. Air Charter was best known for its Channel Air Bridge vehicle ferry service, and for Aviation Traders, an engineering subsidiary which produced the Carvair conversion of the Douglas DC-4.

After the merger, British United acquired the passenger and vehicle carrier, Silver City Airways, and the Jersey Airlines in 1962. By this time the airline was owned largely by two shipping groups, British and Commonwealth Shipping and P. & O., which reorganised the airline under a holding company into a number of specialised operators, including vehicle ferry, Channel Islands, and helicopter operators. This later broke up, with Air Holdings taking the New Zealand-based SAFEAIR, Air Ferry, and British Air Ferries; and later the remaining British United Group was completely disintegrated, with British and Commonwealth retaining the Channel Islands services as British Island Airways, and Bristow Helicopters also becoming independent, while Caledonian acquired B.U.A. itself. By this time, B.U.A. was operating a Vickers VC.10 and B.A.C. One-Eleven fleet on a domestic, European, African, and South American network.

Caledonian Airways was formed in 1961 with a single Douglas DC-7C for charter operations, which was soon joined by further DC-7s and by Bristol Britannias. Although the airline specialised in North Atlantic charter operations, Far East military charter flights were also undertaken. The first jet aircraft, Boeing 707s, were delivered in 1967, and followed by B.A.C. One-Elevens for European charter flights.

British Caledonian Airways introduced North Atlantic scheduled services in 1973 to supplement its existing domestic, European, African, and South American services. A fleet of Boeing 707s, Vickers VC.10s, and B.A.C. One-Elevens is used. Major shareholders include Airways Interests (Thomson), representing the management and employees, the Industrial and Commercial Finance Corporation, Lyle Shipping, Hogarth Shipping, Great Universal Stores, the National Commercial Bank, and Schroeder's Bank.

British Hovercraft Corporation: the British Hovercraft Corporation was formed in 1963 on the amalgamation of the hovercraft activities of Westland Aircraft (q.v.) and Vickers (q.v.), with 65 and 25 per cent shareholdings respectively, the remaining 10 per cent being held by the state-sponsored National Research and Development Council. Westland has since acquired the Vickers shareholding. On its formation, the Corporation acquired the Westland and Vickers experimental designs, plus the Westland SR-N5 (q.v.) and SR-N6 (q.v.), which were put into full-scale production on the world's first hovercraft production line in 1966, and the SR-N4, (q.v.) which is the world's largest hovercraft (or air cushion vehicle, to use the correct term). Two new designs are the BH-7 (q.v.) and BH-8 (q.v.), although the latter is not yet in production. All B.H.C. hovercraft are fully amphibious designs using air effect propellers. Amongst the other activities of B.H.C. can be counted participation in production of the Britten-Norman Islander utility aircraft.

Britten-Norman Aircraft: formed by John Britten and Desmond Norman out of their aerial crop-spraying and air transport operations, the object of Britten-Norman Aircraft was to put into production the BN-2 (q.v.) Islander utility transport aircraft designed by John Britten in 1963. The aircraft first flew in 1965 and entered production in 1968, by which time some

Government support was provided. Orders for the aircraft were such that much of the production had to be undertaken by the British Hovercraft Corporation and in Rumania. In 1969 a light aircraft for private owners, the BN-3 Nymph, first flew; it was intended that this aircraft should be produced in kit form for assembly by firms in the developing countries, or by dealers elsewhere, but further work had to be abandoned due to a shortage of finance. In 1970 a military development of the Islander, the Defender, and a stretched three-engined development, the Trislander, appeared and immediately attracted the interest of customers.

Credit difficulties forced the company into liquidation in 1971, but it has since been re-formed and production has been largely unaffected. Fairey Aviation acquired Britten-Norman in 1972, and a new design, the Mainlander, with three Rolls-Royce Dart turboprops, is under consideration.

Bronco, North American OV-10A: *see* OV-10A Bronco, North American.

Broussard, Avions Max Holste (Reims): originally produced by Avions Max Holste, which is now a part of Reims Aviation (q.v.), the Broussard first flew in prototype form in 1952. Some 330 production aircraft for the French armed forces followed throughout the 1950s. Most former French colonies have one or two Broussards for communications duties, and these aircraft are ex-French Army. A high-wing monoplane with twin-fin and six seats, the Broussard uses a single 450 hp Pratt and Whitney R-985 piston engine for a speed of up to 160 mph and a range of up to 750 miles.

Brown, Sir Arthur Whitten (1886–1948): as Lieutenant Arthur Whitten Brown, one of the first two men to fly non-stop across the Atlantic when he navigated the then Captain John Alcock's (q.v.) Vickers Vimy (q.v.) from St John's, Newfoundland, to Clifden, Co. Galway, on 14–15 June 1919.

Buccaneer, Hawker Siddeley: originating as the Blackburn NA-39 design, development of the Buccaneer carrier-borne, low-level strike aircraft started during the mid-1950s, with the first flight taking place in April 1958, followed by deliveries of the production S.Mk.1 models to the Royal Navy from July 1962. The S.Mk.1 used two 7,100 lb thrust Gyron Junior turbojets, but later models of this Mark and the S.Mk.2, which first flew in May 1963, used 11,255 lb thrust Rolls-Royce Spey RB.168 Mk.101 turbofans, giving transonic speeds in level flight at sea level. The South African Air Force's S.Mk.50, with the addition of an 8,000 lb thrust Bristol Siddeley BS.605 rocket engine for assisted take-off from high altitude airfields, is a variation on the S.Mk.2, which has also entered Royal Air Force service. The R.A.F. is receiving the Royal Navy's aircraft in addition to forty-two new aircraft as the Fleet Air Arm is progressively run down during the 1970s. Generally recognised as the most potent low-level strike aircraft in the world, with an ability to penetrate below defending radar, the Buccaneer has a revolving-door bomb bay with accommodation for four 1,000 lb bombs, plus four wing strongpoints for three 1,000 lb bombs or rockets each, and can carry nuclear bombs or Martel or Bullpup stand-off bombs. A range of up to 2,000 miles is provided, and the aircraft has a crew of two in a cockpit with tandem seating.

Buckeye, North American T-2: *see* T-2 Buckeye, North American.

Buffalo, de Havilland Canada DHC-5: *see* DHC-5 Buffalo, de Havilland Canada.

Bulldog, Bristol: the Bristol Bulldog fighter biplane entered service with the Royal Air Force during the mid-1920s, remaining in frontline service for the following eight years as well as being sold to a number of foreign air forces, including those of Denmark and Finland. A 515 hp Bristol Jupiter radial engine provided a maximum speed of 180 mph.

Bulldog, Scottish Aviation: starting life as the Beagle Bulldog, a military trainer version of the Pup light aircraft, the Bulldog first flew in May 1969, and soon gained orders from Sweden, Zambia, and Kenya. Production was transferred to Scottish Aviation in 1970 after Beagle's financial collapse in 1970. A single 200 hp Lycoming IO-360-AIC piston engine gives the aircraft a maximum speed of 153 mph and a range of 600 miles. A twin-seat cockpit with bubble canopy is fitted.

Bullpup: the Maxson Electronics AGM-12 Bullpup air-to-surface guided missile is in production and service with most NATO air forces, including the R.A.F. and Royal Navy. A large conventional or a nuclear warhead can be fitted, and guidance is from the carrying aircraft using flares fitted to the missile which are kept in line with the target. Versions include the AGM-12B, with liquid propellent rocket and high explosive warhead, the AGM-12C, with detail improvements, the AGM-12D with nuclear warhead capability, and the AGM-12E, with a high-fragmentation warhead.

bus stop jet: a term used for the first generation of true short-haul jet airliners, including the Boeing 737 (q.v.) and Douglas DC-9 (q.v.), but with particular reference to the B.A.C. One-Eleven (q.v.). The term is largely derived from the ability of the aircraft to operate viably on short sector routes, and to make do without airfield equipment by having its own built-in landing stairs and auxiliary power unit (q.v.).

Butler, J. W.: one of the two Englishmen (the other being his partner, E. Edwards (q.v.)) who designed in 1867 the first delta-wing aircraft, which bore a strong resemblance to the simplest of paper aeroplanes. It was proposed that the aircraft use a solid-fuel rocket, with control exercised by moving the rocket nacelle along the fuselage. No attempt was made at construction.

Butler Aviation: *see* Mooney Aircraft.

Bv. 107 Sea Knight, Boeing-Vertol: after winning a U.S.M.C. design competition in 1961, Boeing-Vertol commenced development of its private venture Bv. 107 helicopter design as the CN-46A. The first flight was in October 1962, with deliveries of production models starting in 1965.

Bv. 114 Chinook, Boeing-Vertol

A tandem twin-rotor, twin-engined design using two 1,250 shp General Electric T58-G E-8B shaft turbines in U.S.M.C., U.S.N., and C.A.F., service, or two Bristol Siddeley Gnome H1200s in Royal Swedish Air Force and Royal Swedish Navy service, the CH-46A can carry up to twenty-five fully-equipped troops or a 4,000 lb payload, and has a maximum speed of 186 mph and a range of 300 miles.

Bv. 114 Chinook, Boeing-Vertol: the Boeing-Vertol Bv. 114 Chinook, or CH-47, won a U.S. Army design competition in 1959 as a development of the Bv. 107 (q.v.). The Chinook first flew in September 1961, with deliveries to the United States Army beginning in August 1962, replacing H-21, H-34, and H-37 helicopters. Two 2,650 shp Lycoming T55-L-7 shaft turbines power the twin-rotor helicopter, which can carry up to 44 fully-equipped troops, and has a maximum speed of 180 mph and a maxium range of 250 miles with a 10,500 lb payload. The Chinook is also under production in Italy for the Italian and Iranian armed forces.

B.W.I.A. International–British West Indian Airways: originally formed in 1940, British West Indian Airways was acquired by British South American Airways in 1947, ownership passing to B.O.A.C. (q.v.) shortly afterwards when the Corporation also acquired B.S.A.A. Another B.O.A.C. acquisition, British Caribbean Airways, was merged into B.W.I.A. in 1949. Since independence has been granted to most of the former British colonies in the Caribbean, British West Indian Airways has been owned by the Government of Trinidad and Tobago, for which it operates an extensive Caribbean network, plus services to the United States, Canada, and Guyana, using a fleet of Boeing 707s.

Byrd, Rear-Admiral Richard Evelyn, U.S.N. (1888–1957): the then Commander Richard E. Byrd became the first man to fly over the South Pole on 28–29 November 1929, using a Fokker F.VII Trimotor aircraft, repeating a 1926 performance over the North Pole in the same aircraft.

C

C-: U.S.A.F. and U.S.N. designation for transport aircraft, but it is sometimes used by other countries, and appears on such aircraft as the N.A.M.C. C-1 (q.v.), C.T.A. C-95 (q.v.), and C-160 (q.v.).

C-1, N.A.M.C.: a high-wing, medium range, S.T.O.L. military transport developed for the Japanese Air Self-Defence Force as a C-46 (q.v.) replacement. Two 14,500 lb thrust Pratt and Whitney JT8D-9 turbofans give a maximum cruising speed of 450 mph and a range of 750 miles with the maximum payload of 17,600 lb, and a ferry range of 1,800 miles. The first flight was in 1970, and production of up to sixty aircraft for the J.A.S.D.F. is under way.

C-1A Trader, Grumman: the C-1A was developed from the Grumman S-2 (q.v.) Tracker carrier-borne anti-submarine aircraft for carrier-onboard-delivery duties. First flight, using the designation TF-1, was in January 1955, and the current designation was not adopted until 1962. Two 1,525 hp Wright R-1820-82 piston engines provide a maximum speed of 265 mph and a maximum range of 1,500 miles. Eleven passengers or up to 3,500 lb of freight can be carried. On the U.S.N.'s larger carriers Traders have been replaced by C-2 (q.v.) Greyhounds.

C-2 Greyhound, Grumman: a development of the E-2A Hawkeye airborne-early-warning aircraft, the C-2 Greyhound is a carrier-onboard-delivery aircraft designed to replace the C-1A (q.v.) Trader on the larger U.S.N. carriers. First flight was in November 1964. Two 4,050 shp Allison T56-A-8 turboprops give a maximum speed of 330 mph and a maximum range of up to 1,500 miles. Accommodation is provided for thirty-nine passengers or up to 15,000 lb of freight.

C-5 Galaxy, Lockheed: currently the world's largest aircraft, the C-5 Galaxy has been in U.S.A.F. service since 1970, and is now the backbone of Military Airlift Command strength. Four 41,000 lb General Electric TF-39 turbofans give a maximum speed of 600 mph and a range of up to 6,000 miles, while the maximum payload is 265,000 lb. A feature of the aircraft, in spite of its size, is its relatively short take-off characteristic, which can be as little as 5,000 feet. A civil version has been proposed, and this would probably use six up-rated Rolls-Royce RB.211 engines. The main application as foreseen at present would be as a car transport on either a transatlantic or trans-United States basis.

C-9A Nightingale, McDonnell Douglas: an aero-medical version of the McDonnell Douglas DC-9 (q.v.) airliner, but equipped for up to forty stretchers. In U.S.A.F. service only.

C-45 Expeditor, Beech: a military communications version of the Beech 18 (q.v.).

C-46 Commando, Curtiss: originally conceived as an airliner, the C-46 first flew in March, 1940, and before long the first of many different variants for the then U.S.A.A.F. was entering service. The C-46 also flew during and after World War II with many Allied air forces, and has since also enjoyed considerable popularity amongst South American and Asian air arms. More than 3,000 aircraft were built. The C-46 can carry more than fifty troops or up to 15,000 lb of freight. Two 2,000 hp Pratt and Whitney R-2800-51 piston engines give this rather bulbous low-wing transport a maximum speed of 240 mph and a range of up to 1,800 miles.

C-47 Dakota, Douglas: undoubtedly the most famous and the most successful transport aircraft of all time, some 13,000 Douglas DC-3 (q.v.) and C-47 Dakota transports were produced mainly during World War II, but the aircraft had first appeared in 1936 as a successor to the

C-54 Skymaster, Douglas

DC-2 (q.v.). Few air forces have not used the aircraft, which was also produced under licence in the Soviet Union as the Lisunov Li-2. A refurbished and modernised version of the C-47 is the C-117 Skytrain for the U.S.N., while the AC-47 is a gunship version for use in the Vietnam War. The basic aircraft uses two Pratt and Whitney R-1830-90C piston engines giving a maximum speed of 220 mph and a maximum range of 1,500 miles, while the usual seating capacity is thirty. Two 1,535 hp Wright R-1820 piston engines are used in the Skytrain.

C-54 Skymaster, Douglas: another aircraft originally intended for airline service (as the DC-4 (q.v.)) but pressed into military service on the advent of World War II, the Douglas C-54 remained in production throughout World War II and for some years afterwards, serving with several Allied air forces. Up to fifty troops or 22,000 lb of freight can be carried, while the maximum speed is 274 mph and the maximum range is 1,500 miles. Early production versions had a triple fin, but a single fin was soon substituted.

C-95 Bandeirante, C.T.A.: a light transport and communications aircraft in production for, and service with, the Forca Aérea Brasileira, the C-95 was designed by Max Holste as a Beech C-45 (q.v.) successor. A low-wing aircraft, the first flight was in October 1968, and two United Aircraft PT6A-20 turboprops of 550 shp provide for a maximum speed of nearly 300 mph and a range of up to 1,000 miles.

C-118 Liftmaster, Douglas: the prototype of the Douglas C-118 Liftmaster first flew in February 1946, and more than 160 were built for the U.S.A.F. and U.S.N. Basically a development of the C-54 (q.v.) Skymaster, but with stretched fuselage and extended range, the C-118 is also a military version of the DC-6 (q.v.) Cloudmaster. Four 2,500 hp Pratt and Whitney R-2800-52W piston engines give a maximum speed of 360 mph with a range of up to 3,800 miles, while either eighty passengers or 27,000 lb of freight can be carried. A number of NATO

air forces and the R.N.Z.A.F. have also used the aircraft.

C-119 Packet, Fairchild: developed from an early twin-boom transport design, the C-82 of the last years of World War II, the C-119 Packet, or 'Flying Boxcar', first flew in 1947 as a modified C-82 with up-rated engines and stretched fuselage, plus some other modifications. A considerable number of aircraft were eventually produced, using either two Pratt and Whitney R-4360 or two Wright R-3350 piston engines, giving the final versions a maximum speed of 300 mph and a maximum range of 2,250 miles, while the maximum payload is either sixty troops or 30,000 lb of freight. The aircraft saw service with a number of NATO air forces, plus those of Brazil, India, and Nationalist China. Many of the remaining aircraft have been fitted with auxiliary jet engines, usually the Westinghouse J34-WE-36 or Bristol Siddeley Orpheus, to give enhanced take-off performance. A number of the U.S.A.F. aircraft in Vietnam were AC-119 gunships.

C-121 Constellation, Lockheed: a development of the Lockheed Constellation (q.v.) which first flew in 1939, leading to the civil L-049 version and the military C-69, the C-121 acts in both the transport and airborne-early-warning roles (EC-121), and is comparable to the civil L-749 Constellation and L-1049 Super Constellation which entered airline service during the immediate post-war period. Four Wright R-3350 piston engines allow for a maximum speed of 320 mph and a range of up to 4,500 miles, while up to 100 troops can be carried. The U.S.A.F. and U.S.N. EC-121 versions have outlived the transport aircraft, but will shortly be replaced by Boeing C-135 developments as a part of the airborne warning and command system (AWACS) proposal. A few C-121s remain in service with the Indian Air Force and the Israel Defence Force/Air Force.

C-123 Provider, Fairchild: originally developed by Chase Aircraft as a glider, the C-123 first flew in its powered form as the Chase Avitruc in October 1949. Production was transferred to Fairchild in 1953. A number of variants of the basic design have been built, incorporating detail modifications, but the most popular remains the C-123B, with two 2,300 hp Pratt and Whitney R-2800-99W piston engines giving a maximum speed of 250 mph, a maximum range of 1,500 miles, and a payload of either sixty troops or 24,000 lb of supplies. Apart from service with the U.S.A.F., the aircraft is also used by Saudi Arabia, Thailand and Venezuela. Recent modifications have included the fitting of a General Electric CJ-610 turbojet under each wing to boost performance. AC-123 gunship conversions were flown by the U.S.A.F. in Vietnam.

C-124 Globemaster, Douglas: a development of the Douglas C-54 (q.v.) Skymaster, the first Globemaster flew in 1945, and entered U.S.A.F. service as the C-74; but the double-deck which is associated with this aircraft did not appear until the arrival of the C-124, which first flew in November 1949. About five hundred Globemasters were produced for the U.S.A.F. during the late 1940s and early 1950s, and these remain in service with the Air National Guard (the U.S.A.F.'s reserve). Four 3,500 hp Pratt and Whitney R-4360-20W2 piston engines are used in the C-124B, while the 250 or so C-124Cs use 3,800 hp Pratt and Whitney R-4360-63A piston engines. Maximum speed is 300 mph, the aircraft fly 4,000 miles with a 68,500 lb payload or 200 troops.

C-130 Hercules, Lockheed: probably the most successful peacetime military transport aircraft, the Lockheed C-130 Hercules is also used by a few civil operators. First flight of a Hercules prototype was in August 1954, with deliveries of the C-130A to the U.S.A.F. starting in December 1956. A number of minor improvements were incorporated in the C-130B, while the C-130C was a single model of a short take-off version and did not enter production, and the C-130D was a ski-equipped aircraft and only about a dozen were built. The C-130E is one of the latest versions of the standard production model, of which the C-130K is a version

for the R.A.F. using a number of British components. The C-130H is in service on helicopter tanker and air-sea rescue duties. Four Allison T56-A-7P turboprops of 4,050 hp provide a maximum speed of 360 mph and a maximum range of up to 4,700 miles, while either 100 troops or a payload of up to 40,000 lb may be carried. As with other serving U.S.A.F. high-wing aircraft, an AC-130 gunship version was in service in Vietnam.

C-131 Samaritan, Convair: The Convair C-131 Samaritan series consists of military versions of the Convair 240/340/440 series of civil airliners, of which the C-131A approximates to the 240, the C-131B and C-131C to the 340, and the C-131F and C-131G to the 440. There are also a number of T-29A/B navigational trainers. The aircraft are in U.S.A.F., U.S.N. and Luftwaffe service, while Canadair built a small number using Napier Eland turboprops for the then R.C.A.F. Two 2,500 hp Pratt and Whitney R-2800-99W piston engines provide the standard aircraft with a maximum speed of 300 mph and a range of 500 miles, and there is accommodation for fifty-eight persons.

C-133 Cargomaster, McDonnell Douglas: the C-133 was developed as a result of Douglas winning a U.S.A.F. design competition for a long-range transport capable of carrying heavy and bulky items of equipment, including guided missiles – a similar requirement to that which resulted in the Short Belfast (q.v.) transport for the R.A.F. First flight of the C-133A was in April 1956, using four Pratt and Whitney T34-P-7W turboprops of 6,500 shp each, and this was followed by a production run of thirty-five aircraft; while the first of fifteen C-133Bs, with modified rear doors and 7,500 shp Pratt and Whitney T-34-P-9W turboprops, flew in October 1959. The maximum speed is 360 mph and the maximum range 4,000 miles, and either 200 troops or 110,000 lb of cargo can be carried.

C-135 Stratolifter, Boeing: in common with the Boeing 707 (q.v.) and 720 airliners, the C-135 Stratolifter was evolved from the private-venture Boeing 367-80

83

jet transport, which first flew in July 1954, and incorporated Boeing's experience with the B-47 Stratojet bomber. Initial U.S.A.F. orders were for the KC-135 Stratotanker, for operation with Strategic Air Command. Its first flight was in August 1956, the first of 700 C-135 and KC-135 aircraft being delivered in June 1957. Developments included the EC-135 flying control-room for S.A.C., of which seventeen were built, the EC-135J and RC-135C/E electronics-reconnaissance aircraft, the VC-135A V.I.P. transport, and the C-135A/B Stratolifter transport for the Military Air Transport Service. A number of KC-135F tankers are in service with the Armée de l'Air, while the Luftwaffe operates four Boeing 707-320s. The performance of the different versions varies, including the use of turbojets and turbofans respectively in the C-135A and C-135B, but the standard specification includes four Pratt and Whitney J57-R59W turbojets of 13,750 lb thrust each, giving a range of 1,200 miles fully loaded, and a maximum speed of 585 mph.

C-140 JetStar, Lockheed: first flight of the prototype JetStar was in September 1957, using two Bristol Siddeley Orpheus turbojets, but production models of both civil and military aircraft have been built with four Pratt and Whitney JT12A-6A turbojets of 3,000 lb thrust, making the aircraft the smallest four jet-engined aeroplane in the world. The U.S.A.F.'s aircraft are used for radar calibration work, although a number are VC-140 V.I.P. transports. Maximum speed is 575 mph and the range is 2,185 miles.

C-141 Starlifter, Lockheed: the C-141 Starlifter was built as a result of Lockheed winning a 1961 U.S.A.F. design competition for a long-range strategic transport aircraft. First flight was in December 1963, and deliveries to the only operator, the U.S.A.F., started in 1965. Four 21,000 lb thrust Pratt and Whitney TF33-P-7 turbofans give a maximum speed of 560 mph and a maximum range of 6,500 miles, and up to 130 troops can be carried, or 86,000 lb of freight. The aircraft is specifically designed for transport of Minuteman (q.v.) missiles.

C-169 Transall: the Transall, or Transporter Allianz, project was initiated in January 1959 to meet a joint Armée de l'Air and Luftwaffe specification for a medium-range turboprop transport, Nord Aviation (now Aerospatiale) of France and V.F.W. of Germany being responsible for design, development, and production. The first flight was in February 1963, and deliveries started in 1967. Basically the aircraft, with its high wing and tail doors, was conceived as a Noratlas replacement. Two 6,100 shp Rolls-Royce Tyne R.Ty.22 turboprops, built under licence by M.A.N. of Germany, produce a maximum speed of 330 mph, with a 1,000 mile range when carrying 35,000 lb of freight. Transalls are also in South African Air Force service. Plans for civil versions, including one with a nose door and a raised flight deck position, have not come to anything so far.

C.207 Azor, C.A.S.A.: the C.207 Azor is in Spanish Air Force service on light transport duties. The first flight was in September 1955, at which time this was the largest aircraft of Spanish design. Twenty aircraft were built, all using two Bristol Hercules 730 piston engines of 2,040 hp each, giving a maximum speed of 280 mph and a maximum range of 1,620 miles, and either forty passengers or 8,800 lb of freight can be carried.

C.212 Avicar, C.A.S.A.: a light transport with two 755 shp Garrett TPE 331-201 turboprops, the C.A.S.A. C.212 Avicar first flew in 1970. Eighteen passengers can be carried, or a payload of 4,500 lb. The maximum speed is 230 mph and the maximum range 1,200 miles. The aircraft has replaced the Spanish Air Force's Junkers Ju. 52/3Ms and Douglas C-47 Dakotas.

C.A.A.: *see* Civil Aviation Authority.

C.A.B.: *see* Civil Aeronautics Board.

'Cab', Lisunov Li-2: *see* Li-2 'Cab', Lisunov.

cabin: either the enclosed passenger and pilot compartment of a light aircraft, or the passenger compartment of an airliner. Military aircraft and some light aircraft have a cockpit (q.v.), while the crew section of an airliner is the flight deck (q.v.). Early aircraft were completely open, with the pilot lying on the wing, and later aircraft left the pilot sitting in an open cockpit, even after his passengers were provided with enclosed accommodation. The first cabin aircraft included the Avro (q.v.) and Sikorsky (q.v.) designs of the immediate pre-World War I period, but it was not until the late 1920s that such accommodation could be taken for granted.

Cactus: *see* Crotale, for which 'Cactus' is the South African name.

Calcutta, Short: the Short Calcutta three-engined flying-boat first entered service with Imperial Airways in 1928, being used principally on the Mediterranean leg of the air route to India. The three 540 hp engines provided a maximum speed of 110 mph and a range of up to 650 miles, while this biplane could generally carry about fifteen passengers. A military version for the Royal Air Force was the Rangoon.

Camel, Sopwith: one of the most successful of the World War I fighter biplanes, the Sopwith Camel appeared during the second half of the war, operating on the Western Front and at sea, as the first aircraft to take off from a barge towed by a destroyer. Powerplants varied, from 110 hp Rhône to 230 hp Bentleys. The Camel was succeeded by the Snipe.

Camm, Sir Sydney: one of Britain's leading aircraft designers from the mid-1920s, when he became Chief Designer for Hawker (q.v.), until the mid-1960s, Sir Sydney Camm designed a succession of successful and elegant biplanes for Hawkers, including the Fury fighter and Hart (q.v.) bomber and ground attack series, before designing the Hurricane (q.v.) monoplane fighter, which entered R.A.F. service during the late 1930s and fought during the first part of World War II. The Hurricane was followed by the Typhoon and Tempest fighters and, during the early 1950s, by the Sea Hawk (q.v.) and Hunter (q.v.) jet fighters. The Hunter was one of the most successful post-World War II military combat aircraft, and many

examples are still in service throughout the world. Camm's last design was the P. 1127 Kestrel, from which the Hawker Siddeley Harrier (q.v.) vectored thrust V./S.T.O.L. fighter was developed for service with the R.A.F. and the U.S.M.C.

Campania, Fairey: the Fairey Campania biplane seaplane was introduced in 1916 for service from H.M.S. *Campania*, a converted merchant liner, from which ship the aircraft flew in 1917, using trolley gear jettisoned on take-off. The Campania had an endurance of up to three hours at a cruising speed of 80 mph. A feature of the aircraft was the ability to fold the wings back for storage in the confined hangars of a ship, although neither this feature nor the trolley take-off from the ship's deck can be counted as unique.

Canadair: a subsidiary of the American General Dynamics Corporation (q.v.), Canadair has been largely engaged on licence-production for the Canadian Armed Forces, or on subcontract work. In recent years, however, the company has developed a number of its own designs, including the CL-41 Tudor jet trainer, which entered R.C.A.F. service as the CT-114 (q.v.), the CL-84 (q.v.) V./S.T.O.L. light transport, and the CL-215 (q.v.) amphibian. Earlier Canadair licence-production programmes have included Avon-powered versions of the North American F-86 (q.v.) Sabre, and developments of the Bristol Britannia (q.v.), including the CP-107 Argus, the CC-106 Yukon, and the CL-44 (q.v.). Convair C-131 (q.v.) Samaritan transports were also built for the then R.C.A.F., and more recently the Northrop F-5A/B (q.v.) Freedom Fighter has been built as the CF-5A/B for the C.A.F., and the NF-5A/B for the Royal Netherlands Air Force.

Canadian Pacific Airways: *see* C.P.-Air.

canard: in aviation, the term canard refers to an aircraft with a tail-first layout. Such aircraft included the Wright brothers' (q.v.) designs, but the concept fell into disfavour before World War I and has only recently been revived, with such aircraft as the SAAB-37 (q.v.) Viggen. The Tupolev Tu-144 (q.v.) supersonic airliner

Canberra, B.A.C.

has a small retractable canard aerodynamic surface.

Canberra, B.A.C.: originally an English
Electric (q.v.) design, the Canberra jet
bomber first flew in prototype form in
May 1949, deliveries of the B.Mk.2 for
the R.A.F. starting in January 1951. This
was followed by the B.Mk.6, with up-
rated engine and extra fuel tankage, and
the B(I) Mk.8 interdictor bomber with
offset cockpit. A version of the B(I) Mk.8,
the B(I) 12, was sold to New Zealand and
South Africa, while the Indian Air Force
received the B(I) Mk.58. Although basically a medium jet bomber, or interdictor
bomber, the Canberra was also used as a
night fighter and reconnaissance-bomber,
including the P.R.Mk.7, based on the
B.Mk.6 and the P.R.Mk.9. There were
also ground attack and target tug conversions, with trainer versions. Production
was mainly in the United Kingdom, with
licence-production in Australia, and in
the United States by Martin, which produced the Canberra as the B-52 for the
U.S.A.F. Although few Canberras are
now with their original owners, the aircraft is still in widespread use.

Typical Canberra performance using
two Rolls-Royce Avon 109 turbojets
(B(I) Mk.8 and P.R. Mk.9) of 7,4000 lb
thrust includes a maximum speed of 540
mph and a range of 3,000 miles, or 800
miles with a full warload of 8,000 lb, of
which 6,000 lb can be carried internally.

cantilever wing: a wing without any
external bracing. External bracing was a
feature of many of the early monoplanes,
and indeed most of the monoplanes built
up to and including the early 1930s. One
of the first aircraft with a cantilever wing
was the Junkers J.1 (q.v.) 'Blechesel' or
Flying Donkey, which was also the first
all-metal aeroplane and appeared in 1915.
A related concept was the monocoque
(q.v.) aeroplane.

capacity ton-miles: a measure of production. The C.T.M. of an airline or a route
indicates the number of tons of capacity
produced multiplied by the miles flown. It
does not, of course, indicate the amount of
capacity sold, the measure of consumption

being load ton-miles (q.v.). The metric
equivalent of the C.T.M. is the A.T.K., or
available tonne-kilometre.

Caproni: an Italian aircraft manufacturer
of some prominence during World War I
and until the end of World War II,
Caproni's first significant aircraft were a
series of heavy bombers produced between
1916 and 1918. Between the wars a number of transport and reconnaissance aircraft were produced, including the Ca.97
reconnaissance aircraft, the Ca.101, Ca.
111, and Ca.133 transports, and the Ca.73
and Ca.101 bombers. The Caproni-
Campini (q.v.) was an early jet aircraft.

Caproni-Campini: the Caproni-Campini
jet aircraft design used a single engine
and bore a resemblance both to the
Heinkel He. 178 and the Gloster E.28,
but was much less successful than either
of these; development was soon abandoned after the first flight in 1941. The aircraft does have the distinction of being
the first jet aircraft to make a cross-
country flight.

Caquot balloon: one of many different
tethered observation balloons in use
during World War I, this French design
by a Captain Caquot, with its roughly
airship appearance and bulbous fins,
proved itself to be the most stable. The
barrage balloons (q.v.) of World War II,
and paratroop training balloons since,
owe much to Caquot's design.

Caravelle, Aerospatiale: although a civil
aircraft, the Caravelle was built as a
result of Sud Aviation winning a Government design competition for a short- and
medium-range jet airliner in 1953. First
flight of the prototype Caravelle was in
May 1955, and the aircraft first entered
service in May 1959, to become the first
French airliner to enjoy any real success.
A number of versions have been produced,
starting with the Caravelle I with Rolls-
Royce Avon 522 turbojets; the Caravelle
III used up-rated Avon 527s and had a
number of detail improvements, the VIN
used Avon 531s, and the VIR Avon 533s.
The main production models were the III
and the VIR.

A fuselage stretch in 1964 produced the

Super Caravelle, which has a number of aerodynamic improvements, and Pratt and Whitney JT8D-1 turbofans of 14,000 lb thrust. It is also known as the 10R. The 11R has an even longer fuselage stretch. The 10R has a maximum speed of 518 mph and a range of up to 1,700 miles with a payload of 20,000 lb. Passenger accommodation varies from 68 to 104 passengers, depending on class-mix and pitch.

Cargomaster, McDonnell Douglas C-133: *see* C-133 Cargomaster, McDonnell Douglas.

Caribou, de Havilland Canada DHC-4: *see* DHC-4 Caribou, de Havilland Canada.

carrier-onboard-delivery: a carrier-borne transport aircraft which enables supplies and personnel to be flown onboard the ship. Usually such aircraft are conversions of anti-submarine or airborne-early-warning aircraft, such as the Royal Navy's Gannet (q.v.) and the United States Navy's C-1A (q.v.) Trader or C-2 (q.v.) Greyhound.

C.A.S.A.: a Spanish company established in 1923, C.A.S.A. has produced a number of its own designs during recent years, including the C.204, C.207 (q.v.), and C.212 (q.v.) light transport aircraft, as well as undertaking licence production of other designs for the Spanish Air Force. Amongst the more famous of C.A.S.A.'s licence production programmes can be included the Heinkel He. 111, built with Rolls-Royce Merlin engines during the post-war period, while more recently Northrop F-5A/B Freedom Fighters have been built as SF-5A/Bs. Northrop is a major shareholder in the company. Currently, C.A.S.A. is also producing components for the German H.F.B. 320 (q.v.) Hansa executive jet and for the European airbus, the A.300B (q.v.).

CAT: *see* clear air turbulence.

Catalina, Convair PBY-5/6: originally designed as a flying-boat during the late 1930s, the Convair Catalina has survived longest in its PBY-5A amphibian form, which is still in active service in Latin America. Most NATO air forces have used the aircraft in one form or another, although service with the R.A.F. was wartime only, and strictly speaking pre-NATO. The name Catalina was originally given to the aircraft by the R.A.F. In common with the DC-3, licence-production of the Catalina was allowed to the Soviet Union. Some 1,200 flying-boats and 944 amphibians, including the PBY-6 development, were built. Two Pratt and Whitney R-1830 radial engines of 1,200 hp give a maximum speed of more than 220 mph and a range of up to 2,500 miles.

catapult: steam catapults are used on modern aircraft carriers to launch naval bomber, fighter, and strike aircraft which, with the high-wing loadings of modern combat aircraft, could not take off without catapult assistance. However, one of the first aircraft to be catapulted was the Langley Aerodrome (q.v.) of 1903, although it fouled the catapult, which was on the roof of a houseboat.

The first successful launching of an aircraft from a catapult was in November 1915, when an AB-2 flying-boat was catapulted from the stern of the U.S.S. North Carolina. It was not long before most battleships and cruisers were fitted with catapults for launching seaplanes for spotter duties. During World War II, British merchant vessels were equipped with catapults to launch Hawker Hurricane fighters as a defence against enemy air attack.

category, weather: for purposes of convenience when certifying airports, aircraft, and equipment for poor weather operation, I.C.A.O. weather categories are used and clearance is provided or sought for operations in weather down to a certain category. The categories are:

Category 1: operations down to the decision height of 200 feet and with visibility of not less than 2,600 feet.

Category 2: operations down to the decision height of 100 feet with visibility of not less than 1,300 feet.

Category 3a: operations down to and along the runway with a minimum visibility of 700 feet.

Category 3b: operations down to and along the runway with a minimum

Cayley, Sir George, Bt

visibility of 150 feet, which is sufficient for taxiing only.

Category 3c: operations down to and along the runway in nil visibility.

Cayley, Sir George, Bt (1773–1857): Sir George Cayley is widely accepted as the father of the science of aerodynamics (q.v.). His model glider of 1804 was the first real aeroplane with mainplane, or wings, and adjustable tail surfaces with a fin, giving both control and stability. A paper, 'on Aerial Navigation', published in 1809, contained Cayley's findings and laid the basis for all subsequent studies of aerodynamics, breaking away from the ornithopter concept and pointing towards the idea of a mainplane, tailplane, fuselage, and undercarraige as essentials for an aeroplane. Sir George Cayley's work was later taken further by another Englishman, Horatio Phillips (q.v.).

Later, Cayley produced the first practical helicopter design, with laterally offset contra-rotating rotors, vertical rudder, and a steam engine to drive a pusher propeller. It is generally considered that an efficient powerplant, had such been available, could have made this design a feasible proposition. Many of the early practical helicopter research aircraft of the 1930s, including the Focke-Achgelis (q.v.), borrowed from Cayley's work. Cayley's helicopter design appeared in 1840, but he had earlier, in 1796, produced a model helicopter based on the Launoy (q.v.) and Bienvenu (q.v.) design of 1784.

In 1852, a ten-year-old son of one of Cayley's servants made the first controlled gliding flight in a glider designed by Sir George Cayley, which had been towed downhill into a light breeze. A coachman made a flight later, possibly the following year, and nearly resigned! These were the first manned heavier-than-air flights in history.

Cayuse, Hughes 500 (OH-6): see Hughes 500.

CC-: international civil registration index mark for Chile.

CC-106: Yukon, Canadair: a licence-built development of the Bristol Britannia (q.v.), the CC-106 Yukon uses Rolls-Royce Tyne 515 turboprops of 5,730 shp, and also has a stretched fuselage. First flight of the prototype was in November 1959, and a total of twelve were built for the then R.C.A.F. The aircraft remains in service. Side-loading cargo doors were fitted at the front and rear of the aircraft, although the civil CL-44 (q.v.) Britannia-development which followed immediately after CC-106 production ended had a swing tail. Range of the CC-106 is 2,400 miles with a 60,500 lb payload or 135 troops. Maximum speed is 400 mph. An earlier Canadair development of the Britannia was the CP-107 (q.v.) Argus.

CCCP-: international civil registration mark for the U.S.S.R.

CENTO: the Central Treaty Organisation, usually known as CENTO, was first formed in 1955, and for a while was known as the Baghdad Pact. Members were the United Kingdom, Iran, Iraq, Turkey, and Pakistan, with the United States as an associate member. The present title was adopted after Iraq withdrew from the alliance following a revolution in 1958. Pakistan has gradually ceased to play any active part in CENTO, and now looks towards China for support. A formal withdrawal from the Organisation by Pakistan took place in 1972.

CENTO is solely concerned with countering a Communist attack on one of the member states, and would not be activated by a threat from any other quarter. There are no commands on the NATO pattern, but control is exercised through the Council of Military Deputies. There is also an Economic and Counter-Subversion Committee.

Certificate of Airworthiness: the British document certifying that an aircraft, both as a type and as an individual unit, meets the Civil Aviation Authority's airworthiness requirements. It is unusual for a type to lose its C. of A., but this has happened, notably with the de Havilland Comet I (q.v.).

Ceskoslovenske Aerolinee: see C.S.A. Czechoslovak Airlines.

Cessna: although the Cessna Aircraft

Company dates from 1927, its founder, Clyde Cessna, had been building aircraft since 1911, initially producing Blériot-type monoplanes and later going into partnership with Victor Roos to form Cessna-Roos Aircraft, which lasted until 1927. In addition, in 1925 Cessna was one of Walter Beech's partners in Travel Air (see Beechcraft). The Cessna Company was forced to close in 1931 during the depression, but re-opened in 1934, and since has become the world's largest producer of light aircraft.

Amongst the more notable aircraft produced by Cessna can be included the L-19 or, to use the present designation, 0-1 (q.v.) Bird Dog for the U.S. Army and the U.S.A.F., while the Cessna 150, launched in 1957, has become the most popular light aircraft of all time. Although a mass-production manufacturer, Cessna has been responsible for a number of technical innovations, including the progressive introduction of pressurisation to lighter aircraft, and the twin-boom Skymaster series, which combine the benefits of twin engines with the advantages of a single engine in the event of a failure of one unit by having both engines in line, one in front of the cabin driving a puller propeller and the other behind the cabin driving a pusher propeller. The cheapest executive jet today is the Cessna Citation, while the Cessna 340 is the cheapest pressurised aircraft. Cessna's first jet was the T-37 (q.v.) trainer of 1954, which entered service with the U.S.A.F. and many other air forces.

Many Cessna designs are produced by a subsidiary company, Reims Aviation, in France.

Cessna 150: first introduced in 1957, the Cessna 150, a high-wing 2+2 monoplane with a fixed tricycle undercarriage, has become the world's best-selling light aircraft. A single 100 hp Continental 0-200-A piston engine provides a maximum speed of 123 mph and a range of 560 miles. The aircraft is also built in France by Cessna's licensee, Reims Aviation, using Rolls-Royce-built Continental engines. A number of variants on the 150 theme are available, including the 150 Aerobat, and the aircraft can be fitted with floats or ski-equipment.

Cessna 172 and 177 Cardinal: these are respectively fixed and retractable under-carriage versions of the same aircraft. A single 145 hp Continental 0-300 piston engine gives a maximum speed of 138 mph and a range of 700 miles. Four seats are usual, and the aircraft is also the U.S.A.F.'s T-41 trainer. This type is also produced by Reims in France.

Cessna 182 Skylane: a single 230 hp Continental 0-470-R piston engine gives this high-wing monoplane a maximum speed of 167 mph and a range of about 1,000 miles, with four or six seats. A retractable undercarriage is a part of the equipment.

Cessna 310 Skynight: the Cessna 310 has been in production since 1953, although it has been extensively modified since that date. It was Cessna's first twin-engined and first low-wing design. Two Continental I0-470 piston engines of 260 hp each give a maximum speed of 225 mph and a range of up to 1,200 miles. Six seats are usually fitted. A feature of the aircraft is the carriage of most of its fuel in wing-tip tanks.

Cessna 337 Super Skymaster: the Super Skymaster is a developed version of the 336 Skymaster of 1961, which entered production in May 1963. A retractable undercarriage was fitted with the introduction of the Super Skymaster in 1965. A high-wing, twin-boom design, with one engine fore and another aft of the cabin, the aircraft remains easy to control with one engine out. Two Continental IO-360 piston engines of 210 hp give a maximum speed of 200 mph and a range of 1,100 miles. Six seats are fitted. The aircraft is in U.S.A.F. and U.S. Army service as the O-2 (q.v.).

Cessna 340: the lowest-price pressurised aircraft, the twin-engined, low-wing Cessna 340 was introduced in 1971, and uses two Continental GTSIO-520-B piston engines of 320 hp, giving a maximum speed of 250 mph and a range of 1,500 miles, while up to eight seats can be fitted.

Cessna 414: a light twin developed out of the 411, first flown in February 1962. Two 340 hp Continental GTSIO-520-C piston engines give a maximum speed of 250 mph and a range of up to 1,500 miles. Eight seats.

Cessna 421: at one time the cheapest pressurised aircraft, the 421 was introduced in August 1967. Two 375 hp Continental GTSIO-520-D piston engines give a maximum speed of 255 mph and a range of up to 1,800 miles. Eight to ten seats.

CF-: international civil registration index mark for Canada.

CH-3, Sikorsky: *see* S-61, Sikorksy and Westland.

CH-19 Chickasaw, Sikorsky: *see* S-55, Sikorsky and Westland.

CH-34 Choctaw, Sikorsky: *see* S-58, Sikorsky and Westland.

CH-53, Sikorsky: *see* S-65A, Sikorsky.

CH-54 or YCH-54 Skycrane, Sikorsky: *see* S-64 Skycrane, Sikorsky.

Chance-Vought: Chance-Vought Aircraft dated from 1917 and concentrated on building aircraft for the United States Navy, originally under the name of Vought. One of the first designs from the concern to enter production was the UO-1 seaplane biplane, using a Wright 200 hp engine, which was catapulted from cruisers and battleships for fleet-spotting duties. This was the predecessor of the first of the Vought products to carry the Corsair name: the UO-2, also a biplane with floats, although using a more powerful 425 hp Pratt and Whitney radial engine.

The most famous Vought product was the F4K Corsair (q.v.) of World War II, a crank-wing monoplane with a 2,000 hp Pratt and Whitney engine; although basically a carrier-borne fighter, the aircraft provided the backbone of the R.N.Z.A.F.'s strength during the latter part of the war. A number of these aircraft also saw service in the Korean War, for which they were brought out of reserve. During the late 1940s a number of experimental circular-wing aircraft was built and tested, but without any production resulting. The first Chance-Vought jet was the F7U-3 'Cutlass' of the early 1950s, and this was followed by the F-8 (q.v.) Crusader, with a variable-incidence wing.

In 1961 Chance-Vought was acquired by Ling-Temco Electronics, and the resulting group included the airframe division under the name of Ling-Temco-Vought (q.v.). A major item of work since then has been development and production of the L.T.V. A-7 (q.v.) Corsair II for the U.S. Navy, based on the Crusader airframe.

Chanute, Octave (1832–1910): one of the leading glider designers and manufacturers prior to the advent of powered flight, Octave Chanute was a civil engineer, of French birth and American nationality. After experiments with Lilienthal (q.v.) gliders in 1894 and 1895, he sought a better means of controlling the glider than by merely shifting the pilot's weight, but did not, as often suggested, invent control by wing warping (q.v.). The Wright Brothers used Chanute's system of wing rigging on their aircraft, and he was also used as an adviser by the Wrights on their glider and aircraft designs. Chanute's own hang-glider design of 1896 made a large number of successful flights.

Apart from his design and development activities, Chanute also produced the first accurate history of flying, entitled *Progress in Flying Machines*, in 1894, and after his visit to Europe in 1903 was largely instrumental in re-awakening widespread European interest in aviation.

Chaperral: a U.S. surface-to-air guided missile system, rather than an actual missile. The component parts are a modified tracked vehicle and Sperry Sidewinder (q.v.) 1C air-to-air guided missiles modified for ground-to-air work. The launcher is pointed in the general direction of the target, and final guidance is by the missile's own infra-red homing device.

Charles, Professor Jacques A. C.: an eminent professor at the Académie des Sciences in Paris, J. A. C. Charles was commissioned to discover a means of producing hydrogen in sufficient quantities, and to design a balloon, in an attempt to rival the efforts of the Montgolfier brothers, whose balloons were at first thought to contain an unknown gas which was obviously not as light as hydrogen (which had been discovered by the Englishman, Cavendish, in 1766). Charles solved the problem by pouring sulphuric acid on scrap iron, and designed a balloon with the help of the Roberts brothers, using their new rubberised fabric. The balloon which resulted was called a Charlière (q.v.), after Charles.

An experimental balloon was launched on 25 August 1783, but remained tethered until 27 August, when it lifted a weight of 20 lb. Charles and one of the Roberts brothers made the first flight in a hydrogen balloon on 1 December 1783, making a 27-mile flight before landing, when Aine Roberts left the balloon, allowing Charles to make a further 4½-mile flight.

Charlière: the name Charlière was given to the hydrogen balloon invented by Jacques Charles (q.v.), a professor at the Paris Academy of Sciences. The Charlière balloon incorporated many of the features generally associated with the balloon, including the spherical shape, use of hydrogen for lift (although coal gas was sometimes used as a far cheaper alternative), a valve in the crown, a car suspended from a net slung over the balloon, and a supply of ballast which could be jettisoned to extend a flight; but it lacked the rip cord for venting which was a feature of later designs.

charter: the hire of an aircraft and its crew for a flight. Charters are as old as commercial aviation itself, can be divided into several categories, and are strictly regulated. The airline providing the aircraft must, if based in the United Kingdom, have a valid air operator's certificate; the aircraft must have a certificate of airworthiness; and the crew must be properly qualified. Two of the commonest types of charter are the inclusive-tour charter, for which licences are required for each destination served by the airline, unless it is already in possession of a licence for scheduled operation to that destination, and the cost of which to a passenger must also include accommodation for a set period at the destination; and affinity-group charters, in which the aircraft is chartered by an organisation which exists for a purpose other than travel, and the members taking part in the charter party must have been members of the organisation for at least six months. There are also various types of ad hoc charters, including those for movement of ships' crews, and military charter flights.

New British regulations introduced in 1973 have involved a system of advance bookings through licensed travel agents or tour organisers, who are able to charter the aircraft.

The inclusive-tour charter is largely a British innovation, although it is now adopted in the United States and in Europe. The availability of low-cost charter flights can be credited with bringing air travel within the reach of the mass market, as well as stimulating the economy of many resorts with few resources other than labour and good weather. Scheduled airlines have been highly suspicious of charter operations for many years, and at times the International Air Transport Association (q.v.) could be counted as strongly anti-charter. Feeling is perhaps a little less intense now than before, however, with many airlines offering special inclusive-tour party rates for bulk bookings on scheduled flights – this is the I.T.X. concession – and also many I.A.T.A. airlines operate non-I.A.T.A. charter subsidiaries, including B.O.A.C.'s British Overseas Air Charter, B.E.A.'s B.E.A. Airtours, Lufthansa's Condor, and Alitalia's S.A.M. There are, nevertheless, still many airlines operating nothing but charter flights, and airlines which operate both types of air service without any subsidiary arrangement.

Cherokee, Piper: the Piper PA-28 Cherokee was first introduced in 1960 as a low-price, low-wing successor to the successful Piper Tri-Pacer, and has since been developed into a range of aircraft, from the two plus two PA-28-140, with fixed undercarriage and a 150 or 160 hp Lycoming 0-320 piston engine, to the PA-32 Cherokee Six, with six seats and fixed undercarriage, or the PA-28C Cherokee Arrow, with retractable undercarriage. The Piper PA-34 Seneca (q.v.) is a twin-engined, retractable-undercarriage development of the Cherokee Six. Performance varies upwards from the basic aircraft's 160 mph maximum speed and range of 725 miles.

Cheyenne, Lockheed AH-56A: *see* AH-56A Cheyenne, Lockheed.

Chickasaw, Sikorsky H-19: *see* S-55, Sikorsky and Westland.

Chinook, Boeing-Vertol Bv. 114: *see* Bv. 114 Chinook, Boeing-Vertol.

Chipmunk, de Havilland Canada DHC-1: *see* DHC-1 Chipmunk, de Havilland Canada.

chosen instrument: the airline selected for government support in developing and expanding international air services. Preference, and if necessary a subsidy, is given to the chosen instrument in order that its services be developed without suffering undue competition, and the chosen instrument is also the airline nominated for the national share of the services agreed in any air traffic agreement (q.v.). Before World War II the British and American governments were operating chosen instrument policies, respectively favouring Imperial Airways and Pan American World Airways.

Cierva: a company formed to develop and build autogiros or gyroplanes, the former designation being the trade name for Cierva-built versions of the gyroplane. Operations were moved from Spain to England in 1925, with the company's founder, Juan de la Cierva (q.v.), and during the period before World War II a number of designs were put into production. After the war production was hit by the advent of the helicopter, and in 1950 Cierva was taken over by Saunders-Roe (q.v.), a company which built flying-boats and helicopters and was itself merged into Westland Aircraft (q.v.) in 1960. Interest in the gyroplane for sport and pleasure flying was reawakened during the early 1960s, and in 1966 Cierva Rotorcraft

was formed out of Rotorcraft, a company dating from 1962. Gyroplane production is in hand, and development flying with the CR.LTH-1 twin 135 hp Rolls-Royce/Continental 10-360-D piston-engined helicopter with contra-rotating co-axial rotors and five seats is well advanced.

Cierva, Juan de la (1886–1936): the inventor of the gyroplane (in which rotor blades are driven by slipstream from the propeller and provide lift in place of wings), Juan de la Cierva was a Spaniard who had set himself the task of designing an aircraft which would not stall on take-off. Cierva started his work in Spain, where the world's first gyroplane flew in 1923, but moved to England in 1925, and it was in England that his company, Cierva (q.v.), was established, and that most of his subsequent developments took place. Cierva's gyroplanes were given the trade name of autogiro, which has stuck, so that many today use this term instead of the correct one. Unfortunately Cierva died in an aircraft accident before he could develop his work to its logical conclusion.

Citation, Cessna: the first Cessna civil jet aircraft, the Citation has been a logical progression upwards of Cessna's range, utilising military aircraft experience. Two tail-mounted 2,200 lb thrust Pratt and Whitney JT15D-1 turbofans give the Cessna 500 Citation a maximum speed of 450 mph and a range of 1,500 miles, while up to eight passengers can be carried.

Civil Aeronautics Board: The C.A.B. is responsible for regulating America's air transport, including service licensing, but its authority does not extend to safety regulation, for which the Federal Aviation Administration (q.v.) is responsible.

Civil Aviation Administration of China: the Civil Aviation Administration of China is the airline of the Chinese People's Republic, dating from 1964 and operating under the full control of the General Bureau of Civil Aviation. The C.A.A.C.'s predecessors are Skoga, a joint Soviet Union–Communist Chinese airline dating from 1950, and the China Civil Aviation Corporation, a wholly Chinese-owned domestic airline. The Soviet interest in Skoga was taken over by the Chinese in 1964. A very varied fleet is operated on a wide range of duties, as with the Soviet Aeroflot (q.v.), although C.A.A.C. is very much smaller. There are few modern aircraft, due to the tensions between the U.S.S.R. and China in recent years, these being limited to a quantity of Ilyushin Il-62s, Boeing 707s and British Hawker Siddeley Tridents and Vickers Viscounts. Utility aircraft and light transports are built in China, using Soviet designs.

Civil Aviation Authority: the Civil Aviation Authority was formed in early 1972 under the Air Transport Act, 1971, to take over certain of the functions of the British Department of Trade and Industry, the Air Transport Licensing Board, and the Air Registration Board. As a result the C.A.A. is now the regulatory body for air transport safety and competition in the United Kingdom. An Airworthiness Requirements Authority operates within the C.A.A. to deal specifically with safety questions.

CL-28 Argus, Canadair: see CP-107 Argus, Canadair.

CL-41 Tutor, Canadair: see CT-114 Tutor, Canadair.

CL-44, Canadair: the Canadair CL-44 is the civil development of the CC-106 (q.v.) Yukon military transport, itself a derivative of the Bristol Britannia (q.v.). The first flight of the CC-106 was in November 1959, and this aircraft differed from the CL-44, which first flew a year later, in having conventional side loading freight doors at the front and rear of the fuselage; the civil aircraft has a swing tail in which the entire tailplane and rear fuselage swings sideways on hinges. Deliveries started in July 1960, and twenty-seven aircraft were built, in addition to the CC-106 production run of twelve. The last CL-44 had a stretched fuselage, and many of the earlier machines have since been stretched as well, while one has received a wide-body Conroy Guppy conversion. Four Rolls-Royce Tyne 515/10 turboprops of 5,730 shp provide a maximum speed of 400 mph and a maximum range of 5,260 miles, while a payload of 64,000 lb can be carried. The stretched version, known as the Canadair 400 or CL-44J, can accommodate up to 214 passengers.

CL-84, Canadair: the Canadair CL-84 is an experimental V./S.T.O.L. light transport aircraft, using tilt wings for vertical take-off. Two 1,400 shp Lycoming LTC LK-4A turboprops provide for a maximum speed of 330 mph and a range of up to 350 miles. A crew of two and up to sixteen passengers can be carried. Prototypes are currently undergoing trials with the C.A.F.

CL-215, Canadair: the Canadair CL-215 is one of the very few amphibians remaining in production. A high-wing, twin-engined aircraft, of simple but modern design, it was originally intended as a water-bomber for fighting forest fires, scooping up to 1,200 gallons of water on the take-off run, and a number are in service on these duties. It is also offered as a thirty-passenger transport. Two Pratt and Whitney radial engines of 2,110 hp each provide a maximum speed of 220 mph and a range of up to 1,235 miles.

CL 834, B.A.C.: a long-range, helicopter-mounted, air-to-surface guided missile currently under development. The CL 834 is intended to equip the Westland W.G.13 Lynx helicopters of the Royal Navy's frigates and destroyers and to help compensate for the loss of carrier-borne air cover. The CL 834 is supposed to be able to destroy a missile-carrying fast patrol boat before it can come into attacking range of a defending naval vessel or the helicopter.

'Classic', Ilyushin Il-62: see Il-62 'Classic', Ilyushin.

clear air turbulence: clear air turbulence, or C.A.T., is a major problem for modern civil aircraft, largely because the absence of cloud hinders its identification on weather radar. The effect of hitting C.A.T. is a sudden and violent loss of altitude.

'Cleat', Tupolev Tu-114: see Tu-114 'Moss'/'Cleat', Tupolev.

Cloudster, Douglas: a single-seat, single-engined biplane for mail services, the Cloudster was the first Douglas commercial design when it appeared at the start of the 1920s. A 400 hp Liberty engine was used.

CM.170 Magister, Potez: the first flight of the Air Fouga-designed Magister jet trainer took place in July 1952, while the aircraft was a contender for an Armée de l'Air jet trainer requirement. Almost four hundred aircraft were built for the Armée de l'Air during the latter half of the 1950s, and other users included the Luftwaffe, Israel, Austria, Finland, the Netherlands, and Belgium, with licence-production taking place in Israel, Finland, and Germany. The Magister is a twin jet monoplane with a butterfly 'V' tailplane and twin seats in tandem for pilot and instructor. Two Turbomeca Marbore IIA turbojets of 880 lb thrust give a maximum speed of 444 mph and a range of 735 miles. About a hundred of the Armée de l'Air's aircraft are Super Magisters, with 1,058 lb thrust Marbore VI turbojets giving an improved performance. The French Navy uses CM.175 Zephyrs, equipped for carrier operations. The Magister will be replaced during the mid-1970s by the Alphajet (q.v.).

CN-: international civil registration index mark for Morocco.

'Coach', Ilyushin Il-12: *see* Il-12 'Coach', Ilyushin.

coast-to-coast flight, U.S.A.: it is perhaps suprising, in view of the effort devoted to flight across the English Channel, the Mediterranean, and the North Atlantic, that the first coast-to-coast flight across the United States did not take place until 21–24 February 1921, when a U.S. Army Air Service aircraft flew from San Diego to Jacksonville. At this same time the first coast-to-coast air mail flight was also taking place between San Francisco and Long Island. It was not until 4 September 1922 that a one-day crossing of the United States was made by air, when another United States Army Air Service pilot flew a modified D.H.4B. The first non-stop crossing took place on

2–3 May 1923, also with an Army aircraft, a Fokker T-2 transport, flying from Long Island to San Diego.

Placing the trans-U.S.A. services on a more regular footing started in July 1924, when a daily transcontinental airmail service was inaugurated. On 23 October 1930 the first regular passenger service was started by Transcontinental and Western Air, the direct predecessor of today's Trans World Airlines (q.v.).

'Cock', Antonov An-22: *see* An-22 'Cock', Antonov.

Cockerell, Sir Christopher: the inventor of the air cushion vehicle (q.v.), or hovercraft, after conducting experiments with vacuum cleaners. Support was eventually provided by the state-sponsored National Research and Development Corporation for Christopher Cockerell's first full-size experimental hovercraft, the Saunders-Roe SR-N-1 (q.v.), which first 'flew' in 1959, and was the world's first hovercraft. Later knighted for his achievement, Sir Christopher Cockerell now has no direct connection with the hovercraft construction industry.

cockpit: the position of the pilot's seat in an aircraft, although the term is not used when an enclosed flight deck (q.v.) is provided, nor when the pilot's seat is in the passenger cabin (q.v.), as on a light aircraft. A seat for a navigator/observer or instructor may also be positioned in the cockpit. Many early aircraft left the pilot lying on the wing of the aircraft but, starting with the Blériot and Antionette monoplanes, a seat was provided which was partially enclosed by the airframe. It took some years before a canopy was provided for the cockpit, and this feature did not become widespread until the latter part of the 1930s. The R.A.F.'s first aircraft to have a canopy on the cockpit was the Gladiator fighter. During the 1920s it was common for the pilot's position in an airliner to be in a cockpit, even though cabin accommodation was provided for the passengers.

C.O.D.: *see* carrier-onboard-delivery.

'Codling', Yakovlev Yak-40: *see* Yak-40 'Codling', Yakovlev.

Cody, Samuel Franklin (1861–1913): American-born, but a naturalised British subject, Samuel F. Cody had the distinction of building both the first British airship, the 'Nulli Secundus', and the first British aeroplane, the British Army Aeroplane No. 1. Both machines were built at the Royal Engineers' balloon depot at Farnborough, and flown in 1908, with the airship's engine being borrowed for the aeroplane! Cody was at this time a kite instructor at Farnborough, but he continued his work on aircraft until killed while flying one of his machines in 1913.

Cody made no great technical contribution to aviation, but worked at a time when it was an achievement merely to get off the ground, and inspired many of Britain's pioneers.

COIN: see counter-insurgency.

'Coke', Antonov An-24: see An-24 'Coke', Antonov.

'Cold', Antonov An-14: see An-14 'Cold', Antonov.

collaborative aircraft projects: a collaborative aircraft project is one on which the partners share the design, development, and production costs and work, as opposed to a simple licence-construction arrangement, in which the manufacturer who designed and developed the aircraft licenses another manufacturer to produce aircraft in addition to the licenser's own production. Nor should there be any confusion with sub-contract arrangements, under which a manufacturer has component parts of an aircraft built for inclusion in the end product. A number of collaborative ventures have been undertaken in recent years in Europe, and these have included the C-160 (q.v.) Transall, Br.1140 (q.v.) Atlantique, Jaguar (q.v.), Panavia 200 (q.v.) Panther, A.300B (q.v.), Concorde (q.v.), S.A. 330 (q.v.), SA.341, and W.G.13 (q.v.). Design leadership (q.v.) is awarded to one of the partners, and if, as usually happens, the division is on international lines, one of the participating nations receives airframe (q.v.) design leadership, and one design leadership of the engine (q.v.). Seldom is there any duplication of effort on component production, but there may often be a production line in each participating country, although there is only one A.300B production line.

The object of collaboration is to share the costs and not divide the market, and in particular to increase the size of the home market. However, development costs of such projects are officially admitted to be at least 10 per cent above those for a comparable unilateral project, and many in the aircraft industry feel that a true figure is 30 per cent above the costs of a unilateral project. Arguments on prestige, design leadership, orders, work shares, and cost control make such projects highly controversial in many quarters.

Collins, Lt Colonel Michael, U.S.A.F.: one of the crew members of the Apollo XI (q.v.) moon mission which put the first men on the moon, Lt-Colonel Michael Collins had the unenviable task of remaining in the Command Module 'Columbia', which orbited the moon while his two fellow crew members made the landing on 21 July 1969.

'Colt', Antonov An-2: see An-2 'Colt', Antonov.

Comanche, Piper: the Piper PA-24 Comanche was Piper's first low-wing all-metal monoplane when first flown in 1957, and originally used a single 180 hp or 250 hp Lycoming piston engine. Current models use either a 260 hp Lycoming 0-540 or a 400 hp I0-720, giving a maximum speed of 210 mph and a range of up to 1,200 miles in the lower-powered aircraft, while four to six passengers may be carried. *See also* Twin Comanche.

Comet, de Havilland D.H.88: see D.H.88 Comet, de Havilland.

Comet I, de Havilland: the de Havilland D.H.106 Comet I was the world's first jet airliner and made its first service flight with B.O.A.C. on 2 May 1952. Originally de Havilland had considered building a transatlantic mailplane, but rejected this idea as uneconomic and built the medium-range Comet jet instead. Four de Havilland Ghost 50 turbojet engines powered the aircraft, which could carry up to forty-eight passengers, and saw service with Aero Maritime, Canadian Pacific, and B.O.A.C., while many other airlines were interested in the aircraft. Unfortunately a series of accidents due to fatigue cracks around the windows, causing pressurisation failure at high altitude, led to the Certificate of Airworthiness being withdrawn. Two de Havilland Comet IBs remained in R.C.A.F. service, however, and the R.A.F. received a number of Comet IIs, with longer fuselages, oval (instead of rectangular) windows, and Rolls-Royce Avon engines. A further development of the Comet was the Comet III, of which only one was built, but which helped accumulate valuable experience and a clean bill of health before putting the Comet 4 (q.v.) into production.

Comet 4, de Havilland: the de Havilland Comet 4 was the final development of the Comet series of airliners. After a first flight on 27 April 1958, the aircraft was able to make the world's first transatlantic jet flight for B.O.A.C. on 4 October 1958. There were three versions of the aircraft: the 4; the 4B for B.E.A., with a stretched fuselage, clipped wings, and thrust reversers; and the 4C, with the 4B fuselage and thrust reversers, and the 4 wing. Four Rolls-Royce Avon 525 turbojets of 10,500 lb thrust each provided a maximum speed of about 550 mph and a range of up to 4,000 miles, while passenger capacity varied on length of fuselage, pitch, and class-mix for seventy to a hundred seats. Production passed with de Havilland into the Hawker Siddeley Group in 1960. A maritime-reconnaissance development is the Hawker Siddeley HS 801 Nimrod (q.v.).

Commando, Curtiss C-46: *see* C-46 Commando, Curtiss.

Commonwealth Aircraft: originally formed in 1936 by a group of Australian industrialists, Commonwealth Aircraft first produced the North American NA-33 trainer as the Wirraway. A number of other licence-production programmes followed with the advent of World War II, and the nationally-designed CA-2 Wacket trainer was also put into production, and followed by the CA-12 Boomerang ground-attack fighter. After the war the company produced the Australian-designed Winjeel basic trainer, and licence-built the North American F-86 (q.v.) Sabre, fitting these jet fighters with Rolls-Royce Avon turbojets.

competition: air transport competition is severely regulated both by national governments, through agencies such as the Civil Aviation Authority (q.v.) C.A.A. in the United Kingdom and the Civil Aeronautics Board (q.v.) in the United States, and by the International Air Transport Association (q.v.). The extent of such control goes beyond saying which airline may be permitted to fly on certain routes, to cover such items as fares, frequencies, seat pitch, meal service, and in-flight entertainment. In some cases departure times are strictly regulated and even the choice of aircraft is determined, usually by restrictions on capacity. International services are also covered by bilateral air traffic agreements (q.v.), which cover frequency and capacity offered, and a chosen instrument (q.v.) policy sometimes dictates which airline should receive the national allocation of flights on a route.

composite aircraft: the term 'composite aircraft' is applied to those aircraft which in fact consist of two aircraft, the mother plane and a smaller aircraft which is used to complete the journey or mission. It is not usually used to cover airship-borne U.S. Navy aircraft of the 1930s, nor for booster rockets used in launching space rockets or guided missiles.

The first composite aircraft experiment was in 1916, when the Royal Naval Air Service used a Felixstowe Baby tri-motor flying-boat to carry a Bristol Scout fighter, which was released from the patrolling flying-boat when an attacking Zeppelin airship appeared. Commercial use of the concept occurred in 1938 with the Short Mayo (q.v.) composite aircraft, which consisted of a Short S.21 flying-boat which carried a Short S.20 four-engined seaplane to be released en route with a consignment of airmail. Imperial Airways used the Short Mayo experimentally on the North Atlantic and African routes. Militarily, the Luftwaffe 'Mistletoe' bombs consisted of a Messerschmit Bf. 109 (q.v.) fighter and an unmanned Junkers Ju. 88 bomber, which was released near to the target. Many Japanese 'Kamikaze' (q.v.) suicide aircraft were also air-launched.

After World War II, experiments were conducted using a Convair B-36 (q.v.) heavy bomber which carried a Republic F-84F Thunderjet fighter for fighter cover when outside of the range of escort fighters, the F-84 returning to the B-36 after completing its defensive duties. The Bell X-1 and X-2 research aircraft were always air-launched from Boeing B-29 Superfortress bombers, while more recently the North American X-15 was launched from a Boeing B-47 bomber.

Concorde, B.A.C.-Aerospatiale: originating in the B.A.C. and Sud Aviation design studies for a supersonic airliner, which appeared separately during the early 1960s, Concorde became an Anglo-French collaborative aircraft project in 1962, receiving full support from the British and French Governments. Sud Aviation (now Aerospatiale) received design leadership of the aircraft, with production and design work shared with the British Aircraft Corporation, while Bristol Siddeley (now Rolls-Royce) was given design leadership on the engines, with design and production work shared with SNECMA. Although there is no production duplication on the components and sub-assemblies, there are two final assembly lines, in Britain and in France.

The first flight of a Concorde prototype was of a French-assembled aircraft on 2 March 1969, while a British aircraft flew on 9 April 1969. Mach 2·0 was reached in November 1970. The production aircraft have longer fuselages and more powerful engines than the prototypes, and are due to enter airline service during spring 1975. Four Rolls-Royce-SNECMA Olympus 602 turbojets of 38,050 lb thrust each with reheat provide for a maximum speed of almost 1,400 mph and a range of up to 4,000 miles,

while a 130 passenger or 20,000 lb load can be carried. An improved version with up-rated powerplants should enter airline service in 1977.

Condor, Focke-Wulf Fw.200c: see Fw.200c Condor, Focke-Wulf.

Congreve, Sir William: an Englishman, Sir William Congreve effectively re-invented the artillery rocket during the early part of the nineteenth century, starting his work around 1805. Interest in the rocket as an artillery weapon had been aroused by rocket attacks on the British Army in India by forces belonging to the Sultan Tipu. Although even these early rockets made effective weapons if used properly, they never really displaced conventional artillery.

Consolidated Aircraft: see Convair.

Constellation, Lockheed: the Constellation was originally developed as an airliner for T.W.A., starting in 1939, but the project became primarily military with the advent of World War II, and the civil L-049 model was in fact the U.S.A.A.F.'s C-69 transport. After the war a number of C-69s on the production line were modified for civil use, while a development, the L-749 Constellation, entered airline service in 1948. A stretched-fuselage version, with extended range and up-rated engine development, the L-1049 Super Constellation, first flew in October 1950. An even longer-range development, the L-1649 Starliner, followed but failed to enjoy the same degree of success as the L-749 and L-1049. Four 3,400 hp Wright R-3350-CA18-EA3 turbo-compound piston-engines give the L-1049 a maximum speed of 320 mph and a maximum range of 4,800 miles, while up to 100 passengers can be carried. Few Constellations, Super Constellations, or Starliners remain in service today, and those that do are usually EC-121 airborne-early-warning versions for the United States Air Force and Navy (see C-121).

Continental: one of the leading American light aircraft engine manufacturers. Continental engines are used in a variety of light aircraft, although by far the major user is Cessna. Rolls-Royce Motors (q.v.) is Continental's licensee for the European market, including Reims-produced Cessnas. Only piston engines are built, and the latest of these is the Tiara range. Primarily a manufacturer for the civil market, the company's products also appear in a few military aircraft, notably the Cessna 0-1 (q.v.) Bird Dog and 0-2 (q.v.) Super Skymaster.

Continental Airlines: a U.S. domestic trunk airline, Continental Airlines dates from 1934 and the formation of Varney Speed Lines, which operated Lockheed Vegas on a number of routes in California, including San Francisco to Los Angeles. The present title was adopted in 1937, although it was not until 1955 that the first trunk route, Chicago to Los Angeles, was introduced. During 1955 Continental acquired Pioneer Air Lines, a small company dating from 1945. Developments since 1955 have included the addition of many more trunk routes, and services to Hawaii.

Currently, Continental Airlines operates a dense network of scheduled services on the United States West Coast, and from the West Coast to other major centres in the United States, including Hawaii. A 31 per cent interest is held by Continental in Air Micronesia, which operates in the U.S. Trust Territory of the Pacific, and the airline also owns Continental Air Services, a charter and general aviation operator which is primarily active in South-East Asia. Continental's own fleet of aircraft includes Boeing 747s, 720s, 727s, and 707s, with Douglas DC-9s and DC-10s.

contour flying: contour flying is the modern development of the old art of hedge hopping, practised during two world wars as a means of surprising the enemy, flying below the angle of fire of anti-aircraft artillery and, later, also having the advantage of avoiding detection by defending radar. High speed flying at low level requires a radar-computer-auto-pilot system to handle the aircraft and make decisions on avoiding obstructions more quickly than a human pilot could. Most modern bombers use contour flying

techniques, and notable among these is the Buccaneer (q.v.).

Convair: currently a subsidiary of the General Dynamics Corporation (q.v.), Convair was originally formed in 1923 as Consolidated Aircraft, and the first product was a trainer originally designed by the Dayton-Wright concern, itself dating from 1917. Other trainer designs entered production throughout the 1920s, until the company produced its first airliner design, the Fleetster, in 1930. During the 1930s the company produced a monoplane fighter, the PB-2a, and a ground-attack aircraft based on a Vultee design, before developing the first of the Catalina (q.v.) flying-boats and amphibians, which were to prove so successful during World War II, and some of which survive to this day. Possibly even more famous than the Catalina was the B-24 (q.v.) Liberator heavy bomber of World War II, and Convair also produced a maritime-recon-naissance version, the Privateer, for the United States Navy.

After World War II the company experimented with a mixed-power turbo-prop and turbojet fighter, the XP-81, while the giant B-36 (q.v.) bomber, originally designed to be able to bomb Germany in the event of Great Britain collapsing, entered production and for many years served as an important part of Strategic Air Command of the U.S.A.F. A jet-powered version of the B-36, the YB-60, flew but did not enter production. A post-war airliner development was the 240, 340, and 440 range of twin-engined, short-haul airliners, of which a military version was the C-131 (q.v.) Samaritan series.

The first Convair jet to enter production appeared during the mid-1950s. This was the F-102 (q.v.) Delta Dagger fighter for the U.S.A.F., and it was followed by a development, the F-106 (q.v.) Delta Dart, some of which remain in service. The first Mach 2·0 bomber, the B-58 (q.v.) Hustler, also appeared from the Convair stable in 1958, and was soon followed by the 880 and 990 airliners, although these were not a commercial success for Convair. Plans for a short-haul jet airliner during the late

1960s were not proceeded with, and today the company is engaged in sub-assembly manufacture for other General Dynamics products.

co-operative projects: *see* collaborative aircraft projects.

'Coot', Ilyushin Il-18: *see* Il-18 'Coot', Ilyushin.

Corsair, Vought F4U: the Vought, or Chance-Vought, F4U Corsair was a low-crank-wing monoplane fighter for the United States Navy, which first entered service in 1941. Primarily a carrier-borne aircraft, it did however also serve as the backbone of the R.N.Z.A.F.'s wartime strength. The Corsair was active in both World War II and the Korean War, and many of the aircraft found their way to Latin American air forces afterwards. A single 2,100 hp Pratt and Whitney R-2800-18W piston engine provided a 450 mph maximum speed for the F4U-4 version, while a warload of up to 3,200 lb could be carried on under-wing strong-points.

Corsair, Vought 02U: the first aircraft to bear the Corsair name, the 02U first flew in November 1926, and was used for a number of years by the United States Navy as an observation aircraft, equipped either with a wheeled undercarriage or floats. The 02U was a development of the U0-1, which was one of the first production aircraft to be fitted with an arrester hook for carrier operations, in this case aboard the U.S.S. *Langley*.

Corsair II, Ling-Temco-Vought A-7: *see* A-7 Corsair II, Ling-Temco-Vought.

counter-insurgency: a modern concept of considerable importance, counter-insurgency, or COIN, duties are in effect an up-dated version of the old colonial police duties. Recent examples of COIN duties include the action taken by British and Commonwealth forces against Indonesian terrorists in Malaysia during the early 1960s, a considerable amount of the action by the United States and South Vietnamese forces in South Vietnam and Cambodia during the Vietnam War, and operations by the Government of Sri Lanka (formerly Ceylon) against terrorists. Apart from the extensive use of helicopters for rapid reinforcement of positions and supply duties, a number of aircraft types have been either developed specially for, or modified for, counter-insurgency operations, including the North American 0V-10A (q.v.) Bronco and the Douglas A-1 (q.v.) Skyraider.

CP-: international civil registration index mark for Bolivia.

CP-107 Argus, Canadair: development of the CP-107 Argus started in 1954, with the grant of a Britannia (q.v.) licence to Canadair, which had a contract to meet an R.C.A.F. requirement for a maritime-reconnaissance aircraft to replace its Avro Lancasters. Although the Britannia wing, tailplane, and other components were retained, a completely new fuselage was designed, and piston engines were used to extend the low altitude range of the aircraft. Deliveries to the R.C.A.F. started in 1957. Four Wright R-3350-EA-1 piston engines of 3,700 hp produce a maximum speed of 315 mph and a range of up to 4,000 miles, while 8,000 lb of weapons can be carried internally, with a further 3,800 lb on underwing strong-points. A replacement aircraft is under consideration.

C.P.-Air: Canadian Pacific Airways, or C.P.-Air, is the air transport subsidiary of a major group, originally formed to operate a railway, which now also operates shipping, road transport and hotels, and undertakes oil, gas, and timber production and property development.

A plan by the Canadian Government during the 1930s to form a joint private and state enterprise airline with Canadian Pacific came to nothing, and it was not until 1942 that Canadian Pacific Airlines was formed on the amalgamation of ten small independent airlines, the largest of which was Canadian Airways, in which Canadian Pacific had held the controlling interest for some twelve years. The railway company itself had held authority to operate aircraft since 1919, and in 1940 had assisted in the formation of the North Atlantic ferry service for delivery of war-planes to the R.A.F. and R.C.A.F. in Great Britain. A number of flying schools were also operated during the war period. After World War II military charter flights were operated, with this traffic proving to be particularly important during the Korean War.

The airline initiated trans-Pacific services to Sydney and to Hong Kong in 1949, while in 1955 the Quebec local services were traded with the then Trans-Canada Airlines for a route to Mexico City, which was subsequently extended to Buenos Aires. During the early 1950s, the airline was one of the operators of the de Havilland Comet I, the world's first jet airliner. A polar route from Vancouver to Amsterdam was inaugurated in 1955, and in 1959 the remaining third level operations were exchanged for trans-continental rights, and Bristol Britannia turboprop airliners ordered.

The present title was adopted in 1968, and today the airline operates across the Pacific, to Latin America, and to continental Europe, as well as being allowed to compete with Air Canada on trans-Canada services. A fleet of Douglas DC-8s and Boeing 727s and 737s is operated.

CR-: international civil registration index mark for Portuguese Overseas Provinces (mainly in Africa).

'Crate', Ilyushin Il-14: see Il-14 'Crate', Ilyushin.

'Creek', Yakovlev Yak-12: see Yak-12 'Creek', Yakovlev.

crop dusting/spraying: the aircraft is an effective means for dusting or spraying crops as a protection against pestilence, the advantages being those of speed and the area which can be covered in one pass, as well as the important point of not damaging crops in large fields while attempting to spray those in the middle. One of the earliest operators of crop-spraying aircraft was Huff Daland Crop Dusters, formed in 1925, which eventually became Delta Airlines, one of the main American domestic airlines. A number of aircraft manufacturers, including Piper

(q.v.), consider this market so important that a special range of aircraft exists, and of course helicopters are also useful for this work.

Crotale: the French Crotale surface-to-air guided missile system has been developed for the French and the South African armed forces (the latter use the term Cactus) by Thomson-C.S.F. and Engins Matra. It will also be used by the Lebanon. Crotale is an all-weather system designed to destroy aircraft flying as fast as Mach 1·5, and as low as 180 feet. A single-stage solid fuel rocket is used, and the missile is radar-guided onto its target.

Crusader, Ling-Temco-Vought F-8: see F-8 Crusader, Ling-Temco-Vought.

Cruzeiro do Sul: the Brazilian airline Cruzeiro do Sul was originally formed in 1927 by the German Condor Syndicate, backed by Deutsche Lufthansa (q.v.) and shipping interests, largely as a feeder to air and sea services across the South Atlantic. The present title was adopted in 1942 when German control in the airline was replaced by Brazilian interests. The airline has since become an important Brazilian domestic airline, with some international services to neighbouring states. A fleet of Boeing 727s, Aerospatiale Caravelles, N.A.M.C. YS-11s, and Douglas DC-3s is operated.

CS-: international civil registration index mark for Portugal.

C.S.A.-Czechoslovak Airlines: C.S.A. dates from 1923, when a nationalised air transport organisation was formed with the title of Czechoslovak State Airlines, using military aircraft and personnel on a domestic route network which was built up over the years, starting in 1924. Another airline, C.L.S. inaugurated services to other major European centres from Czechoslovakia during this period. Between them the two airlines had an extensive network, with C.S.A. operating a few international services, at the time of the German occupation of Czechoslovakia in 1939. C.L.S. was operating Douglas

DC-2s and DC-3s in its fleet during the late 1930s.

After World War II operations restarted in 1946 using ex-Luftwaffe Junkers Ju. 53/3Ms, and these were soon joined by ex-military C-47s. During the latter part of the decade Czechoslovakia was drawn firmly into the Soviet Bloc, and its routes primarily linked Prague with other East European capitals. The fleet consisted of Ilyushin Il-12 and Il-14 piston-engined airliners, with the first jets, Tupolev Tu-104s, arriving in 1957. On the usual Communist pattern, the airline undertakes every kind of air transport activity. The present fleet includes Ilyushin Il-62s, Il-18s, and Il-14s, and Tupolev Tu-104, Tu-124, Tu-134, and Tu-154 airliners.

CT-114 Tutor, Canadair: the first flight of the Canadair CL-41 Tutor was in January 1960, and the following year the R.C.A.F. ordered 190 Tutors under the designation CT-114. Deliveries started in October 1963. A single 2,633 lb thrust General Electric J85 turbojet gives a top speed of 500 mph, with a maximum range of 950 miles. A counter-insurgency version, the CL-41G, is in service with the Royal Malaysian Air Force, and uses an up-rated J85-24 turbojet with a thrust of 2,950 lb. Underwing strongpoints allow or a warload of up to 3,500 lb.

C.T.A.: the abbreviated name of the Brazilian Centro Tecnico Aerospacial, or the Aerospace Technical Centre, a government-owned aircraft design and development organisation. Actual production of C.T.A. designs usually passes to Embraer (q.v.), a joint state and private-enterprise concern. C.T.A.'s current design is the C-95 (q.v.) Bandeirante light transport.

C.T.M.: *see* capacity ton-miles.

'Cub', Antonov An-12: *see* An-12 'Cub', Antonov.

Cubana: the Cuban national airline, Empresa Consolida Cubana de Aviacion, was formed in 1929 as the Compania Cubana de Aviacion S.A., and the present title was not adopted until nationalisation by the revolutionary Castro régime in 1959. At the time of nationalisation the airline was operating services throughout the Caribbean with a new fleet of Bristol Britannia turboprop airliners. Currently, the airline operates domestically and to a limited number of European destinations, using a fleet of Bristol Britannias, Ilyushin Il-14s and Il-18s, Antonov An-24s, and Douglas DC-3s.

Cuckoo, Sopwith: the Sopwith Cuckoo has the distinction of being the world's first purpose-built carrier-borne torpedo bomber. After the first flight in June 1917, the aircraft entered service aboard H.M.S. *Argus* in October 1918 – too late for action in World War II. A large number of Cuckoos were built, and the work had to be subcontracted to other manufacturers, including Blackburn. A single-engined biplane, the Cuckoo was powered by a 200 hp Hispano-Suiza engine.

Curtiss, Glenn Hammond (1878–1930): One of the most notable of the early American pioneers of aviation, Glenn Curtiss was the next American to fly, after the Wright brothers. A motorcycle manufacturer, he formed the aircraft firm which bore his name in 1908, and produced its first design, the June Bug, that same year. His entry at the First International Aviation Meeting at Reims in 1909 was amongst the more important, and included victory in one of the air races.

He is frequently criticised for his re-design of Langley's Aerodrome (q.v.), which he undertook to investigate the aircraft's flying capabilities before the outbreak of World War I.

Curtiss Aircraft: one of America's first aircraft manufacturers, Curtiss was formed in 1908 by Glenn Curtiss (q.v.), and produced its first design, the June Bug, that same year. After some success at the Reims International Aviation Meeting of 1909, the company produced a number of aircraft for racing and training, including the JN-2 and JN-4 biplanes which were used by the American Army for flying training during the war. After the war the Curtiss NC flying-boats were the first aircraft to fly across the Atlantic (*see* Atlantic flights) in 1919, although they used the southerly route via the Azores and took some time, due to stops *en route*. A number of the designs which followed were also flying-boats or seaplanes, and amongst these was the CR-3 of 1924, using a Curtiss D-12a engine and establishing a world speed record of 227 mph. A return to training aircraft marked the late 1920s and the 1930s, although there was also the Sparrowhawk fighter, which participated in experiments in flying to and from U.S. Navy airships.

The mid-1930s saw Curtiss produce the last American biplane airliner, the Condor; this aircraft also had the distinction of being the first sleeper plane. The Hawk fighter saw limited service during the late 1930s and during the first part of World War II, as did its successor, the P-40 Warhawk (R.A.F. Tomahawk). The Hawk itself was the first Curtiss all-metal monoplane. Training types for World War II included the AT-9 Jeep and the SNC-1, while the SB2C Helldiver carrier-borne dive-bomber saw extensive use in the Pacific. However, the most important Curtiss wartime product was the C-46 (q.v.)

Commando transport, some of which remain in service to this day.

CV-: international civil registration index mark for Cuba.

CX-: international civil registration index mark for Uruguay.

Cyprus Airways: Cyprus Airways dates from 1947, when it was formed with the Cyprus Government and B.E.A. each holding a 40 per cent interest, and private Cypriot interests holding 20 per cent. Initially aircraft were leased from B.E.A., but the airline obtained aircraft of its own in 1948, reverting to leased B.E.A. aircraft in 1955 and 1956 after two very bad years due to terrorist activity on the island. Currently, five Hawker Siddeley Trident airliners on lease from B.E.A. operate on services in the Eastern Mediterranean and to Western Europe. Ownership today is divided between the Government of Cyprus, with 53·2 per cent, B.E.A. with 22·7 per cent, and private interests, which hold the remainder.

D

D-: international civil registration index mark for Federal Germany.

D.VII, Fokker: undoubtedly the most successful and most potent German warplane of World War I, the Fokker D.VII biplane did not appear until 1918, replacing the Albatros D.V and other fighters, and being built by both Fokker and Albatros. A 185 hp B.M.W. engine powered the aircraft, some of which were built in the Netherlands by Anthony Fokker after the Armistice, using smuggled parts.

D.XII, Pfalz: a German biplane fighter of the late World War I period, the Pfalz D.XII was an Albatros D.V replacement, and was powered by a 160 hp Mercedes engine.

Daedalus: from Greek mythology, the story of Daedalus and his son, Icarus, is one of the oldest concerning men and flight. Daedalus and Icarus were prisoners of King Minos on the island of Crete, and their only possible means of escape to the mainland of Greece was by flight. Daedalus made wings from bird-feathers and wax, enabling the two men to fly like birds away from Crete. Icarus ignored the advice of his father and flew too near to the sun, which melted the wax in his wings, and he crashed to his death.

Daimler Benz: *see* Mercedes Benz.

Dakota, Douglas C-47/DC-3: *see* C-47 Dakota *and* DC-3, Douglas.

Daland, Huff: Huff Daland founded the company which bore his name, Huff Daland Crop Dusters, in 1925, at which time it was the world's first aerial cropspraying company. The main reason for the company's formation was the boll weevil, a persistent pest which seriously affects the cotton crop in the southern United States. Passenger operations started in 1929, and the company has since developed into the Delta Airlines (q.v.) of today.

Dassault, Marcel: after changing his name from Marcel Bloch (q.v.) at the end of World War II, Marcel Dassault founded the Avions Marcel Dassault (q.v.) company, which merged with the Breguet (q.v.) concern in early 1972.

da Vinci, Leonardo: *see* Leonardo da Vinci.

DC-1, Douglas: the first of the DC- (Douglas Commercial) series of airliners, only one DC-1 was built, in 1933 for Transcontinental and Western Air, the predecessor of Trans World Airlines. A low-wing monoplane, the DC-1 used two Wright radial engines, giving a maximum speed of around 180 mph. Fourteen passengers could be carried. Both the DC-2 (q.v.) and the DC-3 (q.v.) were developments of the DC-1.

DC-2, Douglas: two 700 hp Wright radial engines powered the DC-2, which first flew in 1934. Unlike the DC-1 (q.v.), which was a one-off aircraft, the DC-2 enjoyed some considerable success in its own right in the United States and Europe, and was far from being just a stepping-stone on the way to the DC-3 (q.v.). Amongst the airlines which operated the DC-2 can be counted T.W.A., K.L.M., and Swissair. A military development was the B-18 (q.v.) bomber.

DC-3, Douglas: the most successful transport aircraft yet, the Douglas DC-3 was the ultimate development in the series which started with the DC-1 (q.v.), and first flew in 1936 as the DST-Douglas Sleeper Transport for trans-U.S.A. services. It could accommodate either fourteen sleeper passengers or twenty-one seated passengers. More than 10,000 DC-3s were built during World War II as C-47 (q.v.) Dakotas for the Allied air forces, and the aircraft was also built in the Soviet Union as the Lisunov Li-2. Some 800 DC-3s were built after the war in addition to the large number of military C-47s (q.v.) which were converted for civil duties. B.E.A.'s Pionairs were modified DC-3s with thirty-two seats or, in the freight version, space for up to 6,620 lb of cargo. Many DC-3s still remain in service, mainly using two Pratt and Whitney R-1830-90C piston engines of 1,200 hp each, giving a maximum speed of 200 mph and a range of up to 1,500 miles.

DC-4 Skymaster, Douglas: the DC-4 was the first four-engined Douglas airliner design, and first flew in early 1939. The first production models had a triple fin and were intended for American domestic airlines. However, with the advent of World War II the DC-4 was soon conscripted into U.S.A.A.F. service as the C-54, with a number of design modifications which included a single fin and uprated engines. After the war some limited production was undertaken by Douglas for civil users, but most of the aircraft which entered airline service were converted military aircraft, while Douglas concentrated on the DC-6 (q.v.) development. Canadair also built, during the postwar period, the DC-4M North Star with Rolls-Royce Merlin engines, and this was also B.O.A.C.'s Argonaut class. Most DC-4s use four 1,450 hp Pratt and Whitney R-2000-5D piston engines, giving a maximum speed of 210 mph and a range of up to 2,000 miles. Payload is 20,000 lb or up to eighty passengers.

DC-5, Douglas: the only DC- airliner, apart from the basically prototype DC-1, not to achieve success, and the only high-wing aircraft of the range, the DC-5 was first flown in early 1939. Altogether only a dozen DC-5s were built, and most models of this twin-engined, tricycle undercarriage airliner were bought by the U.S.A. and U.S.A.A.F.

DC-6 Cloudmaster, Douglas: a post-World War II development of the Douglas DC-4 (q.v.), the Douglas DC-6 incorporated a pressurised and stretched fuselage with up-rated engines giving an all-round improvement in performance. The U.S.A.F.'s C-118 (q.v.) was the military transport version of the DC-6, and civil developments included the DC-6A Liftmaster freighter and the DC-6B. Altogether some 700 of the aircraft were built. The DC-6B uses four 2,400 hp Pratt and Whitney R-2800 CB-16 piston engines to give a maximum speed of 320 mph and a range of more than 3,000 miles, while up to 108 passengers can be carried. A further development of the basic DC-4 (q.v.) was the DC-7 (q.v.).

DC-7, Douglas: the fastest piston-engined airliner built, the DC-7 was the final development of the DC-4 (q.v.)/DC-6 (q.v.) family, with a considerably extended range. In production during the mid-1950s, the aircraft used four 3,400 hp Wright R-3350-18EA4 Turbo Compound piston engines for a maximum speed of more than 360 mph and a range of up to 6,000 miles, while a maximum payload of 22,000 lb or 100 passengers can be carried. A few remain in service.

DC-8, McDonnell Douglas: the Douglas-designed DC-8 project was started in June 1955 as a rival to the Boeing 707, Convair 880 and de Havilland Comet 4 jet airliners. A number of variations were built almost from the introduction of the aircraft to airline service in 1959, including the DC-8 Srs. 10 with 13,500 lb thrust Pratt and Whitney JT3C-6s, and the Srs. 20 with 15,800 lb thrust JT4A-3s, both of which were meant for U.S. domestic airlines. The longer-range DC-8 Srs. 30 used the Srs. 20's engines, but had a much-increased fuel capacity, while the Srs. 40 used Rolls-Royce Conway 509 turbojets of 17,500 lb thrust, and the Pratt and Whitney 18,000 thrust JT3D-3 turbofan powered the Srs. 50, which was also available as a convertible passenger-freight aircraft, the Jet Trader.

The first of a new range of DC-8s, designed the Srs. 60 and sometimes known as the 'Super Sixty', flew in 1966. All the Srs. 60 aircraft use four 18,000 lb thrust Pratt and Whitney JT3D-3D turbofans, but the Srs. 61 has a thirty-six foot fuselage extension and accommodation for up to 250 passengers. The Srs. 62 has only a small fuselage stretch compared with the standard DC-8s, but the aircraft has a much extended range with extended wing tips, while the Srs. 63 has the fuselage of the Srs. 61, with the range and the wings of the Srs 62.

A DC-8 Srs. 50 can carry a maximum of 190 passengers or up to 34,360 lb of freight, and has a maximum cruising speed of 580 mph and a maximum range of 5,750 miles. The Srs. 63 can accommodate up to 250 passengers, and has a maximum cruising speed of 580 mph and a maximum range of 8,000 miles.

DC-9, McDonnell Douglas: work on the Douglas DC-9 short-range jet airliner started in April 1963, and the first flight was in February 1965, with deliveries starting in the November. The basic model was the DC-9 Srs. 10, and this was soon followed by the Srs. 20, the Srs. 30, with stretched fuselage and high-lift devices on the wings, and the Srs. 40, with up-rated engines and a further fuselage stretch. The Srs. 30 uses two 14,000 lb thrust Pratt and Whitney JT8D1 turbofans, giving a maximum speed of up to 600 mph and a range of 1,500 miles, while up to 115 passengers can be carried, according to class mix, seat pitch, and fuselage length. DC-9s in U.S.A.F. service for Medivac duties are known as the C-9 Nightingale.

DC-10, McDonnell Douglas: the McDonnell Douglas DC-10 trijet airbus for medium- and long-range routes first flew in 1971, with deliveries starting in early 1972. Three General Electric CF-6-6D advanced technology turbofans of 40,000 lb thrust each provide for a maximum speed of up to 640 mph and a range of 3,400 miles in the standard DC-10-10 version, while up to 380 passengers can be carried. Extended range versions of the basic aircraft are in production, including the CF-6-50-powered DC-10-30, with a 4,250 mile range, and the DC-10-40, with a 4,500 mile range, or up to 6,500 miles with reduced payload; the latter aircraft uses a Pratt and Whitney JT9D turbofan. Twin and four jet variants for short- and very-long-range services respectively are under consideration.

Defender, Britten-Norman: the military version of the BN-2 (q.v.) Islander aircraft, using two Lycoming piston engines and fitted with underwing strongpoints for stores. The aircraft is in service with a number of air arms and can undertake counter-insurgency, ground attack, and light transport and communications duties.

de Havilland, Captain Sir Geoffrey (1882–1965): after producing a series of designs while working at the Army Aircraft Factory (later the Royal Aircraft Factory), Sir Geoffrey de Havilland (q.v.) joined the Aircraft Manufacturing Company (*see* Airco) in 1914 as Chief Designer. De Havilland's Airco designs included the D.H.1, D.H.2 (q.v.) fighter, and D.H.4 (q.v.) bomber biplanes. After World War I he established his own company, and during the inter-war period produced a series of successful light and light transport aircraft. During World War II the company, de Havilland Aircraft (q.v.) produced the D.H. 98 (q.v.) Mosquito. After the war the world's first jet airliner, the Comet I (q.v.), was built by de Havilland Aircraft.

de Havilland Aircraft: founded after World War I by Captain Sir Geoffrey de Havilland (q.v.), de Havilland Aircraft produced a series of airliner, light transport and light aircraft designs, including the D.H.60 (q.v.) Moth series of light aircraft and trainers, the D.H.84 (q.v.) Dragon series of light biplane transports, and the D.H.66 Hercules trimotor airliner of the 1920s. In 1934 the de Havilland D.H.88 Comet twin-engined monoplane racer appeared. D.H. aircraft dating from before de Havilland Aircraft was founded were those built by the Aircraft Manufacturing Company, of which Geoffrey de Havilland was the Chief Designer. World War II interrupted the further development of two promising airliner designs, the Albatross (q.v.) and the Flamingo, which had entered service with Imperial Airways shortly before the outbreak of World War II. Wartime aircraft included the D.H.82 Tiger Moth (*see under* D.H.60 Moth) basic trainer, D.H.89 Dragon Rapide (*see* D.H.84 Dragon) light transport, and the D.H.98 (q.v.) Mosquito light bomber, and its successor, the Hornet (q.v.).

After World War II de Havilland rapidly

de Havilland Canada

put a range of jet-powered aircraft into production, many of which used the company's own engines. The world's first jet trainer was the de Havilland Vampire, (q.v.), which was also available as a fighter and fighter-bomber, as was its successor, the Venom (q.v.)/Sea Venom; both aircraft were built under licence in Europe. The final development of these twin-boom jet fighters was the Sea Vixen (q.v.). Unfortunately the success of these military aircraft was marred by failure of the world's first production jet airliner, the Comet I (q.v.). Other civil projects enjoyed considerable success, however, notably the Dove (q.v.) and Heron (q.v.) light transports, while the Canadian subsidiary, de Havilland Canada (q.v.) produced a range of aircraft starting with the DHC-1 (q.v.) Chipmunk trainer. A number of experimental aircraft were also built, including the tailless Swallow.

The Comet eventually surmounted its earlier problems, and de Havilland Comet IIs entered service with the R.A.F., while the Comet 4 (q.v.) became the first jet airliner to operate transatlantic scheduled services. Successors to the Dove and the Comet 4B were evolved in the D.H. 125 executive jet and the Trident (q.v.) airliner, but these projects entered production after de Havilland was merged with other companies to form Hawker Siddeley Aviation (q.v.). The guided weapons side of de Havilland's varied aviation interests became a part of Hawker Siddeley Dynamics (q.v.); the aero-engine activities, which were not much younger than the airframe business, became a part of Bristol Siddeley Engines (q.v.); while de Havilland Canada retained its identity, and de Havilland Aircraft of Australia was merged with another Hawker Siddeley predecessor company, Hawker, to become Hawker de Havilland.

de Havilland Canada: de Havilland Aircraft of Canada was formed in 1947 out of an earlier de Havilland assembly plant to produce a range of indigenous aircraft designs. The company enjoyed immediate success with the DHC-1 (q.v.) Chipmunk trainer, chosen by many air forces, and built in Canada and Great Britain. Later

aircraft, such as the DHC-2 (q.v.) Beaver and the DHC-3 (q.v.) Otter, helped to establish the company's reputation for building short take-off utility aircraft, and these were followed by the larger DHC-4 (q.v.) Caribou and DHC-5 (q.v.) Buffalo. A very successful feeder liner is the DHC-6 (q.v.) Twin Otter development. Currently the company is partnering America's Boeing (q.v.) Corporation in a NASA-sponsored project for a S.T.O.L. civil airliner based on the DHC-5 Buffalo, with jet engines in place of the basic aircraft's turboprops. The company also has its own plans for a civil S.T.O.L. airliner, the D.H.C.-7 (q.v.), which is being developed with Canadian Government backing.

The Canadian Government took an option in the purchase of de Havilland Aircraft of Canada in 1972.

de-icing equipment: an invention of the inter-war period, de-icing equipment did not become commonplace on civil and military aircraft until well into the 1930s, and early examples of aircraft so equipped were easily identifiable by black strips on the leading edges of the wings. The need for such equipment requires little explanation, since the rapid icing up of an aircraft at altitude, helped by condensation on the wings, had led to many fatal accidents due to the extra weight placed upon the airframe. The basic theory of de-icing equipment is that a de-icing fluid, such as alcohol or some other substance with a low freezing point, is sprayed on to the aerodynamic surfaces of the aircraft as and when necessary.

Delfin, Aero L-29: see L-29 Delfin, Aero.

Delta 2, Fairey: the Fairey Delta 2, or F.D.2, was a high-speed research aircraft, using a single Rolls-Royce Avon turbojet with reheat and having a delta wing configuration and no tailplane. On 10 March 1956 the Delta 2 set a world air speed record of 1,132 mph near Chichester, Sussex. Earlier trials with the Delta 2 and its predecessor, the Delta 1, are sometimes claimed to have influenced Marcel Dassault when designing his successful Mirage (q.v.) series of fighters.

Delta Airlines: originally formed in 1925

as Huff Daland Crop Dusters (*see* Daland, Huff), the company started passenger operations in the state of Georgia in 1929. The airline activities have developed extensively since that time, and have included the acquisition of Chicago and Southern Airlines (which dated from 1934) in 1953. Basically Delta, which is one of the major U.S. domestic airlines, operates from the southern United States, with services to the major northern, western and eastern cities, and throughout the Caribbean area. In August 1972 Delta acquired Northeast Airlines, which dated from 1933, when it was formed as Boston-Maine Airways. Currently a fleet of Boeing 747s and 727s, Lockheed TriStars, Douglas DC-8s, DC-9s, and DC-10s, Convair 880s, and Lockheed Hercules aircraft is operated.

delta wing: the first known delta wing design was that of two Englishmen, Butler (q.v.) and Edwards (q.v.), in 1867, and this also envisaged jet propulsion, first included in a practical aircraft design by a Frenchman, Charles de Louvrié (q.v.), two years earlier. In appearance, the Butler–Edwards design resembled a paper dart, and the definition of a delta wing is that it should resemble an isosceles triangle in appearance. Many delta wing designs have omitted a tailplane, using elevons instead of separate elevators and ailerons; but the presence of a tailplane in itself has nothing whatsoever to do with the definition of an aircraft's wing as a delta.

Practical delta wing designs appeared after World War II in Great Britain and the United States, and involved a number of manufacturers, including Avro, Boulton Paul, Convair, and Gloster. The first delta wing aircraft to enter operational service was the Gloster Javelin (q.v.), which first flew in November 1951, and entered service with the R.A.F. as an all-weather fighter in 1952. Other aircraft followed, including the Avro Vulcan (q.v.) nuclear bomber, and the Convair F-102 (q.v.) Delta Dagger and F-106 (q.v.) Delta Dart. A world speed record was gained in March 1956 by a British delta wing research aircraft, the Fairey Delta 2 (q.v.), although a

Convair F-106 exceeded this in December 1959.

Probably the most famous and successful delta wing aircraft has been the French Dassault Mirage (q.v.), versions of which undertake fighter, interceptor, fighter-bomber, reconnaissance, bomber, and conversion training duties, although since the addition of the Mirage G and Mirage F.1, no longer are all Mirages delta wing aircraft. The Anglo–French Concorde (q.v.) supersonic airliner does not, strictly speaking, employ a delta wing, since the leading edges are curved. Nor is the other variant, the so-called double delta, strictly speaking a delta wing either, since it entails a change of angle partway along the leading edge of the wing – typical aircraft are the Swedish SAAB-35 (q.v.) Draken and SAAB-37 (q.v.) Viggen.

Demoiselle, Alberto Santos-Dumont: an expatriate Brazilian living in Paris, Alberto Santos-Dumont (q.v.) was well to the fore in pioneering aviation in Europe. The Demoiselle was a very light aircraft of high-wing monoplane layout, built and flown in 1908, using a 28 hp Darracq engine. Unusually for the aircraft of the period, the pilot perched inside the fuselage, which was largely built of bamboo, and uncovered. Technically the design was a success.

Deperdussin: a French aircraft manufacturer in existence prior to the outbreak of World War I, Deperdussin specialised in a series of monoplanes, largely destined by Bechereau. The company's real fame lies in building the world's first monocoque monoplane in 1912. Known as the Monocoque Deperdussin (q.v.), it was built of plywood and used a 160 hp Gnome rotary engine.

Derringer, Wing: a compact light twin-engined aircraft with side-by-side seating for two in an enclosed cabin, the Wing Derringer is currently in production in the United States. A maximum speed of 250 mph and a 1,000 miles range come from two 160 hp Lycoming piston engines.

design competition: a design competition, as the name implies, is the competition

design leadership

which ensues after an air arm or a government issues a specification for a new aircraft, the winner of the design competition getting the contract for construction of the aircraft. At one time it was usual for two or more designs to be built, and a final decision to depend on the results of the 'fly-off', but this is seldom the case today, due to increased knowledge about aerodynamic theory and the vastly inflated price of modern technology.

design leadership: only found in a collaborative (q.v.) aircraft project, design leadership is awarded to one of the companies involved on the airframe, and to another company – usually in a different country if the project is international – for the engine. It implies overall control of the design, although detailed design work on various parts of the project are left to the partner responsible for those parts, e.g. Hawker Siddeley is responsible for design and production of the wings of the A.300B (q.v.) airbus, while Aerospatiale is responsible for the overall design since design leadership has been vested in the company.

D.E.T.A.–Direccao de Exploracao des Trasportes Aereos: formed in 1936 by the provincial government of the Portuguese territory of Mozambique, D.E.T.A. originally operated local services. Since that time the route network has been increased to include services to neighbouring countries in Southern Africa with whom Portugal has good relations – the political situation preventing services to East and Central Africa. Currently the fleet includes Boeing 737 and Fokker F.27 airliners.

deterrent: the weapon or weapon system capable of inflicting unacceptable damage on a potential enemy is usually termed a deterrent, and it is an essential feature of the balance of terror (q.v.), but not necessarily of the balance of power (q.v.). Today, the term is used in terms of nuclear capability, the theory being that the Soviet Union would not dare attack the West because of the retaliation which could be expected from British and American nuclear weapons; and certainly overall peace has survived a Cold War and

several crisis points. A deterrent does not require to be nuclear, however, since it is generally considered that German use of poison gases during World War II was prevented by the fear of retaliatory action by Britain.

An essential of any deterrent is the will and the means to use it, although it is also true that once used, the deterrent effect is ended.

Deutsche Lufthansa: *see* Lufthansa.

Devon, Hawker Siddeley: the R.A.F. versions of the Dove (q.v.) light transport and communications aircraft, from which there are no major differences.

Dewoitine: a French aircraft manufacturer of the inter-war period, Dewoitine built civil, military, and carrier-borne naval aircraft. Amongst the more notable of the company's designs were the D.14, a single-engined high-wing monoplane air liner which first appeared in 1924, the D.332 trimotor fourteen-passenger all-metal airliner of 1936, and its development, the D.338 of 1937, which also had a retractable undercarriage. It also built the D.500, D.510, and D.520 fighters for the French Navy and Air Force which were in service during the late 1930s, although a poor match for the Luftwaffe's Bf. 109s.

D.H.2, Airco: designed by Geoffrey de Havilland (q.v.) during World War I while Chief Designer for the Aircraft Manufacturing Company, the D.H.2 pusher-propeller biplane was one of the Allied aircraft credited with ending the Fokker scourge, resulting from Fokker's advantage of using a propeller-synchronised machine gun which could fire straight ahead through the propeller disc. Some 400 D.H.2s were built, starting in 1916, for the Royal Flying Corps. The aircraft used a single 100 hp Gnome engine.

D.H.4, Airco: another de Havilland design for the Aircraft Manufacturing Company, the D.H.4 day bomber first appeared in 1917. A biplane of conventional appearance, it used a 375 hp Rolls-Royce Eagle engine. A distinction for the aircraft was that licence-built versions were the first American-built aircraft to see active service in World War I. After the war a number of D.H.4s were modified for civil airline and airmail duties in Europe and the United States.

D.H.9, de Havilland: one of de Havilland's first designs as an aircraft manufacturer in his own right, the D.H.9 biplane was something of a disappointment and it was in fact the D.H.9A which entered R.A.F. service in large numbers as a day bomber after the end of World War I. A single 400 hp Liberty engine was used. A number of D.H.9s were converted for civil duties.

D.H.50, de Havilland: a successor to the D.H.4/D.H.9 series, the D.H.50 was used during the 1920s by a number of airlines as a four-passenger aircraft or mailplane. A single-engined biplane, the powerplant was an Armstrong Siddeley Puma.

D.H.60 Moth, de Havilland: in many ways the Moth was more than a single aeroplane type, since it gave rise to a closely-related series of light and training aircraft from de Havilland's during the inter-war period. The D.H.60 Moth biplane used a single 100 hp de Havilland Gipsy engine and provided open tandem seating for two when it entered production in 1925. The D.H.80 Puss Moth of 1929 used a 120 hp Gipsy engine and was a three-seat, high-wing, cabin monoplane. There were also the D.H.87 Hornet Moth and the Fox Moth biplanes of the early 1930s. A later aircraft was the D.H.94 Moth Minor of 1937, a low-wing monoplane with two tandem seats. Most famous of the Moths was the D.H.82 Tiger Moth biplane, which became the standard R.A.F. and Fleet Air Arm basic trainer for many years. A post-war crop spraying derivative of the Tiger Moth was the Thruxton Jackaroo. Speeds of the Moth aircraft varied considerably in the 85–115 mph mark, according to model, and ranges varied from 200 to 600 miles.

D.H.66 Hercules, de Havilland: the D.H.66 Hercules tri-motor airliner entered service with Imperial Airways in 1927, and was used on services to the Middle East initially, later flying to India and to South Africa. The 450 hp Bristol Jupiter engines gave the fourteen-passenger biplane a maximum speed of 130 mph. Although cabin accommodation was provided for the passengers, the pilot sat in an open cockpit.

D.H.84 Dragon, de Havilland: in common with its smaller cousin, the D.H.60 Moth, the D.H.84 Dragon was also the parent of a family of related aircraft, although all Dragons were biplanes and, with one exception, light transport aircraft. The eight-passenger D.H.84 Dragon first flew in 1933, with deliveries starting the following year, and used two de Havilland Gipsy Major engines for a cruising speed of 110 mph. The four-engined, sixteen-passenger D.H.86 Dragon Express appeared in 1934, while the most famous of the Dragons, the twin-engined D.H.89 Dragon Rapide, also made its first flight that year. The eight-passenger Dragon Rapide remained in production throughout World War II and for a short time afterwards, serving with civil and military users throughout the world, and being named the Dominie in R.A.F. service. Baby of the family was the D.H.90 Dragonfly, a five-seat business and private-owner aircraft which first appeared in 1936.

D.H.88 Comet, de Havilland: the D.H.88 Comet of 1934 bore no resemblance to the post-World War II de Havilland aircraft of that name, being a twin-engined racing aeroplane with piston engines. The D.H.88 Comet won the 1934 MacRobertson England to Australia air race.

D.H.98 Mosquito, de Havilland: the de Havilland D.H.98 Mosquitos started to enter R.A.F. service in 1941 on light bomber and photographic-reconnaissance duties, but were soon employed as long-range escort fighters, night fighters, pathfinders (dropping marker flares on to targets for a following night bomber force), and armed-reconnaissance aircraft. The aircraft was also used as a fast transport for wartime service with B.O.A.C. between Scotland and Sweden. A mid-wing monoplane of wooden construction, the Mosquito used two Rolls-Royce Merlin piston engines and was known as a strong and highly-manoeuvrable aircraft.

D.H.110 Sea Vixen, de Havilland

Maximum speed was not far short of 400 mph, and a post-war development, the Hornet, reached 480 mph on one occasion, to become the fastest piston-engined aircraft. After the war Mosquitos found their way into a number of air forces, including those of the Netherlands, Belgium, and France.

D.H.110 Sea Vixen, de Havilland: *see* Sea Vixen, Hawker Siddeley.

DHC-1 Chipmunk, de Havilland Canada: the de Havilland Canada DHC-1 Chipmunk was the first design by de Havilland's Canadian subsidiary and first appeared during the late 1940s, as a D.H.82 Tiger Moth replacement. An all-metal, low-wing monoplane with tandem seating for two in a covered cockpit, the Chipmunk was built in Canada and in Britain, and entered service with a number of air forces and air arms, apart from those of Britain and Canada. A single 145 hp de Havilland Gipsy Major 8 piston engine gives a maximum speed of 138 mph and a range of 280 miles. R.A.F. Chipmunks are being replaced by Scottish Aviation Bulldogs at present.

DHC-2 Beaver, de Havilland Canada: the second de Havilland Canada design, the Beaver first flew in August 1947, as a high-wing utility transport with a single 450 hp Pratt and Whitney R-985-AN-1 piston engine. Most of the aircraft were ordered by military users, and many remain in service today. A six-place aircraft with freight-carrying capability and a short take-off characteristic, the DHC-2 Beaver can fly at up to 140 mph and has a range of up to 800 miles. A development, the Turbo-Beaver, has enjoyed some limited production. A larger aircraft, the DHC-3 Otter, is based on the Beaver.

DHC-3 Otter, de Havilland Canada: essentially a larger version of the DHC-2 (q.v.) Beaver, the DHC-3 Otter first flew in December 1951, and although less of a success than the smaller aircraft, still enjoys fairly extensive use. A single 600 hp Pratt and Whitney R-1340 piston engine gives the eleven-seat Otter a cruising speed of 160 mph and a range of up to 1,000 miles. As with the Beaver, the standard non-retractable undercarriage can be replaced by skis or floats. The DHC-6 (q.v.) Twin Otter is a much-modified development.

DHC-4 Caribou, de Havilland Canada: the first de Havilland Canada twin-engined design, the Caribou S.T.O.L. military transport first flew in July 1958, with deliveries to the U.S. Army starting in October 1959. A number of other air forces have ordered the aircraft, including the C.A.F., R.A.A.F., and India, Ghana, Kuwait, and Zambia. The DHC-4 Caribou is a high-wing monoplane with two 1,450 hp Pratt and Whitney R-2000-7M2 piston-engines, and can carry thirty-two passengers for up to 300 miles. The maximum speed is 216 mph. A turboprop development led to the DHC-5 (q.v.) Buffalo.

DHC-5 Buffalo, de Havilland Canada: the de Havilland Canada DHC-5 Buffalo won a U.S. Army design competition in 1962. First flight of a Buffalo was in April 1964, although a turboprop Caribou flew in 1961. Deliveries to the U.S. Army started in the spring of 1965, and orders from the then R.C.A.F. and from Brazil followed. Two 2,850 shp General Electric T64-GE-10 turboprops give the Buffalo a maximum speed of 271 mph and a range of more than 500 miles. A high-wing monoplane with a 'T' tail, the Buffalo can carry forty passengers, or freight.

A joint de Havilland Canada-Boeing project, with NASA sponsorship, is aimed at developing a jet-powered short take-off airliner based on the Buffalo. Development flying with two prototypes is under way.

DHC-6 Twin Otter, de Havilland Canada: the DHC-6 Twin Otter first flew in May 1965, and deliveries to users started in Spring 1966. Basically a twin-engined development of the DHC-3 (q.v.) Otter, the Twin Otter differs from the basic Otter in having a tricycle undercarriage, a different tailplane and, of couse, two engines. The aircraft is in production and is proving to be popular with military and civil users in many countries. Two 578 shp Pratt and Whitney PT6A-6 turboprops give the eighteen-seat aircraft a maximum speed of 180 mph and a range of up to 900 miles. Freight can be carried instead of passengers, and floats or skis fitted instead of wheels.

DHC-7, de Havilland Canada: a design project for a four-engined, forty-passenger S.T.O.L. airliner, a decision to develop the DHC-7 was taken in 1972, and the Canadian Government is supporting the project. The aircraft will use turboprop engines.

dihedral: an upward inclination of the wings, from the root to the tip, giving a 'V' appearance from a front or rear view. The advantage of this is to give increased stability in roll. *See also* anhedral.

Dinfia: an Argentinian aircraft manufacturer, Dinfia specialises in limited production runs of military and civil aircraft for communications duties and agricultural operations. The more notable aircraft in recent years include the I.A.35 (q.v.) Huanquero light transport for the Argentinian Air Force, its successor aircraft, the I.A.50 (*see under* I.A.58) Guarani II, and the I.A.53 crop-spraying aircraft.

dirigible: the term dirigible is used to describe a steerable airship (q.v.). The first dirigible was that built by Henri Giffard (q.v.), which made a flight in 1852. Other significant dirigible manufacturers have included Count Ferdinand von Zeppelin (q.v.), and important craft have included the German Hindenburg (q.v.) and the British R-100 (q.v.) and R-101 (q.v.).

dive-bomber: as the term implies, a dive-bomber dives towards its target before releasing its bombs at the lowest possible altitude and climbing away. Perhaps the shortest-lived class of warplane, the dive-bomber did not appear until the late 1920s, one of the first being a Martin biplane for the U.S. Navy. Earlier introduction of the type probably was prevented by the high standard of structural integrity required in view of the stresses to which a dive-bomber could expect to be subjected. Primitive bomb racks are also often unacceptable in a dive-bomber, which fre-quently requires the bomb to be swung clear of the propeller.

At the outset of World War II the dive-bomber was perhaps at the peak of its existence. Amongst the aircraft in service could be counted the Fleet Air Arm's Blackburn Skua, which had first flown in 1937; the United States Navy's Curtiss Helldiver and Douglas Dauntless, both of which saw extensive action in the Pacific; and the Luftwaffe's famous Junkers Ju.87 (q.v.) Stuka. A later arrival on the scene, although non-combatant, was the SAAB-17 (q.v.). However, by this time the basic weaknesses of the concept were showing through, such as the relatively slow speeds and vulnerability of the aircraft themselves, and the proven superiority of heavy strategic bombing and low level interdiction bombing.

DM-: international civil registration index mark for the German Democratic Republic (East Germany).

Do.X, Dornier: the largest aircraft at the time of its first flight in 1929, the Dornier Do.X flying-boat used twelve radial engines; initially these were 550 hp Siemens, but later models used 615 hp Curtiss Conquerors, giving a maximum speed of 140 mph. The engines were mounted on top of the wing, in six back-to-back groups with tractor and pusher propellers. A number of prestige flights were made by the aircraft across the Atlantic in 1930–2.

Do.17, Dornier: the Dornier Do.17 was originally designed as an airliner, although the long and narrow fuselage, which earned the aircraft the name of the 'flying pencil', rendered it unsuitable for its allotted task, and it was soon adopted as a medium bomber during the German rearmament programme of the 1930s. After first entering Luftwaffe service in 1937, the aircraft gave very satisfactory service and paved the way for its successor and development, the Do.217. Two 1,000 hp Bramo radial engines powered the Do.17.

Do.24, Dornier: a Dornier flying-boat design which was also built in the Netherlands by Aviolanda for use by the Royal Netherlands Navy in the East Indies, the

Do.24 used three 880 hp B.M.W. radial engines mounted in a high wing. The aircraft was intended as a Wal replacement. A turboprop development is planned.

Do.27, Dornier: the Dornier Do.27 was originally designed and built in Spain – with the production work undertaken by C.A.S.A. – after the end of World War II imposed restrictions on aircraft design and development in Germany. Production was transferred to Federal Germany in 1956, more than two years after the first flight of a prototype in Spain, and the aircraft was delivered in large numbers to the German armed forces. Orders have been received since from a number of European and African air arms. A single 275 hp Lycoming GO-480-B1A6 piston engine gives this five-seat, high-wing monoplane a S.T.O.L. performance, a maximum speed of 140 mph, and a range of some 600 miles.

Do.28D Skyservant, Dornier: the Dornier Do.28 originally evolved as a twin-engined version of the Do.27, with the engines mounted beside the fuselage, and a number of aircraft of this type were produced. The Do.28D Skyservant retains the layout of the smaller aircraft, but has a crew of two and upwards of twelve passenger seats. The two 380 hp Lycoming IGSO-540 piston engines give the aircraft a maximum speed of 200 mph and a range of 1,000 miles. It is in quantity production for the German armed forces.

Do.31E, Dornier: an experimental V./ S.T.O.L. transport, the Do.31E started as a joint project with Hawker Siddeley Aviation, but full responsibility reverted to Dornier after the British Government withdrew support for the project. Flights as a conventional and as a V./S.T.O.L. aircraft have been conducted successfully. A high-wing aircraft, the Do.31E uses two 15,500 lb thrust General Electric turbojets for forward movement, and eight 4,400 lb thrust Rolls-Royce lift jets in two wing-tip pods for lift. The maximum speed is in excess of 400 mph, the range is up to 1,000 miles, and up to thirty passengers can be accommodated. Much of the information gained from the Do.31E

programme, which has now been discontinued, has been applied to the Do.231 V./S.T.O.L. airliner project, the future of which is uncertain.

Do.335, Dornier: a tandem-engine fighter which was in production and service with the Luftwaffe during the closing days of World War II. One of the fastest piston-engined fighters of the war, the twin-engined Do. 335 could fly at more than 450 mph, using two Daimler Benz DB 603 piston engines of 1,900 hp each, one mounted conventionally in the nose, driving a tractor propeller, and the other mounted in the tail, driving a pusher propeller. A tricycle undercarriage was fitted, and the fin extended downwards as well as upwards.

dog fight: a name given to aerial combat between two opposing fighter units, and probably coined during World War II. The combat became a dog fight after the initial formations were broken up by the intensive fighting.

Dominie: currently the name given to the R.A.F.'s Hawker Siddeley HS 125s (q.v.), in use as navigational trainers, but differing little from the civil version. The term was also used for the R.A.F.'s de Havilland D.H.89 Dragon Rapides (*see under* D.H.84 Dragon).

Doodlebug, McDonnell: the first McDonnell aircraft, the Doodlebug, first flew in 1929, and was unusual for the time in being fitted with both a slotted leading edge and flaps, giving a far lower stalling speed than the then conventional aircraft. A 110 hp Warner radial engine powered the Doodlebug.

doppler: a navigational aid in widespread use, particularly for civil transport aircraft navigation, giving the pilot the groundspeed and heading of his aircraft, as opposed to the airspeed (q.v.). Radio signals of a known frequency are sent to the ground and their frequency recorded on the rebound, using the doppler system. The principle behind the doppler is that the frequency of a radio wave appears to shift in proportion to the velocity of the observer, and, therefore, providing one starts from a known point and notes the differences in radio wave frequencies, ground speed and heading can be deduced.

Dornier: originally founded in 1922 by Claude Dornier, the Dornier concern soon became famous for its series of large flying-boats during the inter-war period, including the Wal, the Do.X (q.v.), and the Do.24 (q.v.), which was also built by Aviolanda in the Netherlands. Dornier also produced a diesel-powered flying-boat, the Do.18, during this period. Re-armament in Germany during the 1930s affected the company, which had its Do.17 (q.v.) airliner modified for service as a medium bomber, later blossoming into the Do.217 of the middle years of World War II. Although mainly a flying-boat and bomber manufacturer, towards the end of World War II Dornier produced the Do.335 (q.v.) tandem-engined fighter.

After the war Dornier found itself in a repeat of the post-World War I situation, with aircraft design and manufacture severely restricted within Germany. The first post-war Dornier, the Do.27 (q.v.) light transport, made its début in Spain, although production was soon transferred to Germany. The Do.28D (q.v.) soon followed the Do.27, and meanwhile the company produced Lockheed F-104G (q.v.) Starfighters and Fiat G.91s (q.v.) for the Luftwaffe while Germany's armed forces were once again rebuilt. The Starfighter production programme was followed by the Bell 204/205 (q.v.) Iroquois helicopter. Dornier partnered Hawker Siddeley on the Do.31E (q.v.) V./S.T.O.L. transport project until Britain withdrew, and was also a member of the consortium building the European A.300B (q.v.) airbus, until withdrawing in 1971. Dornier is currently the German partner in the Franco–German Alphajet (q.v.) advanced jet trainer project.

Douglas, Donald: founder of the aircraft company which today bears his name, in spite of a merger with McDonnell (q.v.), Donald Douglas was with Martin (q.v.) before founding his own company in 1920. Although the company achieved fame with the World Cruiser of 1924, it was not until the DC-2s (q.v.) and DC-3s (q.v.) of the 1930s that real commercial success and recognition came to Douglas Aircraft, which was then firmly under the control of its founder.

Douglas Aircraft: founded by Donald Douglas (q.v.) in 1920, Douglas Aircraft started production with the Cloudster (q.v.), which first flew in 1920, and was followed in 1921 with the DT-1 (q.v.) torpedo-bomber. Fame came to the company in 1924 with the World Cruiser, which was the first aircraft to fly around the world. A succession of aircraft entered production during the late 1920s, including the 0-2 observation aircraft for the U.S.A.A.C. and the M-2 mailplane. In 1932 the first flight of the DC-1 (q.v.) marked the start of the Douglas Commercial series, which continues to this day, although on a more immediate level it led to the commercial success of the DC-2 (q.v.) and DC-3 (q.v.). A military development of the DC-2 was the B-18 bomber, and by the outbreak of World War II, followed by America's subsequent involvement, the DC-3 was in production as the U.S.A.A.F.'s C-47 (q.v.), while Douglas also built the C-54 (q.v.) transport (originally the civil DC-4 (q.v.)), the Dauntless dive-bomber, the A-20 (q.v.) Havoc attack-bomber, and the A-26 Invader bomber.

The end of World War II saw some limited DC-4 production, although the company was preoccupied with the DC-6 (q.v.) pressurised development of the earlier aircraft, and the A-1 (q.v.) Skyraider attack aircraft. A few experimental aircraft were in hand during the late 1940s, including the Skystreak and the Skyrocket, but it was not until the early 1950s that the first production jets, the Skyray and the A-4 (q.v.) Skyhawk fighters and the B-66 (q.v.) Destroyer bomber, were flown. Before this, however, the giant C-124 (q.v.) Globemaster was built for the U.S.A.F., and later there were C-133 Cargomasters, also for the U.S.A.F., and the DC-7 (q.v.), the fastest piston-engined airliner, for the world's major airlines.

The jet transport stage reached Douglas

with the DC-8 (q.v.) in 1959, and during the 1960s the company strengthened its position in this field with the addition of the DC-9 (q.v) short-haul airliner, and an agreement with Piaggio (q.v.) to market the PD-808 (q.v.) executive jet as a joint venture. In 1967 cash-flow problems led to a merger with McDonnell, to form McDonnell Douglas, but the company's product range has been maintained since with the DC-9, the new DC-10 (q.v.) tri-engined airbus and its planned developments, and the A-4 Skyhawk.

Dove, de Havilland: the first post-World War II British transport aircraft to fly, the Dove made its first flight in September 1945, as a Rapide replacement aimed at the executive, light transport, and military communications markets. In fact the figure of 540 built was the highest total for any predominantly civil British aircraft during the post-war period. A larger four-engined development, the Heron (q.v.), also enjoyed some success, while military versions of the Dove, with only minor differences from the civil specification were known as the Devon and Sea Devon. Two de Havilland Gipsy Queen piston engines of 380 hp each give the eight-to-eleven passenger Dove a maximum speed of 240 mph and a range of up to 900 miles. Many remain in service, particularly in the United States.

Dr.1, Fokker: the famous Fokker Dr.1 triplane fighter first appeared in 1916, and gained most of its notoriety as the mount of the famed 'Red Knight of Germany', Baron Manfred von Richthofen (q.v.). A 110 hp rotary engine powered the aircraft, which was somewhat lacking in structural integrity and owed much to skilled flying.

Dragon, de Havilland D.H.84: *see* D.H.84 Dragon, de Havilland.

Draken, SAAB-35: *see* SAAB-35 Draken.

DT-1, Douglas: a military development of the Douglas Cloudster (q.v.) of 1920, the Douglas DT-1 was the first military aircraft to be built by the company. A folding-wing biplane torpedo-bomber, the DT-1 entered service with the U.S. Navy in 1921, and was one of the aircraft types operated from the aircraft carrier U.S.S. *Langley*.

Dumont, Alberto-Santos-: *see* Santos-Dumont, Alberto.

Dunning, Commander E. H., R.N.: on 2 August 1917 Commander E. H. Dunning landed a Sopwith Camel on the fore-deck of the converted battle cruiser H.M.S. *Furious* – the first time an aeroplane had landed on a ship under way. A second attempt, a few days later, ended in tragedy when the aircraft's engine stalled, and Dunning drowned before he could be rescued by a ship's boat.

du Temple, Félix (1823–1890): a Frenchman, Félix du Temple had the distinction of designing and building in 1874 the aircraft which made the first powered take-off. However, the aircraft's take-off was assisted by a downramp run, and it did not sustain the flight long enough for it to be counted as such. A steam engine was used in the aircraft. A similar performance can be credited to the Russian, Mozhaiski (q.v.).

E

E-1B Tracer, Grumman: an airborne-early-warning development of the S-2 (q.v.) Tracker and C-1A (q.v.) Trader series, the Grumman E-1B Tracer first flew in March 1958, following which more than sixty aircraft were built for the United States Navy. A high-wing monoplane with search radar in a saucer-shaped radome mounted on top of the fuselage, and a triple fin in place of the single fin of the Tracker and Trader, the E-1B uses two 1,525 hp Wright R-1820-82 piston engines for a maximum speed of 265 mph and a seven-hour endurance. On the larger United States Navy attack carriers, the E-1B has been replaced by the E-2A (q.v.) Hawkeye.

E-2A Hawkeye, Grumman: unlike the E-1B Tracer, which it has replaced on the larger U.S. carriers, the E-2A Hawkeye was conceived from the start as an airborne-early-warning aircraft, with the C-2 (q.v.) Greyhound as a carrier-onboard-delivery development. The first flight of a Hawkeye prototype took place in October 1960, followed by a pre-production model in April 1961, with deliveries to the sole user, the U.S. Navy, starting in January 1964. A high-wing monoplane with a search radar radome mounted on top of the fuselage, the E-2A has a quadruple fin, and uses two 4,050 shp Allison T56-A-8 turboprops for a maximum speed of 300 mph and an endurance of eight hours.

E.III, Fokker: the Fokker E.III monoplane fighter's fame springs not from any attributes of the aircraft itself, but from the distinction of being the first aircraft to use a synchronised machine gun, allowing the pilot to fire through the propeller disc without the need for deflector blades, or the need to stand up in the cockpit to fire over the propeller disc. Fokker had offered his invention to the Allies, who turned it down, but so effective was it to prove after the advent of the E.III in 1915 that the name 'Fokker Scourge' was coined for the devastating attacks on Allied aircraft which resulted. The 'Fokker Scourge' was not ended until the advent in 1916 of the Airco D.H.2 (q.v.) pusher-engine biplane fighter. An 80 hp Oberursal rotary engine powered the E.III.

E.28/39 Whittle, Gloster: the first Allied turbojet aircraft to fly, the Gloster E.28/39, often known as the Gloster Whittle, first flew on 15 May 1941. A Whittle (q.v.) centrifugal flow turbojet powered the aircraft, which was used solely as a test-bed for the Whittle engine, first used operationally on the Gloster Meteor (q.v.) jet fighter.

EA-: U.S.N. designation for electronics countermeasures (q.v.) aircraft, usually modified versions of attack aircraft – see A-1 (q.v.), A-3, and A-6 (q.v.).

early warning system: a radar system for advanced warning of attacking or intruding enemy aircraft and missiles. See airborne-early-warning *and* ballistic missile early warning.

East African Airways: the East African Airways Corporation was formed in 1946 by the Governments of Kenya, Uganda, Tanganyika, and Zanzibar – the last two now forming the Republic of Tanzania – with assistance from B.O.A.C. (q.v.). The first aircraft were six de Havilland Dragon Rapide biplanes for local services, and these were later joined by de Havilland Doves, Lockheed Lodestars, and Douglas DC-3 Dakotas. International operations started in 1957, using Canadair DC-4M Argonauts on services to London, Salisbury, and destinations in the Indian sub-continent. The Argonauts were soon replaced by Bristol Britannias, and in turn these were replaced by de Havilland Comet 4s in 1960. Vickers VC.10 airliners were introduced in 1966, and the fleet now includes these aircraft, plus McDonnell Douglas DC-9s, Fokker F-27 Friendships, Douglas DC-3 Dakotas, and de Havilland Canada DHC-6 Twin Otters.

Eastern Air Lines

An extensive African, Asian, and European network is operated, in addition to domestic services. A subsidiary is Simbair, which operates third-level services to the Seychelles.

Eastern Air Lines: Eastern Air Lines dates from 1926 when Harold Pitcairn formed Pitcairn Aviation to operate domestic mail services on behalf of the U.S. Post Office, starting with an Atlanta–New York service, and introducing further services in 1927 and 1928 before selling the airline to North American Aviation. The airline was renamed Eastern Air Transport in 1930, when Fokker trimotor airliners were introduced, and Curtiss Condor biplane airliners were also added to the airline in 1931. General Motors acquired Eastern in 1933.

The present title was adopted in 1935. A ruling by the C.A.B. in 1938 that manufacturers must not have a financial interest in airlines resulted in Eastern Air Lines breaking away from General Motors. Douglas DC-2 airliners entered service in 1938, and the following year DC-3s were introduced. During World War II, about half of the E.A.L. fleet and personnel served with the U.S.A.A.F.

After World War II, Douglas DC-4s, Lockheed Constellations and Super Constellations, and Martin 404s were introduced, followed by DC-7s in 1953. A tentative order for de Havilland Comet I jet airliners had to be abandoned when the aircraft was grounded, but the first turboprops for the airline, Lockheed Electras, entered service in 1959, followed by Douglas DC-8 jets in 1960, and Boeing 720s, for the airline's shuttle service between Washington Boston, and New York, in 1961. The shuttle service, with no advance booking and a guaranteed walk-on seat, was an air transport innovation by E.A.L. when introduced in 1961. The airline's route network had by that time expanded to include Canada and Mexico.

A merger with American Airlines (q.v.) was considered during the early 1960s, but was not proceeded with.

Currently E.A.L. operates a fleet of Boeing 747s and 727s, Douglas DC-8s and DC-9s (including the Super-Sixty DC-8s), and Lockheed TriStars.

Eastern Provincial Airways: a Canadian airline, Eastern Provincial Airways was founded in 1949 and commenced operations during the following year. In 1963 Eastern Provincial acquired Maritime Central Airways. Eastern Provincial now operates a network of scheduled services in the eastern Canadian provinces, including Newfoundland, and from Montreal, using a fleet of Boeing 737-200s, Handley Page Heralds, Douglas DC-4s and DC-3s, Carvairs, and de Havilland Canada DHC-6 Twin Otters, DHC-3 Otters, and DHC-2 Beavers, with a Beech 18 and a Cessna 180B.

East European Mutual Assistance Treaty: *see* Warsaw Pact.

EC-: U.S.A.F. designation for airborne early warning aircraft, as in EC-121 (q.v.).

EC-: international civil registration index mark for Spain.

EC-121 Constellation, Lockheed: a development of the C-121 (q.v.) Constellation transport, the EC-121, which first entered U.S.A.F. service as the RC-121 with Air Defence Command in 1953, has outlived its transport counterpart. Used for airborne early warning duties, the EC-121 is easily distinguishable from the C-121 or standard Constellation by the large radomes protruding from above and below the fuselage. The same four 3,250 hp Wright R-3350-91 piston engines are used as in the C-121, giving the aircraft a maximum speed of 320 mph and a range of 4,500 miles. A development of the Boeing C-135 (q.v.) will replace the EC-121 in the near future as a part of the AWACS (q.v.) aircraft programme.

E.C.M.: *see* electronics countermeasures.

Edwards, E.: E. Edwards was, with J. W. Butler (q.v.) one of the originators of the first delta wing (q.v.) aircraft design in 1867. The aircraft was never built, but the intention was that rocket power should be used.

Egyptair: the Egyptian state airline, Egyptair, was formed in 1932 as Misr

Airwork, receiving assistance from a British company, Airwork (q.v.). The first scheduled service, between Cairo and Alexandria, started in 1933 with de Havilland Dragon biplanes, and these were later supplemented by Dragon Rapides and Dragon Expresses. A number of international services were being operated throughout the Middle East by 1939. Operations were suspended during World War II, but after the war Vickers Vikings were obtained, and the name changed to Misrair, which was until recently the name of the separate Egyptian domestic airline. Vickers Viscounts were introduced during the late 1950s, and in 1960 the airline became the first in the Middle East to operate jet airliners, de Havilland Comet 4Cs. The title of United Arab Airlines was taken in 1961, and the present title was adopted in 1971.

Today a fleet of Boeing 707s, de Havilland Comets, Ilyushin Il-62s and Il-18s, Antonov An-12s and An-24s, Tupolev Tu-154s, and Yakovlev Yak-40s is operated on a domestic and international route network which includes Europe, the Middle East, Africa, and major points in the Far East.

EI-: international civil registration index mark for the Republic of Ireland.

8R-: international civil registration index mark for Guyana.

EJ-: international civil registration index mark for the Republic of Ireland.

ejector seat: higher aircraft speeds with the advent of the turbojet meant increased difficulty for a pilot attempting to bail out from an aircraft in trouble, with the danger of being swept into the tailplane. The only means at the time of avoiding this was the ejector seat, invented by Sir James Martin (q.v.) at the end of World War II. Initial experiments with ejector seats were made on piston-engined aircraft, and indeed many turboprop types were fitted with ejector seats in due course, so the system was not confined to the pure jet.

The basic principle of the ejector seat remains that an explosive charge beneath the seat ejects the seat clear of the aircraft, leaving the pilot to parachute to earth in the normal way, except that in some cases barometric pressure devices open the parachute, in case the pilot has lost consciousness in a high-speed or high-altitude ejection. Early models required the cockpit canopy to be opened or blown free first, but later models have contact-breaker devices on the back of the seat so that time may be saved by the seat smashing through the canopy. Ejector seats are usually provided for both pilot and navigator, and in certain cases other crew members also have them, but this is only possible in certain types, such as the Gannet anti-submarine aircraft, which has the whole crew in cockpits in which ejection devices can be fitted without major structural alterations.

Naval aircraft usually have ejector seats capable of underwater ejection in addition to inflight ejection. Special seats are available for low altitude ejection, or for ejection from V./S.T.O.L. combat aircraft.

Certain aircraft, such as the General Dynamics F-111 (q.v.), have an ejectable crew module which is parachuted to earth, and for the future a limited flying capability is planned for crew modules to allow the crew to escape to friendly territory.

Development of the ejector seat is also dealt with under Sir James Martin (q.v.).

EL-: international civil registration index mark for Liberia.

El Al Israel Airlines: El Al (the Hebrew for 'To the Skies' or 'Onward and Upward') was formed in 1948, and operations began in 1949 with Douglas DC-4s flying to Rome and to Paris. These were soon joined by Lockheed Constellations for African and transatlantic services. Bristol Britannia turboprop airliners were received in 1957, and in 1961 Boeing 707-420 jet airliners were placed in service. These have since been supplemented by additional aircraft of that type and by Boeing 720s, while Boeing 747s were introduced in 1971.

El Al is a purely international airline, owned by the Israeli Government, with more than 50 per cent of the share capital, the state-owned Z.I.M. Shipping Company, and the Histadrut Federation of Labour. In turn, El Al has a 50 per cent interest in Arkia, the domestic airline formed in 1950, which operates Handley Page Heralds.

Electra I, Lockheed: the Lockheed Electra first appeared in 1934, as a twin-engined, ten-passenger, all-metal mono-plane airliner. Two 420 hp Pratt and Whitney radial engines gave a maximum speed of 220 mph. Later versions of the aircraft had a slight fuselage stretch to allow twelve passengers to be carried. The Electra saw service with airlines throughout the world, and was one of the first commercial aircraft to be fitted with de-icing equipment. A retractable under-carriage was fitted.

Electra II, Lockheed L-188: development of the Electra turboprop medium-range airliner started in 1956, and the aircraft first entered airline service in January 1959, with a total of 165 being built for airlines in the United States and the Pacific area, and for K.L.M. in Europe. A few remain in service, primarily on freight duties. Four 3,750 shp Allison 501-D13 turboprops provide a maximum speed of 405 mph and a range of 2,500 miles, while a maximum payload of 26,400 lb or up to 99 passengers can be carried.

electronics countermeasures: the object of electronics countermeasures equipment is to minimise the effect of defensive radar and tracking apparatus on the ability of an intruding aircraft to penetrate these defences undetected. E.C.M. techniques include jamming of defensive radar, cluttering and confusing radar systems in an attempt to hide the intruding aircraft or attacking missile, and producing false echoes to simulate targets against which defensive measures can be applied. High-power transmissions on the frequency of a defending radar system will jam the defending radar and confuse the system until the defending radar's frequency is changed, when equipment is available which picks up the new frequency (or the basic fequency if it is unknown in the first place), and high-power repeat transmis-sions provided for jamming. False echoes, more powerful and therefore more attrac-tive to defending radar than the echo of the intruding aircraft, can also be pro-vided, subject to the restrictions of pay-load and space in the intruding or attack-ing aircraft or missile.

Aircraft themselves can be made less easily detectable on radar screens by the use in their construction of materials which make detection more difficult through having poor reflective qualities, and the shape of the aircraft is also important. Special paints can also help, but for these the frequency of defending radar needs to be known.

In order to avoid many of the space, payload and power shortcomings of E.C.M. devices in combat aircraft, special E.C.M. aircraft are used by the U.S.A.F. and U.S.N. These are known to be very effective indeed, and it is certain that the United States leads in this very important field.

elevator: an aerofoil control surface in an aircraft, usually in the tailplane, move-ment of which puts the aircraft into a climb or a dive. Tailplanes are not always a feature of delta wing aircraft, which often combine the elevator and aileron (q.v.) functions in elevons (q.v.).

elevon: an aerofoil control surface which fulfils the functions of both elevators (q.v.) and ailerons (q.v.) in an aircraft without a separate tailplane.

Elizabethan: the B.E.A. fleet name for its Airspeed Ambassador (q.v.) aircraft.

Ellyson, Lt. T. G., U.S.N.: Lt Theodore G. Ellyson, U.S.N., became the first U.S. Navy pilot after training with Glenn Curtiss in California in December 1910.

Ely, Eugene B.: the first man to fly from a ship, Eugene B. Ely took off in a Curtiss biplane from a platform built over the bows of the cruiser U.S.S. *Birmingham* on 14 November 1910, before the ship was under way. On 18 January 1911 Ely became the first man to land on a ship when he flew the Curtiss biplane on to a platform built on the stern of the U.S.S. *Pennsylvania* while the ship lay at anchor. Unfortunately, Ely was killed in a flying accident later in 1911.

EMA-124, Agusta-Meridionali: a light helicopter with three seats and a bubble cabin, designed by Agusta but with production licensed to Meridionali, the EMA-124 is powered by a single Lycoming VO-540-B1B3 piston engine. Production so far has been on a very small scale.

EMBRAER: a joint state and private (51:49) Brazilian aircraft manufacturer, Embraer was formed in 1969 and is more properly known as the Empresa Brasileira de Aeronautica, S.A. The company took over the work of C.T.A. (q.v.), the government-sponsored design organisation, and its initial production programme included the C.T.A.-designed C-95 (q.v.) Bandeirante light transport and communications aircraft. The C-95 was soon joined by the EMB-200 agricultural aircraft and by licence production of the Aermacchi M.B. 326 (q.v.) jet trainer.

Empire, Short: one of the most famous civil flying-boats, the Short Empire 'C' class was one of the first aircraft to be ordered in quantity while still on the drawing board – this was an Imperial Airways order for twenty-eight aircraft. Apart from its service on routes to South Africa, India, and Australia with passengers and mail (including the Empire Air Mail Scheme which had enabled Imperial to place such a large order with such confidence), the Empire also flew a number of experimental transatlantic flights in 1937, which involved it in inflight refuelling trials as well. The initial twenty-eight aircraft used four 920 hp Bristol Pegasus radial engines, but a further eleven aircraft used the slightly more powerful Perseus radials. The aircraft could carry about twenty-four passengers, and mail, for up to 760 miles at a cruising speed of 165 mph.

endurance: the period of time, in hours, which an aircraft can spend in the air. Endurance is, strictly speaking, a more accurate assessment of the capabilities of maritime-reconnaissance and airborne-early-warning aircraft than range, since these aircraft types are expected to loiter on station for long periods.

engine: one of the earliest examples of mechanically-powered flight was the Giffard dirigible airship of 1852, which used a steam engine, while a later airship, by Tissandier in 1883, used a battery-powered electric motor. Neither engine, in spite of the intervening period, could really be said to be up to the job. Steam engines were widely used for a number of years, including those in aircraft designed by Stringfellow (q.v.), du Temple (q.v.), Ader (q.v.), Mozhaiski (q.v.), and Maxim (q.v.), and many of these engines were a credit to their designers, but suffered from the relatively low rpm and heavy fuel source which is characteristic of the steam engine. Additional problems from the weight of the water and the difficulties in raising adequate power would have arisen had any of the aircraft concerned reached any real altitude. Elastic motors were used extensively during the latter part of the nineteenth century for powering the models which were so important to the development of aviation.

The Wright Flyer (q.v.) of 1903 used a petrol engine, as did the unsuccessful Langley's Aerodrome (q.v.) of the same year, and before long a number of tolerably reliable powerplants were available from Antionette (q.v.) and Anzani. The Antionette engines were of the in-line type, rather like a modern motorcar engine, while the Anzani engine had a semi-radial arrangement for its three cylinders. Well before the end of the first decade of flight, however, Laurent Sequin had harnessed the rotary engine, invented in 1889 by an Australian, Lawrence Hargrave (q.v.). The rotary engine had the pistons and cylinders arranged in radial fashion round a crankcase, which remained static while the cylinders rotated round it. World War I was fought using rotary engines, such as the Gnome, although in later years and with higher power outputs, the gyroscopic effects of these engines were to prove unacceptable.

A further step which occurred during the heyday of the rotary engine was the 'V' configuration engine, starting with the Rolls-Royce Eagle, which powered the Vickers Vimy (q.v.) bomber, and ending with the Rolls-Royce Merlin, which powered the Spitfire (q.v.), Lancaster (q.v.), D.H.98 (q.v.), Mosquito, and F.51 (q.v.) Mustang of World War II. Supercharging was applied to these engines during World War II to boost power at high altitude. The piston engine was to reach its peak of development with the radial engine, developed during the 1920s and 1930s, largely by the Bristol Company.

Work on the turbojet engine was undertaken in Great Britain, Germany, and Italy during the 1930s, but with little recognition in Great Britain of the work of Group Captain Sir Frank Whittle (q.v.). Features of this type of engine were its ability to operate at high speeds and at high altitudes outside the scope of the propeller engine, and it combined these advantages with a relatively simple principle and few moving parts. The early jet aircraft in operational service during World War II were the Messerschmitt Me. 262 (q.v.) and the Gloster Meteor (q.v.). A post-war development was the turboprop, combining the performance of the turbojet with the low-altitude and low-speed performance of the propeller. A development of the turbojet was the by-pass engine or turbofan, with large ducted fans taking the place of the old propeller and offering high subsonic performance. Before this occurred, however, the use of the ramjet, with no moving parts, as a reheat or after-burner on turbojets, provided the power for a rapid climb or a supersonic dash, while still permitting economic cruising for loiter and ferry operations. Reheat was essentially a feature of military aircraft for many years, until the advent of the supersonic Concorde (q.v.) and Tu-144 (q.v.) airliners.

Amongst the most recent developments have been the twin and triple spool engine, with two or three shafts permitting the turbine blades to rotate at the optimum speed for each set of blades, instead of at one speed, and offering higher performance with reduced noise levels. Typical engines include the Rolls-Royce RB.211 and the General Electric CF-6.

Vertical take-off is achieved in civil and military aircraft by the use of lift jets, mounted vertically, or by vectored thrust, as with the Harrier (q.v.), deflecting the exhaust jet for short or vertical take-off.

Rocket motors have only enjoyed a limited, aviation application for high altitude records or as boosters for high altitude airfield take-offs.

English Electric: a British company, mainly involved in diesel and heavy electrical engineering, which entered the aircraft industry at the end of World War II, and found immediate success with its Canberra (q.v.) jet bomber, which was also built in Australia and in the United States, in the latter case as the B-57 (q.v.). The P-1A Lightning (q.v.) interceptor, developed by the company during the 1950s, remained in production throughout the 1960s with the British Aircraft Corporation (q.v.) of which English Electric was one of the founder companies in 1960.

Enola Gay: the name of the Boeing B-29 (q.v.) Superfortress bomber of the U.S.A.A.F. which dropped the first atomic bomb on Hiroshima on 6 August 1945.

Ensign, Armstrong-Whitworth: the Armstrong Whitworth Ensign airliner first flew in January 1938, and at the time was the largest all-metal monocoque aircraft. Accommodation in the four-engined, high-wing monoplane varied from twenty-seven passengers on Imperial Airways' overseas routes to forty on the same airline's European services. The Ensign fleet passed to B.O.A.C. in 1940, and was used during World War II for V.I.P. transport duties. Cruising speed was 170 mph and the range was some 800 miles.

EP-: international civil registration index mark for Iran.

Esnault-Pelterie, Robert (1881–1957): a Frenchman, Robert Esnault-Pelterie built his first glider in 1904, basing his design on the Wright brothers (q.v.) glider of 1902, although with the substitution of ailerons for wing warping, which Esnault-Pelterie considered structurally dangerous. The Esnault-Pelterie glider was not particularly successful, but nevertheless the first design to incorporate ailerons, while Esnault-Pelterie was one of the first designers to devote attention to control in roll. His aeroplane designs, the first of which, the R.E.P. No. 1 of 1907, made only short flights, were tractor monoplanes. The R.E.P. No. 2 of 1908 made a number of successful flights during that year and 1909. Esnault-Pelterie also built aero-engines, and at one time Blériot (q.v.) was a customer for these.

ET-: international civil registration index mark for Ethiopia.

E.T.A.: estimated time of arrival; an inflight as opposed to a scheduled calculation.

Ethiopian Airlines: Ethiopian Airlines was formed in 1945 and commenced international services in 1946 with assistance from Trans World Airlines (q.v.). A domestic and international network of services is now operated, serving major destinations in Africa, Asia, and Europe with a fleet of Boeing 707s and 720s, Douglas DC-6Bs and DC-3s, and a number of light aircraft.

Euler: a German aircraft manufacturer in existence at the outset of World War I, the company's own designs made little headway and it was one of many firms to build the Rumpler-Etrich Taube (q.v.).

Europlane: a consortium formed in 1972 by the British Aircraft Corporation (q.v.), Messerschmitt-Bölkow-Blohm (q.v.) and Svenska Aeroplane A.B. (q.v.), the initial object of Europlane will be to build a short take-off jet airliner with advanced technology engines, if government support is forthcoming.

E.W.S.: *see* early warning system.

F

F-: international civil registration index mark for France.

F-: U.S.A.F. and U.S.N. designation for fighter aircraft. Prior to 1947, the F-designation was preceded by P- (for pursuit) in the U.S.A.A.F. and U.S.A.A.C.

F2F, Grumman: the first single-seat Grumman fighter, the F2F-1 carrier-borne biplane also featured an enclosed cockpit and retractable undercarriage. In U.S. Navy service during the early and mid-1930s, the aircraft used a 650 hp Pratt and Whitney radial engine, and already had the short, tubby outline of the World War II Grumman fighters.

F-4 Phantom II, McDonnell Douglas: the McDonnell Douglas F-4 Phantom II fighter-bomber was originally developed by Douglas for the U.S. Navy, starting in 1953, with the first flight taking place in May 1958, and deliveries of the F-4B starting in December 1960. A U.S.A.F. order followed, the first U.S.A.F. production models, carrying the designation F-4C, entering service in 1963. RF-4B and RF-4C reconnaissance versions for the U.S.N. and U.S.A.F. respectively followed, and developments of the basic aircraft have also included the F-4D and F-4E for the U.S.A.F., with improved avionics and weaponry, and the F-4G and F-4J for the U.S.N., with up-rated engines and other improvements. The Royal Navy and Royal Air Force respectively operate F-4K and F-4M versions, both of which use Rolls-Royce RB.168 Spey 202 turbofans with reheat and 20,515 lb thrust in place of the standard aircraft's General Electric engines. Many of the Royal Navy's aircraft are being transferred to the R.A.F. with the phasing out of the Fleet Air Arm's fixed-wing carrier-borne

operations. The U.S.M.C. operates F-4B and RF-4B Phantoms.

Phantoms have also been sold to a number of other air forces, including those of West Germany, which uses a single-seat version of the Phantom, Japan, Iran, and South Korea. Two 16,500 lb thrust General Electric J79-GE-8 turbojets with reheat power the F-4B/C, giving a maximum speed of Mach 2·5, and a range of up to 2,000 miles, while a 10,000 lb ordnance load can be carried on underwing and under-fuselage strongpoints. A low-wing monoplane, the Phantom normally has a twin-seat cockpit with tandem seating.

F4F Wildcat, Grumman: the Grumman F4F-1 Wildcat was the company's first mid-wing monoplane fighter, and, in common with most of Grumman's designs, was a carrier-borne aircraft. In service with the U.S.N. and the Royal Navy (which named the aircraft the Martlet) during the first part of World War II, the Wildcat was a single-seat aircraft using a single 1,200 hp Pratt and Whitney radial engine.

F.5, Seaplane Experimental Station/Shorts: the last of a series of designs produced during World War I and afterwards by the Seaplane Experimental Station, the F.5 flew in its Short Brothers version in 1924, and was in fact a flying-boat. The aircraft's main distinction lay in being the first flying-boat to have a metal hull.

F-5A/B Freedom Fighter, Northrop: originally conceived in 1954 as a private-venture project, the aircraft obtained the support of the U.S. Department of Defence in 1958 as a means of providing low-cost light fighters under the Mutual Aid Programmes for the less wealthy of America's allies. First flight was in July 1959. Basically, the F-5A is a single-seat aircraft, while the F-5B is a twin-seat version with provision for either a navigator or an instructor to be carried. The M.A.P. F-5A/Bs have been delivered to Ethiopia, Greece, Iran, Libya, South Korea, South Vietnam, and Turkey. The aircraft has also been produced in Canada as the CF-5A/B for the C.A.F. and as the

NF-5A/B for the Royal Netherlands Air Force by Canadair, while C.A.S.A. has assembled the aircraft in Spain as the SF-5A/B for the Spanish Air Force. A few aircraft are also in U.S.A.F. service in South-East Asia. The RF-5A is a reconnaissance version, while the Royal Norwegian Air Force operates RF-5Gs and F-5Gs as well as F-5Bs. Two 4,080 lb thrust General Electric J85-GE-13 turbojets with afterburning are fitted, giving a maximum speed of Mach 1.4 and a range of up to 1,750 miles. Five underwing and under-fuselage strongpoints can carry up to 6,200 lb of ordnance, in addition to wing-tip air-to-air missiles.

Developments of the F-5A/B are the prototype F-5-21, with an up-rated engine, which first flew in March 1969 as a modified F-5B, and the T-38 (q.v.) Talon advanced jet trainer of the U.S.A.F. and the Luftwaffe. Latest production variant is the up-rated F-5E Tiger II, with 23 per cent extra thrust.

F6F Hellcat, Grumman: a development of the F4F (q.v.) Wildcat fighter, the F6F Hellcat was a single-seat, single-engined, low-wing monoplane fighter which served with the U.S.N. and the Royal Navy during the second part of World War II. A single 2,000 hp Pratt and Whitney Wasp radial engine powered this carrier-borne fighter, which was the first low-wing Grumman design.

F.VIIB-3M, Fokker: one of the earliest trimotor civil aircraft, the Fokker F.VIIB-3M was a three-engined development of an earlier single-engined design, and helped to start the fashion for trimotor airliners during the inter-war years. The F.VIIB-3M was first introduced in 1925, and a variety of powerplants were used, although the main types were by Bristol or Wright; and the aircraft offered a slight increase in size and performance on the single-engined F.VII. Developments of the aircraft included the twin-engined F.VIII of 1927, and the twenty-passenger F.IX of 1930.

F-8 Crusader, Ling-Temco-Vought: the Ling-Temco-Vought F-8 Crusader was originally a Vought design which won a U.S.N. design competition in 1953 for a supersonic carrier-borne fighter. The first flight of this high-wing, single-seat monoplane was in March 1955, with deliveries starting two years later. A feature of the aircraft is the variable-incidence wing, which reduces the nose-up position of the aircraft on its landing approach. The basic aircraft was originally designated the F-8A, and further developments included the F-8B, F-8C, and F-8D with improved avionics, and the F-8E with high-performance radar. Aeronavale F-8s are from the F-8E series. Modernisation and modification of the aircraft has resulted in the F-8As becoming RF-8Hs, for reconnaissance duties, while the F-8Bs have become F-8Ls, the F-8Cs are F-8Ks, the F-8Ds are F-8Hs, and the F-8Es in U.S.N. service are now F-8Js.

A single 18,000 lb thrust Pratt and Whitney J57-P-20 turbojet with reheat gives the F-8J a maximum speed of 1,100 mph and a range of up to 1,200 miles, and two 1,000 lb bombs, missiles, or rockets can be carried on underwing strongpoints. A development of the F-8 series is the A-7 (q.v.) Corsair.

F.IX, Fokker: the Fokker F.IX trimotor was one of the larger aircraft of this type when it appeared in 1930, and could carry twenty passengers. A high-wing monoplane, like the F.VII, F.VIIB-3M, and F.VIII, the F.IX used three Bristol Jupiter engines of 500 hp each, giving a cruising speed of 115 mph, and was used by K.L.M. and other airlines, often on long-distance services.

F.13, Junkers: the first all-metal civil aircraft, the Junkers F.13 first appeared in 1919. A low-wing cantilever monoplane, the aircraft was powered by a single 185 hp B.M.W. engine and could carry four passengers. A number of airlines used the aircraft throughout the 1920s, including some floatplane versions.

F-14 Tomcat, Grumman: originally conceived as the U.S.N.'s VFX air superiority fighter project, the prototype F-14B first flew in 1971. Upwards of 400 aircraft are being built for the U.S.N.'s carrier force. A variable-geometry twin-seat aircraft, the F-14 uses two 28,000 lb thrust Pratt and Whitney TF30-P-12 turbofans with reheat for a maximum speed in the region of Mach 3·0. Phoenix, Sparrow, and Sidewinder missiles can be carried. The aircraft has an all-moving twin-fin and rudder units.

F-15, McDonnell Douglas: the McDonnell Douglas F-15 design won a U.S.A.F. air superiority fighter competition and the prototype aircraft first flew in 1972. Entry into U.S.A.F. service will be taking place during the mid-1970s, and more than 100 aircraft are likely to be built. Two 23,150 lb thrust Pratt and Whitney JTF-22 turbofans with reheat give a maximum speed of Mach 3·0. A basically delta wing is fitted, together with a variable-incidence tailplane and twin fins. Phoenix, Sparrow, and Sidewinder missiles can be carried in an internal weapon bay, and quick-firing cannon can also be fitted.

F-27 Friendship, Fokker: the Fokker F-27 Friendship airliner first flew in prototype form in November 1955, and a further prototype, or pre-production model, flew in January 1957. Deliveries to airlines began in August 1958 from the Fairchild production line in the United States, and from the Fokker production line in December of that year. The basic aircraft was designated the F-27 Srs. 100, and was followed by the Srs. 200 with up-rated engines, while the Srs. 300 and 400 which followed were passenger–cargo versions of the two basic aircraft. A military version was the F-27M Troopship. A stretched version of the Friendship was produced as the FH-227 solely by Fairchild in the United States for some time before the Fokker-built equivalent, the F-27 Srs. 500, entered production.

Two Rolls-Royce Dart 532-7 turboprops of 2,255 shp provide a maximum cruising speed of 295 mph and a range of up to 1,200 miles with either up to fifty-two passengers or 13,575 lb of freight.

F-28 Fellowship, Fokker: originally intended as a Friendship replacement, the F-28 Fellowship is in production alongside the earlier aircraft, and has still to imitate its success. Development of the F-28 started in 1962, followed by the first flight in June 1967, deliveries starting in 1968. Plans for a version of the aircraft to be built by Fairchild-Hiller in the United States were abandoned because of the lack of an adequate market. The F-28 uses two Rolls-Royce Spey Junior 555-15 turbofans for a maximum cruising speed of 527 mph and a range of up to 1,235 miles, with either up to sixty-five passengers or 13,560 lb of freight. Part of the aircraft's production is subcontracted to V.F.W. and H.F.B. in Germany, and to Short Brothers in the United Kingdom.

F-47 Thunderbolt, Republic: the Republic F-47 Thunderbolt low-wing, single-seat monoplane fighter appeared early in World War II, with the first production series using a 2,100 hp Pratt and Whitney radial engine and having a hump-backed fuselage and a framed cockpit canopy. The most numerous version of the aircraft, however, was the F-47D, which remained in service in Latin America for some twenty-five years after the end of World War II, and which had a teardrop canopy and a cut-away rear-fuselage. It used a 2,300 hp Pratt and Whitney R-2800-59 radial engine for a maximum speed of 430 mph and a range of up to 1,000 miles; up to 2,500 lb of bombs and rockets could be carried on underwing strongpoints.

F-51 Mustang, North American: the North American F-51D Mustang was another fighter design to appear during the early years of World War II, and to remain in service in Latin America for some twenty-five years after the end of that war. Early models used an Allison engine and had a hump-backed fuselage and a framed cockpit canopy. Later models used a Packard-built Rolls-Royce Merlin engine of between 1,470 and 1,695 hp, according to variant, and had a teardrop canopy and cut-down rear fuselage. Mustangs remained in front-line U.S.A.F. service until the end of the Korean War. Maximum speed of the single-seat, low-wing monoplane, of which the F-51D was the most popular version, was some 437 mph, with a range of up to 1,300 miles and the ability to carry up to 2,000 lb of bombs or rockets on underwing strong-points.

F-80 Shooting Star, Lockheed: the first
U.S.A.F. operational jet fighter, the
Lockheed F-80 Shooting Star first flew in
March 1946, and during the years which
followed saw service with a number of
Latin American countries, usually after
retirement from the U.S.A.F. In addition
to Lockheed's own production, the F-80
was built under licence in Japan and
Canada, the latter work being undertaken
by Canadair, a Rolls-Royce Nene turbojet being substituted for the standard
Allison powerplant. An advanced jet
trainer development, the T-33 (q.v.) saw
service with the U.S.A.F. and with many
NATO air forces as well. A single 5,400 lb
thrust Allison J33-A-35 turbojet provided
a maximum speed of 600 mph and a range
of 1,300 miles.

F-84 Thunderjet, Republic: the Republic
F-84 Thunderjet was another of the early
operational jet fighters of the U.S.A.F.,
and first flew in prototype form in February 1946, although the aircraft did not
reach operational status until after the
F-80 (q.v.). The F-84 served with the
U.S.A.F. and with many NATO air forces.
A single 5,600 lb thrust Allison J35-A-29
turbojet provided a maximum speed of
600 mph and a range of up to 2,000 miles,
while four 1,000 lb bombs could be
carried on underwing strongpoints. The
single-seat, low-wing aircraft had the
distinction of being one of the first swept-
wing fighters. Developments of the F-84
were the F-84F Thunderstreak and the
reconnaissance RF-84F Thunderflash, using a single 7,220 lb thrust Wright T65W-3
turbojet and having a 690 mph maximum
speed.

F-86 Sabre, North American: first flown
in October 1947, the North American F-86
Sabre was undoubtedly one of the most
successful of the early American jet
designs and the first U.S.A.F. swept-wing
fighter. The F-86 played an important
part in the Korean War, and gained an
easy superiority over the MiG-15s (q.v.)
of the Communist forces. Apart from the
U.S.A.F. itself, the aircraft was used by
many other NATO and Allied air forces,
including a short period with the R.A.F.,
and was licence-built in Italy, Japan,

Australia, and Canada – the last two using
Rolls-Royce Avon turbojets in place of
the standard General Electric unit in
order to achieve a higher performance
and also minimise national balance of
payments costs. A number of versions were
built, including the basic F-86A and its
successors in the day fighter role, the
F-86D, F-86E, F-86F, and F-86H, and the
all-weather fighter F-86B and Fiat F-86K.

The F-86 series are still in service, and a
single General Electric J47-GE-27 turbojet of 5,910 lb thrust gives this single-seat
fighter a maximum speed of 687 mph and
a range of up to 925 miles, while two 1,000
lb bombs can be carried on underwing
strongpoints; the aircraft can also carry
missiles and rockets, and is fitted with
cannon. The Canadair and Commonwealth Avon Sabres use a 7,500 lb thrust
licence-built Rolls-Royce Avon 26 turbojet for a 700 mph maximum speed. Few
Sabres remain with their original owners
except in the case of reserve units. A
carrier-borne development was the FJ-113
Fury, although the early versions of this
aircraft had a straight wing.

F-100 Super Sabre, North American:
originally evolved as a supersonic development of the F-86 (q.v.) Sabre, the first
F-100 flew in May 1953, and deliveries to
the U.S.A.F. started in November of that
year. Three basic combat versions of the
aircraft were put into production, the
F-100A interceptor and the F-100C and
F-100D fighter-bombers, the last-mentioned having an up-rated engine. A conversion trainer version, the TF-100F, was
also built and put into U.S.A.F. service.
Many ex-U.S.A.F. aircraft were supplied
to Denmark, France, Nationalist China,
and Turkey, and some remain in service.
A single 17,000 lb thrust Pratt and Whitney J57-P-21A turbojet with reheat provides a maximum speed of 864 mph and a
range of up to 1,500 miles, while missiles
or rockets can be carried to supplement the
four 20mm cannon of the aircraft.

F-101 Voodoo, McDonnell: originally
intended as a long-range escort for the
B-36 (q.v.) bomber, the McDonnell F-101
Voodoo was eventually developed as an
interceptor and fighter-bomber, making

125

F-102 Delta Dagger, General Dynamics

its first flight in September 1954. Main versions of the series were the F-101A and F-101C fighter-bombers and the F-101B interceptor, with reconnaissance versions in the RF-101A and RF-101C. Although the F-101 was basically a single-seat aircraft, the F-101B had twin seats, as did the F-101F supplied to the then R.C.A.F. Two 14,880 lb thrust Pratt and Whitney J57-P-55 turbofans with reheat provide a maximum speed of 1,120 mph and a maximum range of 2,200 miles, while the interceptor can carry air-to-air missiles in an internal weapons bay.

F-102 Delta Dagger, General Dynamics: the Convair-designed F-102 Delta Dagger was the first U.S.A.F. operational delta-wing aircraft, and was the result of a number of experimental delta-wing projects by Convair. First flight of the F-102 was in December 1954, and about 1,000 of the aircraft were subsequently built, most of which are now in reserve. A single 17,200 lb thrust Pratt and Whitney J57-P-25 turbojet with reheat provides a maximum speed of 825 mph and a range of up to 1,100 miles, while missiles or rockets can be carried. A development, initially known as the F-102B, became the Mach 2·0 plus F-106 (q.v.) Delta Dart.

F-104 Starfighter, Lockheed: one of the major production efforts in military aviation after the Korean War, the Lockheed F-104 Starfighter first flew in February 1954. A single-seat, mid-wing monoplane with a flying-tail, the Starfighter was available as an interceptor, fighter-bomber, and multi-mission aircraft. The first production variant was designated the F-104A single-seat interceptor, and was followed by the F-104B two-seat interceptor, and the F-104C and F-104D tactical strike versions, with single and twin seats respectively. A more powerful development was the F-104G, built in Europe as a joint production programme involving Belgium, West Germany, Italy, and the Netherlands, and also in Japan and Canada, for multi-mission duties. This was followed by the still more powerful F-104S for Italy. The conversion trainer version is the TF-104F.

The F-104G, which remains in widespread service, uses a single 15,800 lb thrust General Electric J79-GE-11A turbojet with reheat for a maximum speed of 1,450 mph and a range of up to 1,400 miles, and can carry up to 4,000 lb of rockets, missiles, or bombs or underwing strongpoints.

F-105 Thunderchief, Republic: the first flight of the Republic F-105 Thunderchief supersonic fighter-bomber was in October 1955, and this prototype carried the F-105A designation, while the first production models with more powerful engines were F-105Bs. The F-105Bs were the first of the Thunderchief series to have the type's unique and distinctive reverse sweep engine intakes. A mid-wing, single-seat monoplane, most of the Thunderchief's built have been in the F-105D series, and this type has been much to the fore in the Vietnam War. Only the U.S.A.F. has operated the Thunderchief. A single 26,500 lb thrust Pratt and Whitney J75-P-19W turbojet with reheat provides a maximum speed of 1,485 mph and a maximum range of more than 2,000 miles. A 20mm cannon is fitted, and up to 14,000 lb of ordnance can be carried on under-fuselage or underwing strongpoints.

F-106 Delta Dart, General Dynamics: a development of the Convair F-102 (q.v.) Delta Dagger, the F-106 started its life as the Convair F-102B, but the present designation was adopted before this single-seat delta-wing interceptor made its first flight in December 1956. All of the combat versions were F-106As, and production of this aircraft for the U.S.A.F. totalled 257. The twin-seat conversion trainer version was designated F-106B. A single 24,500 lb thrust Pratt and Whitney J57-P-17 turbojet with reheat provides a maximum speed of 1,525 mph and a range of 1,150 miles, while four air-to-air missiles can be carried in an internal weapons bay.

F-111, General Dynamics: the F-111 was developed by General Dynamics after winning the U.S.A.F.'s TFX two-seat, tactical fighter-bomber design competition in November 1962. First flight of a prototype was in December 1964, and the aircraft has the distinction of being the world's first production variable-geometry aircraft. During the early years of service the aircraft was plagued by teething troubles in the F-111A and F-111D versions, and the F-111E subsequently appeared with modified engine air intakes; the modifications have been incorporated on the earlier versions. The F-111D version has simplified avionics; the F-111K was a version for the R.A.F. which was cancelled before deliveries could start; and the FB-111A is a bomber version for Strategic Air Command. Two Pratt and Whitney TF30-P-3 turbofans with reheat give 21,000 lb thrust each, allowing a maximum speed of 1,650 mph and a range of up to 3,300 miles, while eight air-to-air or air-to-surface missiles can be carried in the weapons bay in the fuselage and on underwing strongpoints which adjust with the sweep of the wings. A feature of the aircraft, which is also in service with the R.A.A.F., is the ejectable crew module.

Fabre, Henri: the inventor of the floatplane, Henri Fabre, designed and flew the first float-plane in March 1910. The aircraft was powered by a 50 hp Gnome rotary engine driving a single pusher propeller. It is not generally regarded as having been the first practical float-plane, this distinction going to a Glenn Curtiss (q.v.) design of 1911.

'Fagot', Mikoyan-Gurevich MiG-15: see MiG-15 'Fagot', Mikoyan-Gurevich.

fail-safe: in aviation the term fail-safe means the provision of a second item of equipment which can take over in the event of failure: most control mechanisms in an aircraft are duplicated or triplicated. Another concept is that of the fail-safe structure, which means that the airframe of an aircraft has sufficient strength to prevent complete disintegration following failure or fracture of a part of the structure, so that the aircraft can operate safely until the damaged section is discovered during a routine maintenance inspection.

Fairchild-Hiller: the Fairchild-Hiller Corporation was formed in 1964 when the Fairchild Engine and Aeroplane Corporation acquired the Hiller Aircraft Company, a manufacturer of light helicopters, including the H-23 light observation and training helicopter. Fairchild itself had a long history of aviation activity dating from the 1920s, with such aircraft as the Fairchild 71, a single-engined, high-wing monoplane land and seaplane transport, followed by the F-24 four-seat, single-engined, light aircraft of the late 1930s, and the PT-23 low-wing monoplane basic trainer of World War II. Although these products enjoyed some success, it was not until near the end of World War II that Fairchild became an important part of the aviation scene, with such aircraft as the C-119 (q.v.) Packet and C-123 (q.v.) Provider military transports, although, of course, the C-123 is very much a post-war aircraft.

During the late 1950s Fairchild became Fokker's licensee for the F-27 (q.v.) Friendship, and Fairchild-built Friendships were in fact in airline service before those from the Dutch company. A development of the F-27 was the FH-227, a stretched version of the basic aircraft. A similar arrangement with the F-28 (q.v.) Fellowship, including a distinctive Fairchild variant, was abandoned because of the low volume of potential sales. Currently, Fairchild-Hiller is engaged on production of the Swearingen-designed Metroliner and on subcontract work, while the Hiller Division produces the Pilatus Turbo Porter under licence. The company acquired another American aircraft manufacturer, Republic (q.v.) in 1965, and this is operated as a separate division. The Fairchild A-10A was a contender in the U.S.A.F.'s ground-attack A-X programme competition.

Fairey Aviation: formed during World War I to build the Campania carrier-borne fighter, Fairey Aviation continued its existence with a variety of military aircraft for shore- and carrier-based operations after the end of the war. Notable amongst these were the Flycatcher (q.v.) of 1923, which was a carrier-borne aircraft, and the Fawn day-bomber for the R.A.F., followed by the Fox (q.v.) day-bomber in 1925, and the general purpose Seal and

Falco, Reggiane

Gordon aircraft for the Fleet Air Arm and for the R.A.F.'s shore-based strength respectively. (At this period, the Fleet Air Arm consisted of R.A.F. aircraft and personnel embarked aboard the Royal Navy's warships.) All of these aircraft were biplanes, as was the Firefly III, an order for which during the early 1930s from Belgium led to the founding of a Belgian subsidiary, Avions Fairey (q.v.).

Fairey's first monoplane was the Hendon bomber, a twin-engined aircraft which was the world's first monoplane heavy bomber, and entered service during the early 1930s. Biplane designs still entered production at this time, and in fact the Fairey Swordfish (q.v.) torpedo-bomber remained in frontline service with the Royal Navy for the first few years of World War II. The Fairey Battle (q.v.) single-engined monoplane light bomber of 1937 was obsolete at the outbreak of World War II, although a development, the Fulmar (q.v.) fighter for the Royal Navy's carriers, was put into full production. Other wartime aircraft included the Barracuda torpedo-bomber and the Firefly (q.v.) fighter, both of which were designed for carrier operations. The Firefly proved to be a particularly successful aircraft of its type.

After the war Fairey continued with Firefly production for a time, before developing the Gannet (q.v.) turboprop aircraft as first an anti-submarine aircraft for the Royal Navy, and later an airborne-early-warning aircraft. Experimental aircraft included the Delta 2 (q.v.) of the early 1950s, which established a world speed record, and during the latter part of the decade the Rotodyne V.T.O.L. transport, which although technically successful was not put into production. In 1960 Fairey's British factories were merged with Westland Aircraft (q.v.) as a part of a Government-inspired plan for reshaping the aircraft industry. The Belgian subsidiary remained, however, as did a Canadian subsidiary manufacturing powered flying controls, with the U.K. parent in the role of a holding company. Fairey returned to airframe production in the United Kingdom in 1972 with the acquisition of Britten-Norman (q.v.).

Falco, Reggiane: the Reggiane Re.2000 Falco I fighter was designed in Italy and first appeared in 1937, but was prevented from immediately proving itself of great value to the Italian armed forces because of powerplant and structural shortcomings, although a few aircraft were exported to Hungary and to Sweden. A development, the Reggiane Re.2001 Falco II, with up-rated engines and other modifications, appeared in 1942 and proved to be a successful fighter.

Falco, Dassault: *see* Mystère 20, Dassault.

Falcon: the Hughes AIM-26 and AIM-4 Falcon air-to-air missile is one of the standard missiles for U.S. and Canadian interceptors, including the F-4, F-101, F-102, and F-106, while versions of the missile are used by the Swiss and Swedish Air Forces, the latter using licence-built versions by SAAB-Scania. A number of different versions exist, including the AIM-26A, using a nuclear warhead, and the AIM-4D/H versions, incorporating modifications which improve close-range performance. Varying in length between seven and twelve feet according to version, the Falcon uses a solid propellant motor and either infra-red or semi-active radar homing. Range is generally within the region of 5-6 miles, and speed is Mach 2·5.

Farman, Henri (1874–1958): born in France of English parents, although becoming a naturalised Frenchman in later years, Henri Farman and his brother, Maurice Farman (q.v.), both worked in France. Henri Farman started by flying and modifying Voisin (q.v.) box-kite aeroplanes, which enjoyed some considerable success, including flying the first circle in Europe and gaining the distance record at the 1909 First International Aviation Meeting at Reims. Henri Farman's own first design also appeared in 1909, and was one of the first aircraft to use the new Gnome rotary engine. A number of Farman designs were used as a basis by other European manufacturers, and some of his products were used by embryonic military air arms for training and reconnaissance duties before the outbreak of World War I.

Farman, Maurice (1877–1964): brother of Henri Farman (q.v.), Maurice Farman had a more practical approach to aviation, as opposed to the somewhat experimental and record-breaking nature of his brother's work. Much of the British Army's heavier-than-air flying before 1914 was on the Farman S.7 or 'Longhorn' biplane, so named because of the extravagant sweep forward of the landing skids, designed to prevent damage to the propeller should the aircraft topple forward on landing. The S.7 was followed by the 'Shorthorn' during the early part of World War I, after which a joint Henri-Maurice product, the Farman F.40, or 'Horace Farman' as it was more generally known, a two-seat biplane fighter with a pusher propeller and the observer-gunner sitting in front of the pilot, entered service in 1915.

Farman-Voisin: from 1907 to 1909 Henri Farman (q.v.) flew, and sometimes modified, Voisin (q.v.) built machines to such good effect that he flew the first circle in Europe on 13 January 1908, and established the distance record at the First International Aviation Meeting at Reims in 1909.

'Farmer', Mikoyan-Gurevich MiG-19: *see* MiG-19 'Farmer', Mikoyan-Gurevich.

Farnborough: possibly the oldest aerodrome in the world, Farnborough's connection with flight began in 1890 when the British Army erected balloon sheds in the area, and in 1907 the first British airship and the first British aeroplane were both built at Farnborough. During the years which followed, Farnborough became famous as the home of the School of Aviation Medicine and of the Royal Aircraft Establishment, the British government-sponsored research organisation, although earlier many of the Royal Aircraft Factory's designs had been built at Farnborough also. The Society of British Aerospace Companies biennial airshows are also staged at Farnborough.

F.E.2. Delta, Fairey: *see* Delta 2, Fairey.

Federal Aviation Administration: the United States authority responsible for air safety regulation, including certification of new aircraft types.

Fellowship, Fokker F-28: *see* F-28 Fellowship, Fokker.

Ferber, Captain Ferdinand (1862–1909): one of the pioneers of aviation in Europe, Captain Ferdinand Ferber, a Frenchman, was a follower of Lilienthal's (q.v.) designs and was working on a Lilienthal-type hang-glider in 1902 when Octave Chanute (q.v.) interested him in the Wright brothers' (q.v.) gliders, which he subsequently built. During the period 1902–4 Ferber was one of the very few men to keep European interest in aviation alive.

FH-227 Friendship, Fairchild-Hiller: licence-built development of the Fokker F-27 Friendship (q.v.) airliner.

Fiat: *see* Aeritalia.

fighter: undoubtedly the second oldest aircraft type, the fighter evolved shortly after the start of World War I in 1914, reputedly when pilots and observers in opposing reconnaissance aircraft started firing at each other with revolvers and rifles, and has since produced a number of offspring in the fighter-bomber, reconnaissance-fighter, interceptor, and ground-attack aircraft.

It did not take long for machine guns to be fitted to the more suitable reconnaissance aircraft during World War I, making these the first true fighters. The machine guns were often fired by the observer, sitting behind the pilot, but in some aircraft the pilot could fire a forward-pointing machine gun, although this usually entailed standing up to fire over the propeller disc. Deflector blades were fitted to the propellers of some aircraft in an attempt to allow the pilot to fire through the propeller disc, giving him better control of the aircraft and better aiming too. A major step forward occurred in 1915 when Anthony Fokker invented the propeller-synchronised machine gun, but this was rejected by the Allies, who then suffered from the so-called 'Fokker Scourge' when the Central Powers accepted the invention, which was fitted to the Fokker E.III (q.v.) fighter. It was some time before the Allies could effectively counter the 'Fokker Scourge', and they did so by using pusher-propeller fighters, such as the Airco D.H.2 (q.v.) and the Horace Farman (*see* Maurice Farman), with the observer sitting in front of the pilot and firing the machine gun ahead, before the first propeller-synchronised Allied designs appeared.

The rate of development slowed down considerably between the wars, but several distinct changes did take place. Monoplanes, of which there had been relatively few during World War I, replaced biplanes during the 1930s, and most fighters became single-seat aircraft, with heavy machine guns, or cannon, mounted in the wings. Canopied cockpits became standard, as did retractable undercarriages, to reduce drag, and since fighters were the fastest of aircraft, they were amongst the first to have flaps and variable pitch propellers. Experiments with jet propulsion were not so far advanced as to produce an operational jet fighter until World War II was well advanced, but experience with racing seaplanes had nevertheless pushed fighter speeds to around the 400 mph mark by 1939. The significant fighters at the start of World War II included the British Supermarine Spitfire (q.v.) and Hawker Hurricane (q.v.), the German Messerschmitt Bf. 190 (q.v.) (or Me. 109) with a cannon mounted in the propeller boss, the Focke-Wulf Fw.190 (q.v.), and the American Curtiss Hawk (q.v.), which was soon followed by the Republic F-47 (q.v.) Thunderbolt and the North American F-51 (q.v.) Mustang.

The major developments of World War II were not confined to the advent of the turbojet engine; with aircraft such as the Mustang, the de Havilland D.H.98 (q.v.) Mosquito, the Bristol Beaufighter (q.v.), and the Lockheed P-38 (q.v.) Lightning, the concept of the long-range escort fighter was born. The Mosquito and Beaufighter also evolved the first night fighters. The Hawker Typhoon, with its anti-tank rockets, was really the first ground-attack aircraft in the modern sense. The Messerschmitt Me.262 (q.v.) jet fighter appeared in 1944, and was soon followed by the Gloster Meteor (q.v.). A rocket-propelled interceptor was the Messerschmitt Me. 163 (q.v.) Komet.

After World War II a succession of jet fighters rapidly appeared, with perhaps the most interesting being the SAAB-21 (q.v.), which had originated as a twin-boom, pusher-propeller fighter, and became the only aircraft to operate both in the piston and turbojet forms. Britain followed the Meteor with the Vampire (q.v.) and Venom (q.v.) from de Havilland, and the Supermarine Attacker (q.v.), which was the first jet aircraft specifically designed for carrier operations. The United States introduced first the Lockheed F-80 (q.v.) Shooting Star, and followed this with the Republic F-84 (q.v.) Thunderjet and the North American F-86 (q.v.) Sabre. Russia built a number of Yakovlev and Mikoyan-Gurevich jet fighters, the most famous of which were the MiG-15 (q.v.) and the MiG-17 (q.v.).

However, much of the post-war effort went into development of interceptors (q.v.), aircraft with a high rate of climb to counter high-flying jet bombers. Ground-attack (q.v.) development was also pushed ahead, sometimes using versions of fighter-bomber or fighter designs, such as the Hawker Hunter (q.v.).

Fighter development in recent years has produced a number of designs, often as an offshoot from interceptor or ground-attack development. Amongst the more notable are the French Dassault Mirage (q.v.) series, successors to the Mystères (q.v.), the British B.A.C. Lightning (q.v.) and Hawker Siddeley Harrier (q.v.), and the American McDonnell F-4 (q.v.) Phantom II and Northrop F-5A (q.v.) Freedom Fighter. Only the last-named really meets the description of a fighter as a highly-manoeuvrable and fast air-to-air combat aircraft, rather than an interceptor, fighter-bomber, or ground-attack type. The Hawker Siddeley Harrier was the world's first operational V./S.T.O.L. aircraft, and brought this concept to fighter development, while the French Mirage G (q.v.) and the American General Dynamics F-111 (q.v.) have introduced variable geometry to fighter construction.

A new American concept, the air superiority fighter, is something of a non-sense, since it is the function of any fighter to gain air superiority. However, basically the air superiority fighter owes more to the interceptor than to the fighter concept, although cannon can be fitted.

Fighter, Bristol F.2B 'Brisfit': the Bristol F.2B Fighter, or 'Brisfit' as it was often called, was a single-engined, twin-seat biplane which first appeared in 1918 in service with the newly-formed R.A.F., and subsequently remained in service throughout the 1920s. A 275 hp Rolls-Royce Eagle engine was used in the Mark 1 version, and later versions used up-rated Eagle engines.

fighter-bomber: the term fighter-bomber is used to describe a fighter aircraft which can be used for light bombing missions, although this role is today complemented by the ground-attack (q.v.) aircraft.

The development of the fighter-bomber is vague, since both the fighter and the bomber evolved from the common ancestry of the reconnaissance aircraft of World War I, and one might possibly describe an aeroplane in which the observer could either fire a rifle or a machine gun, or throw bombs and hand grenades over the side, as a fighter-bomber. However, few of the early fighters were fitted with bomb racks and machine guns for combat, rather than defensive, purposes, so perhaps the evolvement of the fighter-bomber can be said to have started during the 1930s with aircraft such as the Hawker Hart (q.v.) series. Before this, the Fairey Fox (q.v.) of 1925 was a very versatile and manoeuvrable bomber, and the United States Navy's Boeing F4B fighter biplane could carry two 116 lb bombs under its wings.

World War II aircraft, such as the Supermarine Spitfire (q.v.), the Hawker Hurricane (q.v.), Hawker Tempest and Typhoon, the North American F-51 (q.v.) Mustang, the Republic F-47 (q.v.) Thunderbolt, the Bell P-39 (q.v.) Airacobra, and the Focke-Wulf Fw.190 (q.v.), were all available as fighter-bombers, although fame has been more usually accorded to the fighter versions. This tradition of producing fighter-bomber variants of fighters was continued with the jet age, with aircraft such as the de

Havilland Vampire (q.v.) and Venom (q.v.), the Republic F-84 (q.v.) Thunderjet, and the North American F-86 (q.v.) Sabre. Current fighter-bomber aircraft include the McDonnell Douglas F-4 (q.v.) Phantom II and A-4 (q.v.) Skyhawk, the Dassault Mirage III (q.v.) and Mirage 5 (q.v.), the Hawker Siddeley Harrier (q.v.), and the Sukhoi Su-7 (q.v.).

fin: The vertical part of the tailplane, the fin has a stabilising function and includes the rudder, for directional control. Certain aircraft are fitted with an all-moving fin, in which the entire fin is the rudder, e.g. the Mikoyan-Gurevich MiG-23 (q.v.) 'Foxbat' and the McDonnell Douglas F-15 (q.v.) These two aircraft are also fitted with twin fins, due to the need for improved directional control on high supersonic speed aircraft – on other designs this need has resulted in the use of larger fins and strakes.

Finnair: Finnair dates from 1923 when it was formed by Bruno Lucander as Aero, with a single Junkers F.13 seaplane which commenced scheduled operations to Sweden in 1924. Further route development in 1925 by the new airline persuaded the Finnish Parliament to grant a loan towards the cost of further aircraft, and within a short time four additional F.13s and a Junkers G-24 were being operated, while a Ju.52/3M trimotor was acquired in 1931, by which time the airline was operating to most of the major centres in Europe. De Havilland Dragon Rapides were acquired during the late 1930s.

World War II resulted in the almost complete abandonment of services, but during the immediate post-war period services were restarted and Douglas DC-3 Dakotas and Convair 440 Metropolitans introduced. In 1956 Finnair became the first Western airline to operate regular services to Moscow. Finnair's first jets, Aerospatiale Caravelles, were introduced during the early 1960s, and followed in 1969 by Douglas DC-8s for transatlantic air services.

Finnair is owned 73 per cent by the Finnish Government and 27 per cent by private investors. In 1962 Finnair acquired Veljekset Karhumaki, an aircraft repair firm, and through this a 28·9 per cent interest in a charter airline, Kar-Air, with which Finnair also had an operating agreement. Currently the airline operates a fleet of Douglas DC-8s, DC-9s, and DC-10s, Aerospatiale Super Caravelles, Convair 440s, and a Beech Debonair.

'Firebar', Yakovlev Yak-28: *see* Yak-28 'Firebar', Yakovlev.

Firebrand, Blackburn: the last piston-engined combat aircraft to be used by the Fleet Air Arm, the Blackburn Firebrand first entered service in 1945, too late for active service in World War II, and remained with the Royal Navy until replaced by the Westland Wyvern (q.v.) in 1953. A low-wing monoplane torpedo-bomber with a single 2,500 hp radial engine, the Firebrand could also carry bombs or rockets, and was fitted with four 20 mm cannon.

Firefly, Fairey: after a first flight in December 1941, the Fairey Firefly carrier-borne reconnaissance-fighter entered service with the Royal Navy in 1943, replacing the Fairey Fulmar. A twin-seat, single-engined, low-wing monoplane, the Firefly was used extensively in the Far East, often as a night fighter. After the war the Firefly was also used by the Royal Netherlands Naval Air Service.

Firestreak, Hawker Siddeley Dynamics: the Firestreak originated as a de Havilland design, and was the first British air-to-air guided missile to enter service when it was fitted to the R.A.F.'s Gloster Javelin (q.v.) all-weather fighters and the Royal Navy's de Havilland Sea Vixen (q.v.) interceptors in 1958–9. Now superseded by Red Top (q.v.), Firestreak was a clear-weather missile using a solid propellant rocket, and was equipped with two sets of infra-red homing devices, with the nose device locking on to the target, and detonation of the missile controlled by another two devices set back from the nose.

'Fishbed', Mikoyan-Gurevich MiG-21: *see* MiG-21 'Fishbed', Mikoyan-Gurevich.

'Fishpot', Sukhoi Su-9: *see* Su-9 'Fishpot', Sukhoi.

'Fitter', Sukhoi Su-7B: *see* Su-7B 'Fitter', Sukhoi.

5A-: international civil registration index mark for Libya.

5B-: international civil registration index mark for Cyprus.

Five Freedoms of Air Transport: the 1944 Chicago Convention on Air Transport decided the basis of post-World War II international air transport, which was subsequently implemented by the International Civil Aviation Organisation (q.v.) (I.C.A.O.). One of the Convention agreements was the identification of five 'rights to fly', and these duly became known as the 'Five Freedoms of Air Transport'.

The First Freedom allows an airline to operate over foreign territory without landing;

The Second Freedom allows the airline's aircraft to land at a foreign airport for purposes other than the picking up or setting down of passengers, mail or cargo – e.g. refuelling stops are covered by this;

The Third Freedom allows passengers, mail, and cargo taken on board in the airline's own country of origin to be set down at a foreign airport;

The Fourth Freedom allows passengers, mail, and freight to be picked up at a foreign airport and to be flown to the airline's own country;

The Fifth Freedom allows passengers, mail, and cargo from a second country to be set down at an airport in the country of origin of the airline, and then to be flown on to a third country.

In addition, there are three other 'freedoms' which lack official recognition, but which are in fact sometimes referred to and which do exist for practical purposes. These are:

The Sixth Freedom, which is a combination of the Third and the Fourth Freedoms, which are usually gained together by an airline;

The Seventh Freedom is the right to fly passengers, mail, and cargo between two countries using the airline of a third country;

The Eighth Freedom is the right to cabotage traffic, which is the right of an airline to carry traffic between two domestic points in a foreign country. Good examples of this are the B.E.A. and Pan American domestic services from West Berlin to West Germany.

5H-: international civil registration index mark for Tanzania.

5N-: international civil registration index mark for Nigeria.

5R-: international civil registration index mark for the Malagasy Republic.

5T-: international civil registration index mark for Mauritania.

5U-: international civil registration index mark for the Niger Republic.

5V-: international civil registration index mark for Togo.

5W-: international civil registration index mark for Western Samoa.

5X-: international civil registration index mark for Uganda.

5Y-: international civil registration index mark for Kenya.

fix: the ground position of an aeroplane, determined by observations based on the sun or other celestial bodies, or by signals from radio beacons.

fixed-wing: a conventional aeroplane, as opposed to one with wings which rotate or move in flight. The term 'fixed-wing' is also applied to aircraft with wings which fold for ease of storage, i.e. most carrier-borne fixed-wing aircraft.

flag carrier: a concept closely related to that of the chosen instrument (q.v.), although less restricted in practice, since a nation may well have several flag carriers operating on different scheduled routes and only occasionally overlapping. Good examples of the existence of a number of flag carriers are the three British flag carriers, B.E.A., B.O.A.C., and British Caledonian, and the two French flag carriers, Air France and U.T.A.

'Flagon', Sukhoi Su-11: *see* Su-11 'Flagon', Sukhoi.

flaps: flaps can be fitted to either the trailing edge or leading edge (leading edge flaps) of an aircraft wing, and are designed to increase both drag and lift during the landing and take-off sequences by either serving as an extension of the wing or by hinging downwards at such times.

Early aircraft did not require flaps, but as speeds increased and higher wing loadings became commonplace, it proved necessary to resort to some device to reduce stalling speeds. First introduced before World War I, flaps started to be used widely during the early 1930s. Leading edge flaps (*see* slot) are generally of more recent use, and have also been required by a combination of increased speeds and wing loadings, while wing sizes have been reduced to minimise drag at high speeds; or leading edge flaps have augmented slots and trailing edge flaps to increase lift and reduce take-off distances.

'Flashlight', Yakovlev Yak-25: *see* Yak-25 'Flashlight', Yakovlev.

flight deck: the normal term for the control cabin of an airliner, a flight deck differs from a cockpit in having accommodation for pilots, navigator, and flight engineer, with space for movement. Early airliners, such as the de Havilland Hercules and the first Ford Trimotors, used cockpits, but the advent of the modern airliner during the early 1930s saw the pilots' positions brought inside the aircraft.

flight information region: a flight information region is an area within which a flight information service is available from the air traffic control (q.v.) authorities.

flight level: the height above sea level at which an aircraft is flying.

flight plan: advance notice given by an aircraft's captain to the air-traffic control authorities, including details of his aircraft, proposed route, times, intended cruising altitude, and airspeed.

flight time: the flight time is the period of duration of a flight, reckoned from the time of take-off to the time of landing.

float-plane: basically a predecessor of the seaplane (q.v.), the float-plane began with the float-glider, first built and flown in 1905 by Gabriel Voisin (q.v.) for Archdeacon and Blériot (q.v.), using a combination of the Wright brothers and the Hargrave designs, which was successfully towed-off the River Seine in Paris by a motor-boat. The first powered float-plane was built and flown by Henri Fabre (q.v.) in 1910, and this was followed by the first truly practical float-plane, built in 1911 by Glenn Curtiss (q.v.) in the United States.

The original float-plane was soon developed into the seaplane, flying-boat (q.v.), and amphibian (q.v.), but the term was retained for conversions of landplanes, such as the Junkers F.13 (q.v.), the de Havilland DHC-6 (q.v.) Twin Otter, and the Cessna 150 (q.v.). While most such aircraft were civil in character, a few military aircraft, such as the Fairey Flycatcher (q.v.) and Swordfish (q.v.), also appeared as landplanes and float-planes.

Flycatcher, Fairey: one of the first aircraft to be specifically designed for operation from ships at sea (including not only aircraft carriers, but also battleships with take-off and landing platforms over the gun turrets), the Flycatcher was a single-seat fighter biplane using one 400 hp Armstrong-Siddeley Jaguar engine. The Fairey Flycatcher first appeared in 1923, and remained in service until the mid-1930s in landplane, float-plane, and amphibian versions.

Flyer, Wright: the early Wright brothers (q.v.) aeroplanes were known as the Flyer series, and the first of these, appropriately enough known as the Wright Flyer I, was the first aircraft to fly when it made four flights on 17 December 1903. Almost a year later the Wright Flyer II made the first of eighty flights, which were to include the first circle, but it was not until the first flight of the Flyer III in September 1905 that the first truly practical aeroplane could be deemed to have arrived. The Flyer III was fully manoeuvrable, and was the first aeroplane to fly for more than half an hour at a time, while a modified Flyer III made the first passenger-carrying flight on 14 May 1908.

flying-boat: the flying-boat, like the sea-plane (q.v.), was a development from the early float-plane (q.v.), but differed in having a hull which settled in the water and acted as an undercarriage. One of the early pioneers in this field was the American, Glenn Curtiss (q.v.), who built the world's first flying-boat in 1912. Development continued during World War I on both sides of the Atlantic, and amongst the more notable flying-boats of the war period and immediately after were the Curtiss HS-1 and the NC-4 (q.v.), the latter having the distinction of making the first flight across the Atlantic (*see* Atlantic flights) in 1919. The Seaplane Experimental Station at Felixstowe produced a number of designs, as did the Italian Aermacchi (q.v.) concern, and it was a Seaplane Experimental Station design, the F.5, built by Short Brothers, which had the first metal hull.

The flying-boat had its heyday during the inter-war period, when a variety of manufacturers produced a succession of civil and military designs which did much to open up the air routes of the world and make air travel a reality. Notable aircraft included the Dornier Wal and its successors, the giant Do.X (q.v.) and the tri-motor Do.24 (q.v.). The United States produced several successful designs by Sikorsky, Curtiss, and Martin before Boeing produced its Boeing 314 (q.v.) for transatlantic services, just before World War II broke out in Europe, and Consolidated produced the most successful flying-boat ever, the PBY-5 Catalina (q.v.), for maritime-reconnaissance and air-sea rescue duties. In the United Kingdom two of the major manufacturers were Supermarine (q.v.) and Short Brothers (q.v.), who were eventually joined by Saunders-Roe (q.v.), producing such aircraft as the Supermarine Walrus, the Short Calcutta and Rangoon, and the long-range Empire (q.v.) flying-boat.

Flying-boats were used extensively during World War II for maritime-reconnaissance and transport duties, the most famous including the Short Sunderland (q.v.) and the Blohm und Voss (q.v.) Bv.138 and Bv.222 designs, while both Japan and the United States produced a number of flying-boats.

The flying-boat went into a sharp decline at the end of World War II, although Martin (q.v.) in the United States and Beriev (q.v.) in the Soviet Union built flying-boats for a number of years after the war, and the Catalina remained in production in the United States and the Soviet Union for a period. Short Brothers produced the Sandringham and the Solent as developments of the Sunderland, although other Short designs were abandoned. Saunders-Roe built the giant Princess (q.v.) flying-boat, with ten engines, but this was abandoned after four aircraft were built as being too big for the prevailing air transport market; the Saunders-Roe SR.A1 (q.v.) jet flying-boat fighter was also abandoned after successful trials. In the United States the Hughes Hercules (q.v.), the largest wooden aeroplane built, and still the aircraft with the largest wingspan to date, shared the fate of the Princess, although only one Hercules was built and it made only one flight.

The only flying-boat to have been designed and put into production in recent years has been the Shin Meiwa PS-1 (q.v.), a turboprop flying-boat in service with the Japanese Maritime Self-Defence Force.

flying-bomb: the flying-bomb preceded the surface-to-surface missile (q.v.) and the stand-off-bomb or air-to-surface missile (q.v.), and was purely a World War II concept. Two types of flying-bomb can be defined:

a) The type used by the Germans, notably the V-1 (q.v.) flying-bomb, which was pilotless and unguided, but was propelled by a single jet engine at about 400 mph for 150 miles. The V-2 was in fact the first surface-to-surface missile.

b) The piloted suicide planes used by the Japanese, and generally known as the Kamikaze (q.v.), were also flying-bombs, although early Kamikaze aircraft were fighter-bombers filled with explosives. Later versions were air-launched and specially-designed flying-bombs.

Neither type of flying-bomb was a real success in strict military terms, since both were wasteful, the German machine often not finding a target of any military importance, and the Japanese using scarce pilots.

flying clubs: a flying club is simply a club, with aeroplanes of its own, for those who cannot afford to own their own aircraft. Some clubs are specialist in nature, concentrating on aerobatics or air touring, for example. Many offer flying training as well.

The most famous, or infamous, use of the flying club concept occurred in Germany during the late 1920s and early 1930s, before re-armament was officially put in hand; the training and flying given in these clubs was entirely military in outlook since they effectively took the place of an air force training school. Interestingly, the post-war East German air force was also born out of the flying club background!

Flying Fortress, Boeing B-17: *see* B-17 Flying Fortress, Boeing.

Flying Tiger Line: the Flying Tiger Line was formed in 1945 in the United States, by a group of war veterans as the National Skyway Freight Corporation, adopting the present title a year or so later to reflect the fact that the founders had fought with the American 'Flying Tigers' unit in Burma and China. From the start the concern was intended as an all-cargo scheduled airline, and the first aircraft, Budd Conestogas, were soon replaced by Douglas DC-3 Dakotas and DC-4 Skymasters, used on military charter flights to Japan, and Curtiss C-47 Commandos. A coast-to-coast U.S.A. scheduled freight service was started in 1949. A trans-Pacific freight service to Tokyo, Hong Kong, and other destinations to Bangkok was started in 1969, by which time the airline was operating Boeing 707-349Cs. The current fleet consists of Douglas DC-8-63F long-range freighters.

Focke-Achgelis: the first practical helicopter was the Focke-Achgelis Fa.61 of 1936, a twin-rotor design with laterally offset rotors mounted on either side of the fuselage. The Fa.61 only appeared in prototype form, but a successor, the Fa. 223 of 1940, was built in limited numbers by Focke-Wulf (q.v.).

Focke-Wulf: a German aircraft manufacturer which came into prominence during the inter-war period. Focke-Wulf initially produced a light training biplane, the Fw.44 (q.v.) Stieglitz, followed by a twin-engined light transport, communications, and navigational training monoplane, the Fw.58 Weihe (or Kite), for the Luftwaffe during the early and mid-1930s. During the mid-1930s Focke-Wulf was associated with the Focke-Achgelis (q.v.) Fa.61, the first practical helicopter, and followed this with production of the Fa. 223 in 1940. Before the outbreak of World War II, the Focke-Wulf Fw.200 (q.v.) Condor, a four-engined airliner, went into service with Lufthansa, and this became the Luftwaffe's standard maritime-reconnaissance aircraft during the war. The Fw.190 (q.v.), a single-engined fighter monoplane, was a successful Luftwaffe fighter during the war, and more than 20,000 were built. A successor to the Fw. 190 was the Ta.152, designed by the company's Chief Designer, Kurt Tank (q.v.). The other main wartime Focke-Wulf product was the twin-boom and twin-engined Fw. 189 reconnaissance aircraft.

After the war Focke-Wulf was a sub-contractor in a number of European joint production programmes based on American designs, before merging with Wesser in 1963 to form V.F.W. (q.v.), which Heinkel (q.v.) joined in 1964.

Fokker: the Royal Netherlands Aircraft Factory, Fokker, was established in 1919 by Anthony Fokker (q.v.), and initially built a number of his wartime fighter designs, using parts which had been smuggled in from Germany. The first notable civil design from the company was the F.VII, a single-engined airliner of the mid-1920s, and considerable success accompanied the development of a tri-motor version, the F.VIIB-3M (q.v.), which was followed by the twin-engined F.VIII, the F.IX and F.XX trimotors, and the F.XXII four-engined airliner, before the outbreak of World War II. One of the few military aircraft built by the firm was the D.XXI fighter, which was in service with the Royal Netherlands Air Force

on the outbreak of World War II, and was in fact one of the few modern aircraft in R.Neth.A.F. service at the time.

German occupation of the Netherlands during World War II brought an end to the production of Fokker designs for the period, but after the war Fokker produced the S.11 Instructor basic trainer, as well as participating in a number of European production programmes of British and American aircraft designs. During the late 1950s great success followed the introduction of the F-27 (q.v.) Friendship light transport, which was also built under licence in the United States by Fairchild-Hiller (q.v.), a jet successor to the turbo-prop F-27 in the form of the F-28 (q.v.) Fellowship has been slow in gaining airline acceptance.

In 1971 Fokker amalgamated with the German V.F.W. (q.v.) concern to make V.F.W.-Fokker, the first of what may well be many similar European aero-industry mergers.

Fokker, Anthony H. G. (1890–1939): a Dutchman, Anthony Fokker started to build aircraft in 1910 at the age of twenty. On the outbreak of World War I in 1914 he offered his services to the Allies, but was turned down and worked instead for the Central Powers. Undoubtedly the most notable of his wartime designs was the E.III monoplane which was fitted with his propeller-synchronised machine gun, and created such havoc amongst Allied aircraft that it was dubbed the 'Fokker Scourge'. Other famous Fokker designs of the war period included the Dr.1 triplane, which was used by many of Germany's leading fighter aces. After the war Fokker established the aircraft factory which bore his name in his native Holland, and largely concentrated on civil aircraft production. *See also* Fokker.

Folland Aircraft: a British company, the most notable achievement of which was the design and production of the Gnat (q.v.) light fighter and advanced jet trainer in the late 1950s, before the company became a part of Hawker Siddeley Aviation (q.v.) in the 1960 British aircraft industry mergers. Production of the Gnat was continued by Hawker Siddeley, and the aircraft has remained in production and under development by Hindustan Aeronautics (q.v.) in India.

Fonck, René: René Fonck was the top-scoring French fighter pilot during World War I, and in fact the top-scoring surviving wartime pilot at the end of the war, with a total of seventy-five confirmed victories and perhaps another fifty unconfirmed, most of which were gained while flying Spad biplanes. After the war Fonck made an unsuccessful attempt at the first New York–Paris non-stop flight, and reorganised French fighter defences during the 1930s.

Ford: the American Ford Motor Company entered aircraft production during the early 1920s with the Flivver, a single-engined, single-seat, low-wing monoplane, and followed this with the Ford-Stout Pullman single-engined airliner of 1924, and the first of the famous Ford Trimotors in 1925. Ford's most successful design was the 5-AT Trimotor of 1930, which could carry up to twelve passengers and used three 450 hp Wright Wasp radials. Although the company undertook some aero-engine production in later years, airframe production ended during the mid-1930s in the face of strong Boeing, Douglas, and Lockheed competition.

4R-: international civil registration index mark for Ceylon.

4W-: international civil registration index mark for the Yemen.

4X-: international civil registration index mark for Israel.

Fox, Fairey: in many ways the predecessor of the fighter-bomber, the Fairey Fox single-engined biplane was designed as a day bomber with a very high performance, and first flew in 1925. History has it that Viscount Trenchard, then Air Marshal Trenchard, R.A.F., was so impressed by a demonstration that he immediately ordered a squadron, and only severe Government financial difficulties prevented further substantial orders. A 480 hp Curtiss D-12 engine powered the Fox, giving a maximum speed of 160 mph, which was higher than many contemporary fighters.

'Foxbat', Mikoyan-Gurevich MiG-23: *see* MiG-23 'Foxbat', Mikoyan-Gurevich.

Freccia, Fiat G.50: *see* G.50 Freccia, Fiat.

Freedom 7: the first American manned spacecraft, the Freedom 7 was launched by a Redstone rocket on 5 May 1961, carrying Commander Alan B. Shepard, U.S.N., to a height of 115 miles before splashdown. The craft did not orbit the Earth.

Freedom Fighter, Northrop F-5A/B: *see* F-5A/B Freedom Fighter, Northrop.

Freighter, Bristol 170: *see* Bristol 170 Freighter.

'Fresco', Mikoyan-Gurevich MiG-17: *see* MiG-17 'Fresco', Mikoyan-Gurevich.

Friendship, Fokker F-27: *see* F-27 Friendship, Fokker.

Fuji: a Japanese aircraft manufacturer, Fuji started operations during the early 1950s by building Beech T-34 (q.v.) Mentor trainers under licence, and developing the LM-1/2 Nikko four/five-seat cabin monoplane from the basic Beech design. An intermediate jet trainer, the T-1 (q.v.), was built between 1958 and 1963 as a North American T-6 Harvard replacement. Since the early 1960s Fuji has been building Bell 204 Iroquois helicopters under licence, and has also put the FA-200 Aero Subaru four-seat cabin monoplane into production.

Fulmar, Fairey: a light-weight development of the Fairey Battle (q.v.) light bomber, the Fulmar served as a carrier-borne fighter-bomber with the Fleet Air Arm during the early part of World War II. A twin-seat, single-engined monoplane with eight Browning machine guns in the wings, the Fulmar was a fairly undistinguished design, and was eventually replaced by the Fairey Firefly (q.v.).

fuselage: the main part, body or hull, of an aeroplane, including the cockpit and/or cabin, but excluding the wings and tailplane.

Fw.44 Stieglitz, Focke-Wulf: the Focke-Wulf Fw.44 Stieglitz was the Luftwaffe's basic trainer throughout the 1930s. A twin-tandem seat biplane, the Fw.44 used a single 150 hp Siemens Sh 14a radial engine.

Fw.190, Focke-Wulf: although slightly less famous than the Me.109, the Focke-Wulf Fw.190 was an excellent fighter and fighter-bomber, and probably superior to the Messerschmitt design. Some 20,000 Fw.190s were built between 1939 and 1945, although this was well below the Me.109's total. A single 1,800 hp B.M.W. radial engine powered the single-seat, low-wing Fw.190. A successor, also designed by Focke-Wulf's chief designer, Kurt Tank (q.v.), was the Ta.152H, which included a pressurised cockpit for high altitude operations, but few of these were operational by the end of the war.

Fw.200 Condor, Focke-Wulf: the Focke-Wulf Fw.200 Condor had a distinguished career both as an airliner and as a World War II maritime-reconnaissance aircraft. The airliner version was first in service during the late 1930s. A four-engined, low-wing monoplane with a retractable undercarriage, and a maximum speed of around 230 mph, the Fw.200 had a considerable range for an airliner of the period, which doubtless led to its adoption by the Luftwaffe for the maritime-reconnaissance role.

G

G-: international civil registration index mark for the United Kingdom of Great Britain and Northern Ireland.

g: symbol for the gravitational force resulting from rapid acceleration or a change in direction. The tolerance of equipment or aircrew is measured in terms of g, with 2g equal to twice the normal force of gravity, 3g equal to three times the normal force, and so on.

G2-A Galeb, Soko: the Soko G2-A Galeb (or Seagull) tandem two-seat basic jet trainer first flew in May 1961, with deliveries to the Yugoslav Air Force starting in 1965. A maximum speed of 505 mph and a range of 770 miles comes from a single 2,500 lb thrust Bristol Siddeley Viper II Mk.22–6 turbojet. Machine guns can be fitted and rockets and bombs carried for training duties and for counter-insurgency operations, although this task is normally left to the J-1 (q.v.) Jastreb development.

G-24, Junkers: the first trimotor Junkers, the G-24 was in many ways a 'blown-up' development of the F-13 (q.v.), retaining a low-wing monoplane layout with all-metal construction. The aircraft pioneered Lufthansa's flights to the Far East during the mid and late 1920s.

G-38, Junkers: the Junkers G-38 first appeared in 1929, and this unconventional aircraft was one of the largest landplanes at the time. The G-38 employed a lifting aerofoil fuselage, with twelve passengers sitting in the wing in cabins in the inboard leading edges, and another twenty-two passengers sitting in the fuselage in the normal way, although on two decks. A total of 2,400 hp was provided by four engines.

G.50, Freccia, Fiat: developed during the late 1930s, the Fiat G.50 Freccia equipped the Italian Aviazone Legionaria in the Spanish Civil War, also playing a prominent role in Italian operations in North Africa before and after the start of World War II. A single radial-engined low-wing fighter monoplane, the G.50 had the hump-backed fuselage typical of many fighter aircraft of the period.

G.91, Fiat: the Fiat G.91 design won a NATO competition for a strike fighter during the early 1950s, and the first flight of a prototype followed in April 1956, a second aircraft being completed in time for a NATO evaluation contest in October 1957. Deliveries of the G.91 to the Italian Air Force started in August 1958, although the bulk of the production aircraft which followed were in fact G.91Rs with cameras mounted in the nose. Fiat built 100 G.91Rs for the Italian Air Force, and another 100 for the Luftwaffe before Dornier (q.v.) started licence-production in Germany of a further 300. Italian aircraft were designated G.91R/1, G9.1R/1A, and G.91R/1B, while the German aircraft were designated G.91R/3 and G.91R/4.

Developments of the basic single-seat aircraft for conversion training purposes were the G.91T/1 and G.91T/3, for Italy and Germany respectively, and these aircraft had a larger fuselage and wing area, in addition to having tandem twin-seating. The G.91T airframe was modified to take two 4,080 lb thrust General Electric J85-GE-13A turbojets with afterburning instead of the standard single Bristol Siddeley Orpheus 803 turbojet of 5,000 lb thrust, and this single-seat aircraft was designated the G.91Y. Range of the G.91R is 400 miles, and of the G.91Y, about 900 miles, while warload has also been increased on the twin-engined aircraft. A number of surplus Luftwaffe G.91Rs have been sold to Portugal.

G.222, Fiat: a tactical military transport aircraft developed in Italy by Fiat for the Italian Air Force, the G.222 first flew in prototype form in 1970, and is now in production and service. Two 3,400 shp General Electric T64-P-4C turboprops give this high-wing monoplane a maximum cruising speed of 260 mph and a range of up to 2,700 miles, while more than 14,000 lb of freight or forty-four passengers can be carried. Tail-doors are fitted for loading heavy equipment.

Gagarin, Flight Major Yuriy Alexeyevich: a Soviet Air Force officer, Flight Major Yuriy Gagarin was the first man to leave the Earth's atmosphere when, on 12 April 1961, he made a single orbit of the Earth in a Vostock I spacecraft. Aged twenty-seven at the time of the flight, Gagarin was killed while flying a jet fighter on 27 March 1968.

'Gainful': more usually known as SA-6 or SAM-6, 'Gainful' is a lightweight surface-to-air missile. The Soviet 'Gainful' is rocket-propelled and about twenty feet in length. Normally 'Gainful' is deployed on a mobile triple launcher.

Galaxy, Lockheed C-5: *see* C-5 Galaxy, Lockheed.

Galeb, Soko G2-A: *see* G2-A Galeb, Soko.

Galland, Generalleutnant Adolf: one of the most successful Luftwaffe fighter pilots during World War II, Galland had learned to fly, along with many of his contemporaries, in gliders before joining the German air force which was being formed secretly in Italy in 1933. Galland also flew with the Legion Condor in the Spanish Civil War, and later in the Battle of Britain using Messerschmitt Bf.109s. Rapid promotion resulted in Galland becoming Commander-in-Chief of the Luftwaffe's fighters at the age of twenty-nine, but a series of policy disputes with the High Command resulted in his dismissal and posting to command of a Messerschmitt Me.262 jet fighter squadron in 1945. Adolf Galland was able to claim 105 victories at the end of the war, and had been awarded the Knight's Grand Cross with diamonds.

'Galosh': 'Galosh' is a Soviet surface-to-air missile for interception of incoming ballistic missiles before they re-enter the atmosphere. An estimated sixty-three 'Galosh' missiles are deployed around Moscow, and each has a range of several hundred miles and can carry a 1–2 mega-ton nuclear warhead with which to neutralise a ballistic missile. It is considered possible that later developments will have a loiter capability.

Gamecock, Gloster: a development of the Grebe (q.v.) fighter biplane of 1923, the Gloster Gamecock first flew in early 1925. A small, single-engined biplane, the Gamecock had a maximum speed of 150 mph.

Gamma, Northrop: one of a number of small, all-metal mailplanes available from American manufacturers during the early 1930s, the Northrop Gamma had a maximum speed of over 200 mph and led to the passenger-carrying Northrop Delta and the A-17 attack aircraft of 1935.

'Ganef': more commonly known as the SA-4 or SAM-4, the 'Ganef' is a highly mobile Soviet surface-to-air missile of about thirty feet in length, and is unusual for a Russian missile in using ramjet propulsion with four jettisonable rocket boosters for take-off, while control is by pivoting wings. Guidance is by radio command associated with radar tracking. Range is believed to be about forty miles, and 'Ganef' first entered service in 1967.

Gannet, Westland: originally developed for the Royal Navy by Fairey Aviation as a carrier-borne anti-submarine aircraft, the Gannet first flew in September 1949, and entered service as the A.S. Mk.1 in the summer of 1953. Initially a single 3,035 shp Bristol Siddeley Double Mamba 101 turboprop engine was fitted to the aircraft, which had the three-man crew in separate cockpits, and a cranked mid-wing. Gannets of this type were supplied to Federal Germany and to Australia, and later second-hand Royal Navy Gannets were supplied to Indonesia. A few anti-submarine Gannets have remained in the Royal Navy on C.O.D. duties, but the A.S.W. function has been taken over by Westland SH-3D Sea King helicopters.

Development of an airborne early-warning version started during the mid-1950s, and the first flight of a Gannet of this type was in August 1958, with deliveries starting in the December. The A.E.W. Gannet differed from the earlier versions in having a re-designed fuselage and tailplane, while up-rated, 3,875 shp Bristol Siddeley Double Mamba 102 turboprops were fitted, giving a loiter time of six hours and a maximum speed of 340 mph. Designated the A.E.W. Mk.3, a few of these aircraft remain in service aboard the Royal Navy's remaining aircraft carrier, H.M.S. *Ark Royal*. Another Gannet version was a conversion trainer modification of the earlier A.S.W. aircraft.

Gardan, Yves: a French aircraft designer, Yves Gardan's most famous design so far has been the GY-80 Horizon (q.v.) light aircraft which was produced by Socata, the light aircraft subsidiary of Aerospatiale. Other Gardan designs include the SITAR GY-90 Mougli, the GY-100 Bagheera, and the GY-110 Sher Khan.

Gargoyle, McDonnell RTV-2: *see* RTV-2 Gargoyle, McDonnell.

Garnerin, A. J. (1769–1823): a Frenchman, André Jacques Garnerin made the first successful parachute descent from a balloon at 3,000 feet over Paris on 22 October 1797. The Garnerin parachute differed from the modern parachute mainly in that the parachutist did not hang from the parachute, but instead travelled in a car slung underneath the parachute, balloon-style. Garnerin's wife was the first woman to make a balloon ascent.

Garrett Airesearch: an American aero-engine manufacturer, Garrett Airesearch generally concentrates on development and production of light turboprop engines, including the powerplants for the North American OV-10A (q.v.) Bronco. Other Garrett interests include Normalair, an aircraft pressurisation equipment manu-facturer with which Westland Aircraft (q.v.) is also associated.

Garuda Indonesian Airways: the history of Garuda Indonesian Airways really dates from the 1930s, which saw the expansion of K.L.M. Royal Dutch Airlines services into the area, and the pre-war operations of K.L.M.'s subsidiary K.N.I.L.M. Gar-uda itself was not formed until 1950, however, when K.L.M. and the Govern-ment of the then newly-independent Indonesia started the airline. K.L.M.'s

interest was taken over by the Indonesian Government in March 1954, and a number of other smaller airlines in Indonesia were acquired during the next decade or so. Currently, Garuda operates an extensive domestic network in an area with limited and poor surface communications, and also services to Rome, Amsterdam, and a number of Far Eastern destinations. The fleet includes Douglas DC-8-50s, DC-9s, and DC-3 Dakotas, Convair 990As, 440s, and 340s, Fokker F-28 Fellowships and F-27 Friendships, and Lockheed Electras – some of which are ex-K.L.M.

gas turbine: basically, a mechanical unit which rotates in reaction to a current of gas passing through or over it. This also defines a turbojet (q.v.) or turbofan, but not a ramjet or rocket motor.

Gazelle, Aerospatiale S.A. 341: *see* S.A. 341, Gazelle, Aerospatiale.

Gemini Programme: the National Aero-nautics and Space Administration's Gem-ini spacecraft carried a two-man crew and was launched by a Titan rocket. The pro-gramme started with Gemini 3 on 23 March 1965, which was the first two-man United States spaceflight and made three orbits of the earth. A total of ten launches were made, those following Gemini 3 being numbered Gemini 4 to Gemini 12 inclusive. Significant Gemini flights in-cluded those of Gemini 4, from 3 to 7 June 1965, when Major Edward Higgins White, U.S.A.F., made the first American 'space walk'; Gemini 6, 15 to 16 Dec-ember 1965 and Gemini 7, 4 to 18 Decem-ber 1965, which made a successful rendez-vous while in orbit; and Gemini 8, on 16 March 1966, which successfully docked with an unmanned Agena spacecraft while in orbit.

General Dynamics: a major United States aerospace group, General Dynamics itself dates from 1880 and the formation of Electro Dynamics, which became General Dynamics in 1952; while the aerospace connection dates from 1908 and the establishment of the Gallaudet Aircraft Company. Over the years a number of aircraft manufacturers have been acquired by General Dynamics, including some famous names. The most important of these have been Thomas, founded in 1909; Dayton-Wright, founded in 1917; Consolidated, founded in 1923 and now General Dynamics' Convair (q.v.) Division; Stinson (q.v.), founded in 1925; Thomas-Morse, founded in 1929; and Vultee (q.v.), founded in 1932. Another important subsidiary is Canadair (q.v.).

General Dynamics is currently producing the F-111 (q.v.) fighter-bomber for the U.S.A.F., and the group also produces the Terrier (q.v.), Tartar (q.v.) and Redeye (q.v.) missiles and the Atlas space launcher.

General Electric: one of the top three aero-engine manufacturers in the world, General Electric has undertaken turboprop, turbojet, and turbofan development and production since the end of World War II, and amongst the company's more notable products can be included the CF-6 advanced technology engine of the DC-10 (q.v.) and A.300B (q.v.) airbuses, as well as the engines for many other civil and military aircraft. If the Boeing 2707 (q.v.) supersonic transport project had not been cancelled, the GE4 engine would have been the most powerful civil powerplant to date. Other aircraft in production today using G.E. engines include the Fiat G.222 (q.v.), the Lockheed S-3A (q.v.) Viking with the CF-34 engine, and the McDonnell Douglas F-4 (q.v.) Phantom II with the J79 engine. Turbines are produced for a number of helicopters, including the heavier Sikorsky products. G.E. has licensed production to SNECMA of France, while the licence originally held by Bristol Siddeley (q.v.) has now passed to Rolls-Royce (q.v.), enabling that company to supplement its own helicopter engine designs with those of the American company.

Ghana Airways: Ghana Airways was formed in July 1958 to take over the former Ghana operations of the West Africa Airways Corporation after the Gold Coast achieved its independence and changed its name to Ghana. Initially the airline was a limited liability company, with B.O.A.C. (q.v.) holding 40 per cent of the capital, and the Ghanaian Government holding the remainder and also underwriting any losses. The first long-distance services to London were operated by B.O.A.C. aircraft in Ghana Airways colours, while Douglas DC-3 Dakotas and de Havilland Herons operated the domestic services.

Full control of the airline passed to the Government of Ghana in 1961, and the title Ghana Airways Corporation was taken. During the 1960s, the airline embarked on an ambitious but commercially unrealistic programme of expansion, with generous supplies of Soviet aircraft in addition to two Vickers VC-10s ordered by the airline from the United Kingdom. A change in government in 1966 resulted in the withdrawal of many services to the Middle East and Eastern Europe, and the return of the Soviet aircraft, while one of the VC-10s was also disposed of.

Currently the fleet consists of a VC-10, two Vickers Viscounts, a Hawker Siddeley HS 748, and several Douglas DC-3s, while the route network includes a small domestic operation and services to the major West African and European destinations.

Gibson, Wing Commander Guy, V.C., D.S.O., D.F.C., R.A.F.: Guy Gibson was made commanding officer of 617 Squadron, R.A.F. Bomber Command, a unit originally formed specially for the raid on the Ruhr Dams in Germany on 16 May 1943, for which action he was awarded the Victoria Cross. He was, however, already an experienced bomber pilot, and held both the D.S.O. and bar and the D.F.C. and bar. Relieved of further operational duties after the Dam Busters raid, Guy Gibson pressed for a further operational mission, which was finally granted; during this, tragedy struck when the de Havilland Mosquito flown by Gibson crashed while returning from a raid on the Ruhr industrial area in 1944.

Gigant, Messerschmitt Me. 323: *see* Me. 323 Gigant, Messerschmitt.

Giffard, Henri (1825–82): the French inventor Henri Giffard devoted his considerable income from the invention of the steam injector to aeronautical development, including what was to be the first successful dirigible airship. The Giffard airship, which not surprisingly used steam propulsion, first flew on 24 September 1852, from Paris to Trappes, a distance of seventeen miles. Unfortunately the Giffard airship lacked sufficient power for more than limited manoeuvres, and could not fly in a circle.

Gladiator, Gloster: the last biplane fighter in R.A.F. service, the Gladiator was also one of the first R.A.F. fighters to have an enclosed cockpit. Gladiators were in production during the mid-1930s, and used a single 830 hp Bristol radial engine. A number of the aircraft survived to see operational service during World War II in Norway and Malta.

Glenn, Lt-Colonel John, U.S.M.C.: John Glenn, a Lieutenant-Colonel in the United States Marine Corps, became the first American to make an Earth orbital flight on 20 February 1962, using a Friendship spacecraft which had been launched by an Atlas rocket.

glide path: the path taken by an aircraft during a normal controlled landing descent.

glider: a glider can be defined as an unpowered fixed-wing aeroplane for gliding or soaring.

The inventor of the glider was the Englishman, Sir George Cayley (q.v.), who designed and built model gliders before building in 1852 a full-sized glider which took off, carrying a ten-year-old boy, after being towed downhill into a stiff breeze. A flight by Sir George Cayley's coachman followed a little later. However, the Cayley gliders had no means of control, and this did not come until a German, Otto Lilienthal (q.v.), produced a series of hang gliders, starting in 1891. The pilot hung by his shoulders from the Lilienthal gliders and controlled the glider by moving his body; altogether Lilienthal built two biplane and five monoplane gliders before being killed in a gliding accident. One of Lilienthal's pupils, Percy Pilcher (q.v.), a Scotsman, also enjoyed some success before meeting the same fate, and had also been working on powered flight.

An Englishman, F. H. Wenham (q.v.), also conducted several experiments using gliders during the nineteenth century, and from these he evolved a theory about weight distribution in aircraft.

The Wright brothers (q.v.), built several gliders before embarking on their famous Wright Flyer; the feature of the Wright gliders was the placing of the pilot within the airframe, instead of his being left hanging from it.

After World War I a number of experiments were made with gliders to discover the feasibility of rocket propulsion, including those by the Rhon-Rossitten Gesellschaft in 1928, and the German Sailplane Research Institute in 1938. Also during the 1920s and 1930s gliding was used extensively in Germany as a means of training future Luftwaffe pilots without infringing the restrictions then in force on German military aviation.

World War II saw the introduction of large troop-carrying gliders, used by the Germans in the invasion of Crete and by the Allies in the Normandy invasion. A post-war development by the Chase Aircraft Company in the United States became a successful powered aircraft, the Fairchild C-123 (q.v.) Provider, after the U.S.A.F. decided not to pursue troop-carrying glider development.

Since World War II gliding has increased its attraction as a cheap means of flying, and all-metal gliders of considerable performance have evolved for the sportsman; increasingly, as a result, there is a tendency for gliding enthusiasts to view the powered aeroplane with the same disdain which yachtsmen retain for motor-boats! However, in an attempt to keep flying costs to a minimum, the wheel has already turned full circle with the introduction of the powered glider, either as a low-cost, low-powered aircraft in the tradition of the early pioneers of flight, or as a means of increasing the range of gliders and enabling gliding enthusiasts to take off without external assistance.

glider bomb: a predecessor of the stand-off bomb or air-to-surface missile, the glider bomb was also sometimes rocket-assisted and also allowed attacking aircraft to launch their bombs some distance from the target and associated defences. The short period between the invention of the glider bomb and the advent of the stand-off bomb meant that the operational applications were few and only within the last couple of years of World War II. The first operational use of the glider bomb was against seven of the Royal Navy's escort vessels on 25 April 1943, when the bombs were launched, unsuccessfully, from Luftwaffe Dornier Do.217s. Two days later, another attack against a British corvette resulted in damage: in each attack, Henschel Hs.293 glider bombs were used.

Gloster Aircraft: a British company, Gloster aircraft first came into prominence in 1923, with the Grebe (q.v.) fighter biplane, and its Gamecock (q.v.) development in 1925. During the late 1920s the Gloster IV seaplane was a runner-up in the 1927 Schneider Trophy race, although the Gloster Gauntlet fighter-biplane of the mid-1930s was rather slower than the IV, with a speed of 240 mph against the seaplane's 270 mph; a reflection of the degree of advancement brought about by the Schneider Trophy races, since the Gauntlet was not considered a slow plane. The last Gloster biplane, the Gladiator (q.v.), distinguished itself over Norway and over Malta during the early years of World War II, in spite of being obsolete at the time and facing overwhelming fighter opposition.

Almost as if to compensate for the Gladiator's obsolete image at the outset of World War II, Gloster built the first Allied jet aircraft, the Gloster E.28/39 Whittle, in 1943, and followed this with the world's second operational jet fighter, the Meteor (q.v.), in 1944. After the war the Meteor established several new speed records. During the early 1950s Gloster produced the R.A.F.'s first all-weather jet fighter and the world's first operational delta wing fighter, the Javelin (q.v.).

Gloster became a part of the Hawker Siddeley Group in 1960, at the time of the British-Government-inspired aircraft industry mergers.

Gnat, Hawker Siddeley: originally evolved by Folland as a lightweight fighter, the first Gnat prototype flew in July 1955, and aircraft of this type were supplied to Finland and India, further aircraft being licence-built in India by Hindustan Aircraft. Development of a tandem twin-seat trainer version for the R.A.F., with a fuselage and wing some 15 per cent larger than on the single-seat fighter model, started in 1958, with a first flight in August 1959; deliveries to the Royal Air Force started in November 1962. The Gnat advanced jet trainer uses a single 4,400 lb thrust Bristol Siddeley Orpheus 101 turbojet, giving a maximum speed of 635 mph and a maximum range of 1,180 miles. Fighter versions use two 30 mm cannon and can carry two 500 lb bombs on underwing strongpoints. The aircraft proved to be so effective in combat with Pakistan Air Force Sabres that it earned itself the title of 'Sabre Slayer' and was put back into production during the late 1960s in India, where a Super Gnat development is now being produced.

Further developments of the Gnat were mooted by Hawker Siddeley during the early 1960s, but rejected by the British Government. Gnats in R.A.F. service will be replaced by the new HS 1184 advanced jet trainer in the mid-1970s.

'Goa': perhaps more commonly known as the SA-3 or SAM-3, the Soviet 'Goa' surface-to-air missile first appeared in 1964. A twenty-foot-long, two-stage solid fuel rocket with radar homing, 'Goa' is guided by elevons on the booster, and is used as a close-range anti-aircraft defence aboard Soviet warships, and also for protection of military installations ashore.

Goddard, Dr R. H.: an American scientist, Dr R. H. Goddard designed and successfully tested the world's first liquid-fuelled rocket in 1926, and opened the way for development of the V-2 (q.v.) in Germany during World War II and for subsequent space explorations, all of which use liquid-fuelled rockets.

Goering, Reichsmarshall Herman: Baron von Richthofen's successor during World War I, Herman Goering ended the war with twenty-two victories and with the Pour le Mérite (or Blue Max), then Germany's highest decoration. After the war he became an enthusiastic follower of

Adolf Hitler and was involved in the 1923 Munich uprising, which was unsuccessful. Nevertheless he succeeded in becoming President of the Reichstag, or Parliament, in 1932.

After Hitler came to power in Germany, Goering became Air Minister, and during World War II he was in supreme command of the Luftwaffe, and one of the most influential members of the National Socialist Party. The available evidence confirms the widely-held belief that Goering, no matter what his accomplishments as a squadron commander, was incompetent in his World War II position. The R.A.F. soon made a nonsense of his extravagant claim that 'no enemy plane will fly over the Reich Territory', and he supported Hitler's policy of using the Me.262 (q.v.) jet fighter as a bomber initially, although it was completely unsuited for the task. Sentenced to death at the Nuremberg War Crimes Tribunal, he committed suicide before the sentence could be carried out.

Goose, Grumman: Grumman's first amphibian design to use a flying-boat (as opposed to a seaplane), type hull and float combination, the twin-radial-engined, high-wing Goose monoplane entered production in 1938, and was extensively used by civil operators before the United States became involved in World War II.

Goupy Biplane: the French-built Goupy biplane appeared in 1909, at the end of the year and too late for entry to the Reims Meeting, which had been held during the autumn. The Goupy established the pattern for biplanes for several decades ahead, with its tractor propeller, wheeled undercarriage, ailerons, long fuselage, and separate tailplane.

Graf Zeppelin: the most successful of Count Ferdinand von Zeppelin's (q.v.) designs, the Graf Zeppelin dirigible airship was built using public subscription funds, and launched in September 1928. The Graf Zeppelin could carry up to twenty passengers and 26,000 lb of mail and cargo for up to 6,000 miles at a speed of around 70 mph. After nine years of highly successful operation, including a round-the-world voyage and numerous transatlantic flights,

the Graf-Zeppelin was withdrawn from service in 1937.

A Graf Zeppelin II was launched in 1939 as a successor to the ill-fated Hindenburg (q.v.), but was scrapped after only some two dozen flights on the outbreak of World War II.

Grebe, Gloster: a light biplane fighter with a single rotary engine, the Gloster Grebe was a Sopwith Snipe replacement when first introduced in 1923. Later, Grebes were used in experiments in operating fighter aircraft from airships.

Gregory, Lt R., R.N.: one of three Royal Navy officers who commenced flying training at Eastchurch, England, in March 1911, and were the first British naval officers to do so. Gregory is sometimes claimed to have made the first flight from a ship under way in May 1912, flying from H.M.S. *Hibernian*, a battleship, in a Short S.27.

Greyhound, Grumman C-2: *see* C-2 Greyhound, Grumman.

'Griffin': More usually known as SA-5 or SAM-5, 'Griffin' is a Soviet long-range surface-to-air missile consisting of a fifty-four foot long, two stage rocket. Primarily intended for anti-aircraft duties, it is considered by some experts to have a secondary anti-missile capability. Radar homing is used.

ground-attack: a development of the fighter-bomber concept, the ground-attack aircraft evolved during World War II as aircraft were developed for this duty possessing more rugged characteristics than are normally required for a fighter. Early examples included the Commonwealth Boomerang and the Hawker Typhoon, both of which were originally designed as fighters; the latter was particularly successful on anti-tank duties. Postwar development has included such aircraft as the Hawker Hunter (q.v.), the L.T.V. A-7 (q.v.) Corsair II, and the Suhkoi Su-7B (q.v.), all of which retain the handling characteristics of the fighter, coupled with a strong airframe and good carrying capacity. Related concepts are counter-insurgency (q.v.) and fighter-bomber (q.v.) aircraft.

Grumman: the American firm of Grumman Aircraft was formed in 1930, and the first product was the FF-1, a tandem twin-seat, carrier-borne fighter biplane with a retractable undercarriage – by no means a common feature at that time. In 1933 the first of many Grumman amphibians, the J2F-1 Duck, appeared, and during the years before the United States entered World War II the company produced the F2F and F3F carrier-borne fighter biplanes, the F4F Wildcat (q.v.) carrier-borne fighter monoplane, and the Goose amphibian. Wartime aircraft included the Widgeon amphibian, the F6F (q.v.) Hellcat fighter and the TBF Avenger – probably the best torpedo-bomber of the war – followed by the F7F Tigercat (the company's first twin-engined fighter) and the F8F Bearcat carrier-borne fighters.

After World War II Grumman continued with Tigercat and Bearcat production for a while, before putting the company's first jet fighter, again for the United States Navy, into production. This was the F9F Panther, followed by its swept-wing development, the Cougar. Meanwhile two more amphibians, the Mallard and the HU-16 Albatross (q.v.), went into production, and work started on a carrier-borne anti-submarine aircraft, the S-2 (q.v.) Tracker, and its A.E.W. and C.O.D. developments, the E-1B (q.v.) Tracer and C-1A (q.v.) Trader. During the late 1950s Grumman produced yet another carrier-borne attack aircraft, the A-6 (q.v.) Intruder, and later the electronics counter-measures development, the EA-6, while also putting the first civil landplane from the company, the turboprop Gulfstream I (q.v.), into production.

More recently Grumman has built a successor to the Tracer in the E-2A Hawkeye (q.v.), with a C.O.D. development in the C-2 (q.v.) Greyhound, and the Gulfstream I has been supplemented by the jet Gulfstream II (q.v.). A S.T.O.L. observation aircraft for the United States Army is the OV-1 (q.v.) Mohawk. Grumman is also testing the U.S. Navy's air superiority fighter for the mid-1970s, the F-14 (q.v.) Tomcat, and plans exist for turboprop re-engining of certain of the company's amphibians, which have stood the test of time remarkably well. In addition to all of this, Grumman designed the lunar landing modules of the Apollo Programme (q.v.).

Guarani II, Dinfia I.A.50: *see under* I.A.58, Dinfia.

'Guideline': more usually known as the SA-2 or SAM-2, the Soviet 'Guideline' surface-to-air missile is standard equipment with the Warsaw Pact forces. 'Guideline' is twenty foot long and has a twenty-two mile range, with solid fuel booster rockets to supplement a liquid fuel rocket. Automatic radar guidance is used, but the missile has proved itself to be relatively ineffectual, particularly at low altitude. Control is by elevons in the booster rocket fins. First introduced during the mid-1950s, 'Guideline' has been extensively developed, and the latest version first appeared in 1969. Some sources credit it with the capability of using a nuclear warhead.

'Guild': the soviet 'Guild' surface-to-air guided missile is about thirty-nine feet in length and uses a dual-thrust solid fuel rocket motor. It is probably a 'Guideline' replacement, and first appeared in 1960, although it is only comparatively recently that 'Guild' has been heard of again.

Gulfstream I, Grumman: the Grumman Gulfstream I was the company's first civil landplane project, and made its first flight in August 1958. A low-wing, twin-engined monoplane for executive, V.I.P., and feeder-liner duties, two 2,185 shp Rolls-Royce Dart 529-8H turboprops give it a maximum cruising speed of 310 mph and a range of up to 2,500 miles, while the pressurised cabin can provide accommodation for up to twenty-six passengers.

Gulfstream II, Grumman: the Grumman Gulfstream II has supplemented rather than replaced the Gulfstream I (q.v.), although it is a turbojet aimed at the same executive, V.I.P. and feeder-liner market. The first flight of a Gulfstream II was in October 1966, and deliveries started in May 1967. Two rear-mounted 11,400 lb thrust Rolls-Royce Spey 25 turbofans give a maximum cruising speed of 520 mph and a maximum range of up to 2,700 miles, while the cabin can accommodate up to thirty passengers.

Gun Bus, Vickers F.5B: most usually known as the Gun Bus, the Vickers F.5B was a development of the earlier F.4B of 1914. A pusher-engined biplane fighter, the observer-gunner in the Gun Bus sat in front of the pilot. Gun Buses first entered operational service with the Royal Flying Corps in France in February 1915, and were withdrawn at the end of 1916. A single 100 hp Gnome Monosoupape rotary engine provided a speed of 70 mph, which was not very fast, even by the standards of the time, but nevertheless the Gun Bus was a popular and effective aircraft.

gunship: a development by the U.S.A.F. during the Vietnam War, the gunship consists of a modified transport aircraft with machine guns and cannon set to fire through the window openings. The U.S.A.F. designation is AC- (q.v.), and the main aircraft types involved are the AC-47, AC-119, AC-123, and AC-130, which are, of course, versions of the C-47, C-119, C-123, and C-130 respectively.

Gurevich, Mikhail: a member of the Mikoyan-Gurevich (*see* Mikoyan) design bureau, which has provided the Soviet Union with the MiG series of fighters, starting during the middle years of World War II.

Guynemer, Captain Georges: one of the leading French fighter aces during World War I, Georges Guynemer had earlier been rejected by an army medical board, but was accepted for flying training in 1915. During the next two years, before his death in September 1917, he flew a Spad fighter biplane and accounted for fifty-three enemy aircraft.

gyroplane: a gyroplane is an aircraft with a rotor blade, or rotary wing, which is driven by slipstream from the propeller. It is popularly known as the autogiro (q.v.), under which heading development is detailed.

H

H-3 Sea King, Sikorsky: *see* S-61, Sikorsky.

H-19 Chickasaw, Sikorsky: *see* S-55, Sikorsky.

H-34 Choctaw, Sikorsky: *see* S-58, Sikorsky.

H-43 Husky, Kaman: development of the Kaman H-43 Husky helicopter for the U.S. Navy started in 1950, with the aircraft first entering service in 1952 as the OH-43D with a 600 hp Pratt and Whitney R-1340-48 piston engine. Later developments for the U.S.A.F. involved the substitution of a turbine engine for the piston engine of the early models, and putting the engine on top of, instead of behind, the cabin, so providing increased accommodation. U.S.A.F. versions are the HH-43E with an 825 shp Lycoming T53-L-1B turbine, and the HH-43F with a 1,150 shp T53-L-11A turbine de-rated to 825 shp to give extra power for operation from hot or high-altitude airfields. The Husky is usually used for airfield crash and rescue duties, and U.S.A.F. machines normally carry a 1,000 lb rescue pack, although up to ten passengers can be carried. Maximum speed is 120 mph and the range is 504 miles.

H-53 Sea Stallion, Sikorsky: *see* S-65A, Sikorsky.

H-54 Skycrane, Sikorsky: *see* S-64 Skycrane, Sikorsky.

HA-: international civil registration index mark for Hungary.

HA-300, Helwan: originally developed by Hispano-Suiza in Spain to the design of Professor Willy Messerschmitt, further development of the HA-300 was undertaken by Helwan in Egypt, largely as a propaganda exercise with which to influence other even less advanced African states. First flight of the prototype, using a 4,850 lb thrust Bristol Siddeley Orpheus B.Or.2 turbojet, was in March 1964. Few production models have been built, and these use a 11,560 lb thrust Brandner E.300 turbojet with reheat, giving a speed claimed to be in excess of Mach 1.0. A lightweight fighter with a single seat, the HA-300 has a mid-wing delta mainplane, with a separate tailplane using all-moving horizontal surfaces.

Halifax, Handley-Page: one of the R.A.F.'s heavy bomber types during World War II, the Handley Page H.P.57 Halifax was a mid-wing aircraft with four Rolls-Royce Merlin in-line engines of 1,280 hp each in the early production models, and Bristol Hercules radial engines in later versions. Nose, tail, and dorsal turrets were fitted to the aircraft, which had a maximum speed of about 300 mph and a range of up to 2,000 miles.

Hampden, Handley Page: a mid-wing, twin-engined medium bomber, the Handley Page H.P.52 Hampden was in R.A.F. service from 1938 to 1942, and was nicknamed the 'Flying Pan-Handle'—a reference to the deep forward part of the fuselage and the very small circumference tail. Two 1,025 hp Bristol radial engines powered the aircraft.

Handley Page, Sir Frederick: founder of the aircraft company which bore his name, Sir Frederick Handley Page was responsible for building a number of aircraft, including his H.P.5 (q.v.) 'Yellow Peril' of 1911, and the world's first successful heavy bomber, the 0/400 of 1917. Amongst the innovations pioneered by Sir Frederick was the Handley Page slot, a modification to the leading edge of an aircraft's wing which helped to reduce landing speeds and thus improve air safety at a time when speeds and wing loadings were increasing.

Handley Page Aircraft: founded in 1909 by Sir Frederick Handley Page (q.v.), Handley Page Aircraft produced a number of aircraft before the outbreak of World War I, including the H.P.5 'Yellow Peril' monoplane of 1911 and its development, the H.P.6 of 1912, and the H.P.7 biplane of 1913. The company came into prominence during World War I with the H.P. 0/400 twin-engined heavy bomber biplane of 1916, which was the first successful heavy bomber aircraft, and was soon followed by the heavier V/1500 of 1918.

After World War I a small number of military aircraft designs were produced, but the first notable aircraft of the period was a civil airliner, the H.P.42 (q.v.) Hannibal of 1930. During the 1930s the company also produced the H.P.50 Heyford bomber and H.P.52 Hampden (q.v.) for the R.A.F. An experimental aircraft, the H.P.75 Manx, preceded the post-World War II Handley Page experiments in tailless aircraft design. World War II aircraft from the company included the Hampden and the Halifax bombers, the latter one of the R.A.F.'s three main heavy bomber types.

The return of peace saw Handley Page developing the Hermes four-engined long-range airliner for B.O.A.C., and the Hastings military transport for the R.A.F. and some Commonwealth air forces. A light transport of the early 1950s was the four-engined Marathon, a high-wing monoplane, and a prototype Marathon-replacement eventually became the Herald (q.v.) twin-turboprop airliner of the late 1950s and the 1960s. The company's military connection was retained during the late 1950s and the 1960s by the production of the Victor (q.v.), a four-engined long-range jet bomber for delivery of Britain's nuclear deterrent, and many of these aircraft remain in service in the tanker aircraft role. Another British manufacturer, Miles Aircraft, was acquired in 1948, and operated briefly as Handley Page (Reading) Ltd.

No new defence contracts were awarded to the company during the 1960s, due to the refusal of its management to merge with one of the three main airframe groups being formed at the British Government's behest in 1960. The difficulties of launching a major aircraft project without Government support led to the decision to develop a small turboprop aircraft for executive and feeder-liner use, the H.P.137 Jetstream (q.v.). Unfortunately, cash flow problems while a production line was being established for the aircraft led to the collapse of the company. Product support for the Herald airliner was sold off to another firm, while Scottish Aviation (q.v.) now builds the Jetstream.

Hansa, H.F.B. 320: *see* H.F.B. 320 Hansa.

Hargrave, Lawrence (1850–1915): an Australian, Lawrence Hargrave is widely believed to have made less of an impact on the development of aviation than might have been the case had he lived and worked closer to the mainstream of activity. However, in spite of the disadvantage of distance (which aviation has since removed), Hargrave managed to conduct numerous experiments with rubber-motored flying models, of which he had built fifty by 1889, and convinced himself (wrongly), of the need to imitate the birds by building aircraft with flappers instead of propellers. More successfully, he evolved the rotary engine, building the first rotary aircraft engine in 1889, and in 1893 invented the box-kite, combining the tandem-wing and biplane configurations. The significance of the latter development lay in the lift provided by the box-kite, which was readily appreciated and adopted in Europe by the Voisin brothers (q.v.).

'Harke', Mil Mi-10: *see* Mi-10 'Harke', Mil.

Harpoon, Lockheed PV-2: *see* PV-2 Harpoon, Lockheed.

Harrier, Hawker Siddeley: originating as a Hawker design using the then new Bristol Siddeley Pegasus vectored thrust turbofan, the aircraft was originally known as the P.1127 or Kestrel. It made its first, tethered, flight in October 1960, and the first of nine aircraft for trials by a tripartite Luftwaffe, R.A.F., and U.S.AF. squadron flew in March 1964. A more powerful engine and various airframe improvements distinguished the production aircraft for R.A.F. service from the six prototypes and nine trials aircraft, and a pre-production model made its first flight in August 1966. Deliveries to R.A.F. squadrons started in April 1969, making the Harrier the first V.T.O.L. fighter to enter service anywhere.

A further up-rating of the aircraft's engine to meet a U.S.M.C. requirement followed. A licence-production agreement with McDonnell Douglas in the United

States was concluded, but never put into action since it was decided that the U.S.M.C. order would be insufficient for economic production of the aircraft under licence, although any further orders could change this picture. The first flight of a two-seat conversion trainer variant, with modified nose and tail units, took place in April 1969, and aircraft of this type have been delivered to the R.A.F. Further development, with wing and engine modifications and search radar, are planned, with particular importance being placed on optimising the aircraft for carrier-borne operations, although trials with the standard aircraft operating from a variety of ships have proved an unqualified success.

The standard R.A.F. aircraft uses a single 19,200 lb thrust Rolls-Royce Pegasus 101 vectored thrust turbofan giving a maximum speed of around Mach $1 \cdot 0$ when flying without any external fuel or stores, and a maximum range of 1,000 miles. Up to 5,000 lb of bombs, torpedoes, mines, missiles, or rockets can be carried from three fuselage and four underwing strongpoints, in addition to two 30 mm cannon pods. Payload varies considerably, depending on whether vertical or short take-off is employed—the disposable weight varies between some 6,000 lb for the former and some 11,000 lb for the latter.

Hart, Hawker: almost certainly the most successful British military aircraft of the inter-war period was the Hart two-seat, single-engined, biplane day bomber, which led to a series of closely related biplanes for a variety of military duties with the R.A.F., and Commonwealth and European air forces and air arms. The first flight of a prototype in 1928 was followed by almost nine years of production of the Hart and Hart variants. Apart from the Hart itself, other aircraft in the series included the Fury and Demon fighters, the Osprey carrier-borne reconnaisance aircraft, the Audax army co-operation biplane, the Hartbees close support aircraft for African air arms, and the Hind, initially developed and put into production as another day bomber variant, but eventually produced and used as a training aircraft. The Fury was the first R.A.F. aircraft to be able to

exceed a speed of 200 m.p.h. in level flight. Although biplanes, with fixed undercarriages and open cockpits, a resemblance in line with the later Hurricane (q.v.) monoplane fighter could easily be traced.

Powerplants varied somewhat, but the most usual were Rolls-Royce Kestrels of 510 hp or 525 hp. Altogether almost 700 aircraft were built.

Hartbees, Hawker: *see under* Hart, Hawker.

Hawk: although in service since 1960, the Raytheon MIM-23A Hawk surface-to-air guided missile has proved to be so effective, particularly at low altitude, that developments are still in frontline service with the United States armed forces and with those of Israel, Japan, Korea, Sweden, Saudi Arabia, Spain, and Nationalist China. A two-stage rocket of some $16\frac{1}{2}$ feet in length, the Hawk has solid-fuel rocket motors and homing radar, and can operate over distances of up to twenty miles.

Hawk, Curtiss: a fighter biplane, the Curtiss Hawk was one of the more important aircraft in the U.S.A.A.C.'s inventory during the late 1920s and early 1930s, while a float-plane version flew with the United States Navy. A 435 hp Curtiss D-12 engine powered the aircraft.

Hawker: a British company formed during the early 1920s, Hawker Aviation took over the former Sopwith (q.v.) works after that company's collapse, and T. O. M. Sopwith played a leading role in the company during its early years, particularly after the death of its founder, Harry Hawker. A number of designs appeared during the early 1920s, including the Cygnet light aircraft, but the company soon settled down to concentrate exclusively on military aircraft production, with the emphasis on the lighter types of aircraft, such as fighters and light day bombers.

It did not take long before the first of the distinctive Hawker biplane fighters arrived, starting with the Hornbill of 1925, followed by the Hornet and the Woodcock, which was also produced in Denmark as the Danecock. The Hart (q.v.) day bomber was so successful that a variety of versions were produced for fighter, observation, ground-attack, and training duties, and one of the fighters, the Fury, became the first aircraft in the R.A.F. to exceed 200 mph. The Hart was one of the first major designs by Sir Sydney Camm (q.v.), who was the company's Chief Designer for many years, and responsible for most of its more successful products.

The famous Hurricane (q.v.) fighter monoplane first flew in 1936, and during the late 1930s and World War II up to 1944 some 14,000 Hurricanes were built, although the aircraft was often outpaced by later and faster German fighters. Hurricane production was supplemented in 1941 by the new Typhoon fighter, until this was superseded by the Tempest fighter near to the end of the war in Europe. Much Tempest production was undertaken by another firm, Gloster Aircraft (q.v.), however, due to the pressure on Hawker's own production facilities. The Tempest was in any case one of the fastest piston-engined fighters.

After World War II Hawkers managed to maintain the momentum of the earlier successes, starting with the Sea Hawk (q.v.) jet fighter for the Royal Navy's aircraft carriers; from this was developed the P.1052 experimental aircraft which preceded the Hawker Hunter (q.v.) jet fighter. The Hunter was produced under licence in Europe and today, some twenty-one years after its first flight, demand continues to run at a high level for refurbished aircraft; this work is an important part of the company's operations. The next development after the Hunter was the Harrier, the world's first vertical take-off fighter aircraft to enter normal service, but much of the development of this aircraft, originally designated the P.1127 by Hawker, took place after the merger of 1960 which created three major private-enterprise airframe groups out of the British aircraft industry. Hawker was in fact one of the major elements in the new Hawker Siddeley (q.v.) Group formed by one of the mergers.

Hawker de Havilland: formed in 1960 from the Australian subsidiaries of Hawker Aviation and de Havilland Aircraft (q.v.), for much of its history the company and its two predecessors produced the designs of the British parent companies under licence

or acted as local agents. Exceptions to this general rule were subcontract work for other manufacturers and the production after World War II by de Havilland of the Australian-designed Drover, a trimotor light transport for the Royal Australian Flying Doctor Service.

Hawker Siddeley Aviation: the airframe design, development, and manufacture subsidiary of the Hawker Siddeley Group, Hawker Siddeley Aviation was formed in 1960 in a government-supported merger of the airframe interests of a number of companies to form three major British airframe companies. The predecessors of H.S.A. included Avro (q.v.), Armstrong-Whitworth (q.v.), de Havilland (q.v.), Gloster (q.v.), Hawker (q.v.), and Folland (q.v.). Current production, which owes much to pre-merger design work, includes the HS 125 (q.v.) executive jet and its BH-200 and BH-600 developments, the HS 748 (q.v.) turboprop airliner, the Trident (q.v.) jet airliner, the Buccaneer (q.v.) strike bomber, and the Harrier (q.v.) V.T.O.L. fighter, while development of the HS 1182 advanced jet trainer for the R.A.F. is well advanced, and H.S.A. is a major subcontractor and design consultant on the European airbus project, the A.300B.

Hawker Siddeley Dynamics: the guided weapons and electronics subsidiary of the Hawker Siddeley Group, Hawker Siddeley Dynamics largely stemmed from the guided weapons divisions of the de Havilland Aircraft Company, now an integral part of Hawker Siddeley Aviation. Products include the Firestreak and Red Top air-to-air guided missiles, with SRAAM-75 and Tail Dog under development for the future, the Blue Steel and Matra air-to-surface missiles, and the Sea Dart and Seaslug surface-to-air guided missiles. The Blue Steel rocket, originally developed as part of a guided missile programme which was cancelled, is frequently used as a launcher in the European ELDO space programme, and has a proven record of outstanding reliability, although the benefits of this have proved to be largely academic, due to unreliable French and German second and third stages.

H.S.D. also produces electronic checkout equipment, TRACE, for aircraft operators, and also for industrial users.

Hawkeye, Grumman E-2A: *see* E-2A Hawkeye, Grumman.

HB-: international civil registration index mark for Switzerland and Liechtenstein.

HC-: international civil registration index mark for Ecuador.

He.51, Heinkel: the last German fighter biplane, the Heinkel He.51 also had the distinction of fighting in the Spanish Civil War with the Legion Condor. Although basically a fighter, the He.51 was a good ground-attack aircraft and was later relegated to training duties.

He.111, Heinkel: officially designed as an airliner during the late 1930s, the Heinkel He.111 was actually intended to be a bomber, and could only carry ten passengers in its transport role. A number of He.111s were already in Luftwaffe service on the outbreak of World War II, and served extensively during the first part of the war, although the twin-engined, low-wing bomber monoplane was poorly armed and an easy prey to Allied fighter attack.

He.115, Heinkel: a twin-engined, mid-wing floatplane for maritime reconnaisance, the Heinkel He.115 was in Luftwaffe service throughout World War II, and aircraft of the type were also used before the war by Norway and Sweden. Bombs, torpedoes, or mines could be carried by the He.115.

He.162 'Volksjaegar', Heinkel: the Heinkel He.162 'Volksjaegar' or Salamander holds the distinction of having been developed in the shortest time of any aircraft, although the Luftwaffe pilots who flew this single-engined turbojet fighter during the closing stages of World War II must have considered this more a curse, and often a fatal one, than a distinction. A maximum speed of 490 mph came from the single engine, which was mounted on top of the fuselage of the aircraft.

He.177, Heinkel: the Heinkel He.177 was the only German attempt at building a heavy long-range bomber during World War II. A mid-wing aircraft, it had the unusual feature of having two engines mounted in each engine nacelle and driving a single propeller – four engines and two propellers altogether! Although the aircraft was pressed into service, mainly on the Eastern Front, it suffered badly from engine overheating problems and earned itself the nickname of the 'Flaming Coffin'.

He.178, Heinkel: the world's first jet aeroplane, the single-engined He.178 first flew in August 1939, although the aircraft was not a practical aeroplane and was solely intended to act as a test-bed. A single-seat, high-wing, monoplane, the aircraft was not unlike the Gloster E.28/39 (q.v.) Whittle in appearance, although the latter aircraft did not fly until 1941.

Heinkel: founded by Ernst Heinkel (q.v.) in 1921, Heinkel was one of the companies violating the terms of the Armistice during the 1920s and early 1930s by building military aircraft in Germany, although many of Heinkel's early sales were to Sweden and to Japan. Amongst the early Heinkels were the H.D.43 fighter biplane of the late 1920s, the H.20 reconnaissance biplane of the same period and, after the Luftwaffe was formed in 1933, the He.45 and He.46 single-engined reconnaissance biplanes. These were followed by the He.51 fighter biplane and the He.60 reconnaissance seaplane, while a larger aircraft was the twin-engined He.59, a large seaplane of biplane construction which was used for a wide variety of duties, including reconnaissance, torpedo-bombing, transport, and ambulance work.

Heinkel monoplanes did not arrive until the late 1930s, before the outbreak of World War II. The first were the He.111, which was initially put into production as an airliner but was in fact intended to be a bomber, and the twin-engined He.115 seaplane for torpedo-bombing duties.

Advanced technological development was not outside of the company's grasp, however. Following Ernst Heinkel's sponsorship of Dr Hans von Ohain's (q.v.) work on the turbojet, the company was able to build in 1939 the world's first jet aircraft, the He.178. Before this a rocket-powered aircraft, the He.176, had also been built. Heinkel was unlucky when the company attempted to build an operational jet aircraft: the He.280 twin-engined jet fighter of 1941 did not enter production, and the He.162 (q.v.), which was developed in the record time of less than three months, proved to be a dangerous machine to fly. A heavy long-range bomber, the He.177, was so prone to overheating troubles that it was nicknamed the 'Flaming Coffin'.

After World War II Heinkel restarted aircraft manufacture once the restrictions on German armament were lifted, and initially the company built Potez CM.170 (q.v.) Magister jet trainers under licence, before merging with V.F.W. (q.v.) in 1964.

Heinkel, Ernst: one of the top four German aircraft designers of the inter-war period, Ernst Heinkel had founded the aircraft company which bore his name in 1921, although his first aircraft design had been produced in 1911 and he had also designed many of the World War I Brandenburg seaplanes. Heinkel produced a number of designs for Japan and for Sweden before the Luftwaffe was formed in 1933, and he provided financial support for the work of Dr Hans von Ohain (q.v.) on the turbojet engine, which enabled Heinkel to build the airframe for the first jet aircraft, the He.178 (q.v.), in 1939. Before this he had also built the He.176 rocket-powered aircraft.

Helicopter: sometimes known as a rotary wing aeroplane, since the rotor blades provide lift and control as well as propulsion, the helicopter was a comparatively late arrival on the aviation scene in practical terms, although helicopter-type aircraft had attracted the attention of many would-be pioneers of flight. The first helicopter designs were those by Leonardo da Vinci (q.v.), and others interested in the helicopter included Sir George Cayley (q.v.) and Horatio Phillips (q.v.). A milestone on the way towards developing a practical helicopter was the gyroplane, which nevertheless differed considerably from the helicopter by having unpowered rotor blades which rotated in the slipstream from the engine's propeller, giving a reduced takeoff. On the other hand, the term 'helicopter' does not figure very prominently at the present time in discussion on V.T.O.L., even though a helicopter is almost by definition V.T.O.L., because the term

V.T.O.L. is reserved for fixed-wing aircraft capable of vertical take-off.

Both Cayley and Phillips produced models of their helicopter designs, but the first helicopter to lift a man off the ground was that built by Breguet in France in 1907, which used a 50 hp Antionette engine. However, the Breguet lacked any means of control and was not allowed to make a free flight. A different criticism, the lack of adequate stability, can be directed at another attempt during the early 1920s by an Argentinian, the Marquis de Pescara. The gyroplane, as already mentioned, cannot be counted a true helicopter, but this type of machine was developed considerably during the 1920s and 1930s. The first practical helicopter was the Focke-Wulf Fw. 61 of 1936, which used two laterally displaced rotors, but this machine was not put into production, although a development, the Focke Achgelis (q.v.) Fa.223 of 1940 was produced in small numbers.

The helicopter as it is known today normally has a main rotor and a tail rotor to provide stability, and this type of helicopter was developed by the Soviet émigré Igor Sikorsky (q.v.) during the late 1930s, culminating in a first flight of his VS-300 in 1939. Wartime development of Sikorsky's designs in the United States and further intensive helicopter development after the war by Bell Aerosystems gave the United States a considerable lead in helicopter development at the end of the 1940s, although since then the Russian and the French designers have made considerable strides, to the extent that the world's largest helicopter is the Russian Mil Mi-12.

The helicopter has proved itself to be an excellent machine for rescue duties, anti-submarine work, assault and supply duties, and as a V.I.P. or executive transport, but has still to make a real impact on passenger services. The extensive helicopter service networks of Sabena and Pakistan International Airways were abandoned some years ago, while city-centre-to-airport services in the United States are dependent upon airline subsidies. The only scheduled passenger helicopter service operating on anything like a viable basis is that of B.E.A. Helicopters between Penzance and the

Isles of Scilly. B.E.A. also has the distinction of being amongst the first operators of helicopter passenger and mail services during the late 1940s.

The major helicopter manufacturers include Sikorsky (q.v.), Boeing-Vertol (q.v.), Bell (q.v.), and Hughes (q.v.) in the United States, the Mil (q.v.) design bureau in the Soviet Union, Westland (q.v.) in the United Kingdom, Aerospatiale (q.v.) of France, and Agusta (q.v.) of Italy.

Helio: an American aircraft manufacturer concentrating on S.T.O.L. light transport and utility aircraft, Helio's main product is the U-10 (q.v.) Super Courier, developed from the Courier of the early 1950s, which is in service with the United States Army and some Latin-American air arms.

Helwan HA-300: *see* HA-300, Helwan.

Henschel: one of the smaller German aircraft manufacturers of the inter-war period and World War II, Henschel produced a number of designs, including the Hs.126 army co-operation aircraft, before starting the development of the Hs.293 rocket-assisted glider bomb. The Hs.293 was in effect the world's first stand-off bomb, and it was first used operationally on 25 August 1943 (*see* glider bomb).

Henson, W. Samuel: as much a visionary as an aircraft designer, W. S. Henson, an Englishman, produced a design for an 'Aerial Steam Carriage' in 1843 which caught the imagination of the public and for many years was reproduced as an indication of the future heavier-than-air flying machines. A high-wing monoplane, steam-powered and with two pusher propellers, the Henson design incorporated a number of 'modern' features. Henson built a model based upon his design in 1847, but this was unable to sustain itself in flight, although it did encourage Henson's friend, John Stringfellow (q.v.), to develop the idea further.

Herald, Handley Page: work started on the Handley Page Herald airliner during the early 1950s, the original intention being that the aircraft should use four Alvis Leonides Major piston engines. Two prototypes flew with the Alvis engines before it

was decided to substitute two Rolls-Royce Dart turboprops, and the prototypes were accordingly modified. The first flight of a Dart Herald was in March 1958, and deliveries of the Series 100 production model started late in 1959. A stretched fuselage version, the Series 200, first flew in 1960, while further developments, the Series 600 with up-rated engines and a further fuselage stretch, and the Series 700 with standard engines, were put into production during the late 1950s. A military version was the Series 400. Herald production ceased with the collapse of Handley Page in 1969, but most of the aircraft produced are still flying. Two 2,100 shp Rolls-Royce Dart 527 turboprops give the Series 200 version a maximum cruising speed of 275 mph and a range of up to 1,800 miles, while up to fifty-six passengers or 11,000 lb of freight can be carried.

Hercules, de Havilland D.H.66: *see* D.H.66 Hercules, de Havilland.

Hercules, Hughes: The Hughes Hercules flying-boat has the twin distinctions of being the aircraft with the largest wingspan so far, and the largest aircraft built of wood. Only one of the high-wing, 320-foot-span, eight-engined Hercules flying-boats was built, and this aircraft made only one flight of 1,000 feet at a height of 70 feet above Long Beach Harbor, California, on 2 November 1947, with Howard Hughes at the controls. The aircraft has remained in store ever since. Undoubtedly the aircraft suffered from the same disadvantage as the British Brabazon and Princess, that of being too big (with accommodation for 700 passengers) for the air transport market of its time.

Hercules, Lockheed C-130: *see* C-130 Hercules, Lockheed.

Heron, de Havilland: a four-engined development of the Dove (q.v.), with a stretched fuselage, the Heron first flew in May 1950. Initial production versions were of the Heron 1, with a fixed undercarriage, but a retractable undercarriage was fitted to the Heron 2, which first flew in December 1952. The Heron is still in service today as an executive aircraft and feeder-liner. Four 250 hp de Havilland Gipsy

Queen 30 Mk.2 piston engines give the Heron 2 a maximum cruising speed of 185 mph and a range of 1,200 miles, while either 3,750 lb of freight or up to fourteen passengers can be carried. About 170 aircraft were built, and Riley in the United States have stretched a number of these and re-engined them with two turboprops in place of the piston engines.

HF-24 Marut, Hindustan: the HF-24 Marut was designed by Professor Kurt Tank, starting in 1956, with the first flight taking place in June 1961. Deliveries to the Indian Air Force started in March 1963, and the aircraft is still in service. Two 4,850 lb thrust Bristol Siddeley Orpheus 703 turbojets give this low-wing fighter a maximum speed of about 700 mph and a range of 750 miles, while four 30 mm cannon are fitted in the nose and four 1,000 lb bombs can be carried under the wings. A development, the Mark 1A, with 6,000 lb thrust Orpheus turbojets with reheat, is also in service. Consideration has been given to developing a Mach 2·0 version using two Brandner E.300 turbojets with reheat, but this is dependent on the unlikely event of further development of this engine for the Helwan HA-300 (q.v.).

H.F.B.: H.F.B., or the Hamburger Flugzeubau, to use the seldom-quoted full title, dates from 1933 and has always been one of the smaller German aircraft manufacturers. Since the war the company has collaborated with other manufacturers in a number of projects, including the C-160 (q.v.) Transall and the Fokker F-28 Fellowship (q.v.), as well as developing and building its own H.F.B. 320 Hansa (q.v.) executive jet.

H.F.B. 320 Hansa: the H.F.B. 320 Hansa is West Germany's contender in the executive jet market. A mid-wing monoplane with the very unusual feature of forward sweep on the wings, the Hansa conforms in having rear-mounted jet engines. First flight was in April 1964. Two 2,850 lb thrust General Electric CJ610-1 turbojets give a maximum cruising speed of 450 mph and a range of up to 1,500 miles. A crew of two and up to eleven passengers can be carried.

HH-: international civil registration index mark for Haiti.

HH-: international civil registration index mark for the Dominican Republic.

high-wing: a monoplane with a mainplane, or wing, running across the top of the fuselage. High-wing layouts are generally considered to provide extra stability.

Hill, Professor G. T. R.: an Englishman, Professor G. T. R. Hill produced a series of tailless (q.v.) aircraft designs which were built by Westland (q.v.) during the late 1920s and 1930s. Hill's work was based on that of John William Dunne, who had built a number of tailless aircraft before World War I. In certain respects Hill's designs, which were named Pterodactyls, were missing the true potential of the tailless aircraft because, like Dunne, he was more concerned with low speed handling and the elimination of the stall than with high speed and low drag.

The first Pterodactyl flew in 1926, and was a single-engined, pusher-propeller monoplane with a swept wing, single seat, and fixed undercarriage. Other designs followed, including a prototype fighter and a prototype cabin aircraft, although neither entered production after test flying.

Hind, Hawker: a day-bomber development of the Hart (q.v.), the Hind retained the same single-engined biplane layout.

Hindenberg: the successor to the famous and highly-successful Graf Zeppelin (q.v.), the Hindenberg was destroyed in the worst airship disaster on record while mooring at Lakehurst, New Jersey, on 6 May 1937, after a flight from Frankfurt. Thirty-three of the ninety-six passengers on board were killed when the airship collided with the mooring mast and caught fire. The Hindenberg disaster was largely due to an American embargo on helium for Germany preventing the airship from using this much safer gas instead of the highly-dangerous hydrogen. Public interest in airships waned after the disaster, and was not revived with the introduction to service of the Graf Zeppelin II shortly before the outbreak of World War II in 1939.

Hindustan Aeronautics: originally founded

155

at Bangalore before the outbreak of World War II to undertake maintenance and assembly work, Hindustan Aeronautics started to produce its own designs after the end of the war. The first Indian-designed aircraft was the HT-2 basic trainer, and this has since been followed by the HAOP-27 Krishnak A.O.P. aircraft, the HJT-16 Kiran jet trainer, and the HF-24 Marut jet fighter. In addition, the company has undertaken licence-production of the Folland Gnat, the Hawker Siddeley HS 748 turboprop airliner, and the Mikoyan-Gurevich MiG-21 interceptor.

'Hip', Mil Mi-8: *see* Mi-8 'Hip', Mil.

Hirondelle, Dassault: first flown during the late 1960s, the Dassault Hirondelle is a twin-engined turboprop feeder liner, executive aircraft, and military navigational trainer. Two 870 hp Turbomeca Astazou turboprops give a maximum cruising speed of 310 mph and a range of up to 1,200 miles, while a crew of three and up to fourteen passengers can be carried.

Hiroshima: the Japanese city which became the first operational target for an atomic bomb (q.v.) on 6 August 1945. Out of a population of 256,000 persons, 68,000 were killed and 76,000 injured.

Hispano Aviacion: a Spanish aircraft manufacturer associated with the Hispano-Suiza aero-engine and armaments firm, Hispano Aviacion has been noted for its post-war manufacture of Messerschmitt Me.109 fighters as the HA-1112-Mil, with Rolls-Royce Merlin engines. The Me.109s preceded the production of the HA-100-El piston-engined advanced trainer during the early 1950s, and the HA-200 Saeta jet trainer during the mid-1960s. All the post-war designs have been the work of Professor Willy Messerschmitt.

Hispano-Suiza: a noted automobile and aero-engine manufacturer prior to World War II, Hispano-Suiza today, as a division of SNECMA (q.v.), produces rocket projectiles and pods for ground-attack duties, and licence-produces aero-engines.

HJT-16 Kiran, Hindustan: work on the Hindustan HJT-16 jet basic trainer started in 1961, and the first flight followed in September 1964, with deliveries to the Indian Air Force starting in late 1966. A side-by-side twin-seat aircraft with a low wing and a pressurised cockpit, the HJT-16 uses a single 2,500 lb thrust Bristol Siddeley Viper II turbojet for a maximum speed of 485 mph and a range of up to 600 miles.

HK-: international civil registration index mark for Colombia.

HL-: international civil registration index mark for the Republic of Korea.

HM-2, Hovermarine: a side-wall hovercraft without amphibious capability, the Hovermarine HM-2 first appeared during the late 1960s and uses a Cummins marine diesel engine and water-immersed propellers. Early development was plagued by teething troubles, but the sixty-five seat hovercraft is now in production, and in service with a small number of operators. Rescue, firefighting, and survey developments are available.

Honest John: a surface-to-surface missile with a range of twenty miles, Honest John is no longer in production, although it remains as standard NATO equipment and is also used by Nationalist China. Developed and produced by Emerson Electric, and designated MGR-1B, Honest John is a twenty-five foot long single stage rocket with a solid fuel motor. Unguided, the missile is stabilised by its motor and by fins. A nuclear warhead or 1,500 lb of high explosive may be carried. The Lance missile (q.v.) is considered to be a likely successor.

'Hook', Mil Mi-6: *see* Mi-6 'Hook', Mil.

Horizon, Aerospatiale GY-80: Designed by Yves Gardan, the Horizon four-seat light aircraft first flew in July 1960, and was then licence-built by Socata, Aerospatiale's light aircraft subsidiary. Production ended in 1969. An all-metal low-wing monoplane, the Horizon has a retractable undercarriage and a single 180 hp Lycoming 0-360-A piston engine gives a maximum cruising speed of 150 mph and a range of up to 780 miles, but 150 hp and 160 hp versions of the same engine were also available when the aircraft was in production.

'Hormone', Kamov Ka-25: *see* Ka-25 'Hormone', Kamov.

Hornet, de Havilland: one of the fastest piston-engined aircraft, and certainly the fastest twin-engined piston-engined aircraft, the de Havilland Hornet was a development of the D.H.98 (q.v.) Mosquito. Two 2,070 hp Rolls-Royce Merlin 130 piston-engines gave the Hornet a maximum speed of 472 mph and a range of 3,000 miles, while four 20 mm cannon were fitted in the nose. A long-range fighter, the Hornet first flew in 1944, but did not enter R.A.F. squadron service until 1945, and was too late for hostilities in World War II.

'Hound', Mil Mi-4: *see* Mi-4 'Hound', Mil.

Hound Dog: an American air-to-surface missile, the Hound Dog was developed by North American Rockwell for the U.S.A.F.'s B-52 (q.v.) Stratofortress bombers and is no longer first line equipment. Hound Dog uses a turbo-jet propulsive unit and can carry a nuclear warhead over distances of up to 600 miles.

hovercraft: *see* air cushion vehicle.

Hovermarine: a British hovercraft manufacturer, Hovermarine developed the HM-2 (q.v.) sidewall hovercraft during the mid-1960s. After encountering persistent teething troubles with the craft, the company found itself in financial difficulties. Hovermarine has since been acquired by an American group, Transportation Technology Inc., and production of the HM-2 is now in progress, with the earlier difficulties overcome.

HP-: international civil registration index mark for Panama.

H.P.5 'Yellow Peril', Handley Page: a high-wing monoplane with a swept leading edge to the wing, the H.P.5 'Yellow Peril', so called because of the colour of the wings, first appeared in 1911 and was one of the first designs by Sir Frederick Handley ·Page (q.v.). Tandem twin-seats were fitted and a 50 hp Gnome rotary engine drove a tractor propeller.

H.P.6, Handley Page: a development of the H.P.5 (q.v.), retaining the high-wing and swept leading edge of the earlier aircraft, but using an 80 hp Gnome rotary engine, the H.P.6 first flew in 1912.

H.P.42 Hannibal, Handley Page: after a first flight in November 1930, the Handley Page H.P.42 Hannibal airliner entered service on the Imperial Airways services to Europe and to India, a total of eight aircraft being built. A biplane with the wings mounted above the fuselage, the H.P.42 used four 475 hp radial engines and could carry up to thirty-eight passengers in a luxurious interior. There was not a single accident during the years of service with Imperial Airways. Cruising speed was about 100 mph and the range was about 300 miles.

HR-: international civil registration index mark for Honduras.

HS-: international civil registration index mark for Thailand.

HS 125, Hawker Siddeley: the de Havilland-designed HS 125 (originally designated the D.H. 125) first flew in prototype form in August 1962. A number of versions of this aircraft have been built, mainly for executive use, but also as a navigational trainer, starting with the 1B and moving up to the current production model, the Srs. 600. More than 300 aircraft have been built to date, including the Royal Air Force's Dominie navigational trainer. Two rear-mounted 3,310 lb thrust Bristol Siddeley Viper 522 turbojets give the Srs. 3B a maximum speed of 500 mph and a range of up to 1,700 miles, while six or eight passengers can be carried in addition to a crew of two. The aircraft is sometimes marketed as the BH-125, due to Hawker Siddeley's marketing arrangement with Beechcraft (q.v.) in the United States.

HS 141, Hawker Siddeley: a Hawker Siddeley project for a V.T.O.L. jet airliner, using two advanced technology turbofans for forward flight and eight or ten wing root-mounted lift jets for take-off and landing. Little has been heard of the project recently.

HS 144, Hawker Siddeley: a Hawker Siddeley project for a light jet transport successor to the turboprop HS 748 (q.v.).

HS 146, Hawker Siddeley

Originally conceived as a Rolls-Royce Trent-powered aircraft, the design is in cold storage at present with the Trent engine, an advanced technology design, shelved. The HS 144 would carry between forty and fifty passengers, if built.

HS 146, Hawker Siddeley: a S.T.O.L. airliner project by Hawker Siddeley, with a high wing and four advanced technology turbofans with a quiet take-off characteristic, the HS 146 would be capable of carrying about 100 passengers.

HS 748, Hawker Siddeley: originally developed by Avro, the HS 748 turboprop light transport was put into production by the Hawker Siddeley Group, and first flew in June 1960, with deliveries starting in Spring 1962. A number of improvements were made to the original production aircraft, resulting in the Srs. 2 version, which itself has undergone further improvement. The aircraft is used by the Queen's Flight of the Royal Air Force, by a number of other heads of state, and as a V.I.P. transport by other air forces. Most users are civil airlines. The R.A.F. uses a special development, the Andover (q.v.), with a stretched fuselage, and modified tail and undercarriage. Hindustan Aeronautics is also building the aircraft under licence in India. Two 2,105 shp Rolls-Royce Dart 531 turboprops give the aircraft a maximum cruising speed of 275 mph and a range of 700 miles with full payload, which can be up to fifty-six passengers or 11,964 lb of cargo

HS 1182, Hawker Siddeley: an advanced jet trainer for the Royal Air Force during the mid and late 1970s, the Hawker Siddeley HS 1182 is under development by the company at the present time. A low-wing aircraft with tandem twin seats, the HS 1182 will use a single Rolls-Royce-Turbomeca Adour turbojet without reheat. The aircraft is a Gnat replacement.

HT-2, Hindustan: the first Indian-designed aircraft, the Hindustan HT-2 basic trainer first flew in August 1951. A tandem twin-seat, low-wing monoplane, the HT-2 uses a 155 hp Blackburn Cirrus Major III piston engine, giving a maximum speed of 130 mph and a range of 350 miles. The aircraft

remains in service with the Indian Air Force and Navy.

HU-16 Albatross, Grumman: *see* Albatross, Grumman HU-16.

Huanquero, Dinfia I.A.35: *see* I.A.35 Huanquero, Dinfia.

Huey Cobra, Bell AH-1G: *see* AH-1G Huey Cobra, Bell.

Huff Daland Crop Dusters: *see* Daland, Huff.

Hughes: the Hughes Tool Company is owned by the millionaire Howard Hughes, who also built and flew the Hughes Hercules (q.v.) flying-boat in 1947. The company's interest in helicopter production dates from the late 1950s when the development of a light helicopter started, eventually resulting in the Hughes 300 (q.v.); this was later supplemented by the Hughes 500 (q.v.). Other aviation-related activities include production of a 20 mm cannon for U.S. Navy aircraft.

Hughes 300: development of the Hughes 300 helicopter dates back to 1955, resulting in the appearance of the 269A twin-seat light helicopter in 1961, from which the 300 has materialised. The Hughes 300 is the United States Navy's TH-55A trainer, and uses a 180 hp Lycoming H10-360-A1A piston engine giving a maximum speed of 90 mph and a range of 300 miles for the three-seat helicopter.

Hughes 500: the first flight of a prototype of the Hughes 500 took place in February 1963, and deliveries to the United States Army of the military version, the OH-6A Cayuse observation helicopter, started in May 1966. Civil versions are available and in fairly widespread use. Powerplant is a single 252 shp or 317 shp Allison T63 turbine, giving the five-seat helicopter a maximum speed of 160 mph and a range of 350 miles.

Hunter, Hawker Siddeley: the first flight of what was to be one of Britain's most successful peacetime military aircraft, the Hawker Siddeley Hunter, occurred in July 1951, with the Hawker P.1067 prototype. The first flight of a production F.Mk.1 followed in May 1953, and deliveries to the

R.A.F. started in 1954. The early aircraft used versions of the Rolls-Royce Avon turbojet, and were followed by the Bristol Siddeley Sapphire-powered F.Mk.2 and F.Mk.5. Exports of the aircraft started with the Avon-powered F.Mk.4, which was delivered to Sweden as the F.Mk.50, to Denmark as the F.Mk.51, and to Peru as the F.Mk.52; while the F.Mk.6 aircraft was sold to India as the F.Mk.56, to Switzerland as the F.Mk.58, and to Iraq as the F.Mk.59, as well as to Jordan, the Lebanon, and Saudi Arabia. A version of the F.Mk.6 optimised for ground-attack duties is the F.(G.A.) Mk.9, with a reconnaissance version in the F.R.Mk.10. Other operators of the Hunter include Rhodesia, Chile, Kuwait, Singapore, and Abu Dhabi. Although the Hunter fighter is a single-seat aircraft, a two-seat trainer version (with side-by-side seating) has been developed, including the T.Mk.7 for the R.A.F. and the T.Mk.8 for the Royal Navy, while export models have included the T.Mk.62, T.Mk.66, T.Mk.66B, T.Mk.67, and T.Mk.69.

A total of some 2,000 Hunters have been built, including 460 produced under licence in Belgium and the Netherlands, while refurbished aircraft are produced for a market in which demand far exceeds supply. The F.Mk.6 uses a single 10,000 lb thrust Rolls-Royce Avon 207 turbojet for a maximum speed of 710 mph and a range of up to 1,840 miles, while four 30 mm cannon are carried in the nose, and two 1,000 lb bombs or rockets can be carried on underwing strongpoints.

Hunting Percival: *see* Percival Aircraft.

Hurricane, Hawker: designed by Sir Sydney Camm (q.v.), the Hurricane was Hawker's first fighter monoplane, and first flew in 1935 in prototype form. A production Mk.I first flew in October 1937, and was followed into production by numerous different versions, including the supercharged Mk. II, the anti-tank Mk.IID, the Canadian-built Mk.X and Mk.XI, the Mk.XII with Packhard-built versions of the Rolls-Royce Merlin engine, and the Sea Hurricane. Most versions of this last were carrier-borne aircraft, but the Mk.IA was a throw-away aircraft for the R.A.F.'s Merchant Service Fighter Unit, whose aircraft were catapulted from merchant vessels for fighter protection of convoys and ditched in the sea afterwards. Some 14,000 Hurricanes were built between 1937 and 1944, using a variety of versions of the Rolls-Royce Merlin engine, which gave the aircraft a maximum speed of at least 315 mph (Mk.I had Merlin II of 1,030 hp). The armament varied widely, including either two 40 mm cannon, four 20 mm cannon, or up to twelve smaller calibre machine guns, while small bombs or rockets could be carried on underwing racks.

Hustler, General Dynamics B-58: *see* B-58 Hustler, General Dynamics.

hydrogen bomb: the hydrogen bomb is a nuclear fusion device or thermonuclear bomb, as opposed to the atom bomb (q.v.), which is a nuclear fission device. A fusion device entails bringing together two very light atoms, and the deuterium and tritium versions of hydrogen, the lightest element, are most commonly used. Fusion is obtained under conditions of considerable heat, for which a trigger fission device is normally used.

HZ-: international civil registration index mark for Saudi Arabia.

I

I-: international civil registration index mark for Italy.

I.A.35 Huanquero, Dinfia: the Dinfia I.A.35 Huanquero training, photographic, and ambulance aircraft first flew in prototype form in September 1953, and deliveries to the Argentinian Air Force started in 1957. Main production versions of this low-wing monoplane, with a twin fin tailplane, included the IA navigational trainer, the IB weapon trainer, the III ambulance aircraft, with accommodation for four stretcher cases and one attendant, and the IV photographic version. Two 620 hp I.A.19R El Indio radial engines power the IA, III, and IV versions, giving a maximum speed of 225 mph and a range of 975 miles, while up-rated 750 hp engines power the IB.

I.A.58, Dinfia: originally developed from the I.A.35 Huanquero (q.v.) as the I.A.50 Guarani I, the aircraft at first used many components from the older aircraft and first flew in April 1963. Subsequent development has produced the I.A.58, with a single swept fin and many other detail improvements. Two 904 shp Turbomeca Bastan VI-A turboprops give the ten-to-fifteen seat aircraft a maximum cruising speed of 300 mph and a range of 1,500 miles. The I.A.58 is in service with the Argentinian Air Force as a communications aircraft.

I.A.I: *see* Israel Aircraft Industries.

I.A.T.A.: *see* International Air Transport Association

Iberia-Lineas Aereas de España: the Spanish airline Iberia was originally formed in 1927 as Iberia Air Transport. Two years later the airline was merged with two other airlines, C.E.T.A. and Union Aerea Española, to form a new airline, C.L.A.S.S.A., in which the majority shareholding was held by the Spanish Government. An all-state airline, L.A.P.E., was established in 1931, and this acquired C.L.A.S.S.A., but by 1938 L.A.P.E. had run into severe difficulties, and the present Iberia airline was formed to take over L.A.P.E.'s fleet, which included Junkers Ju.52/3M trimotors, and a network of domestic and international routes, some of which had been suspended during the Spanish Civil War.

Iberia was initially a mixed-enterprise airline, with the State holding 51 per cent of the share capital and the rest remaining in private hands. However, Iberia also ran into difficulties during World War II, and the Spanish Government acquired the private shareholdings at this time. After World War II a programme of expansion on the international routes started, with a new fleet of Douglas DC-3s and DC-4s, the latter aircraft operating North Atlantic services. During the late 1940s Lockheed Super Constellations supplemented the DC-4s on long-haul routes, and Convair 440 Metropolitans followed for the European routes. Iberia introduced its first jet aircraft during 1961 with the first of a fleet of Douglas DC-8s for the North Atlantic services, while in 1962 Sud Caravelles were obtained for the European services. The DC-3s were not replaced on domestic routes until 1967, when Fokker F-27 Friendship turboprop airliners were introduced.

Currently, Iberia operates an extensive domestic and international route network with a fleet of Boeing 747s and 727s, Douglas DC-8s (including 'Super-Sixty' versions), DC-9s, and DC-10s, Fokker F-27 Friendships and F-28 Fellowships, and some Convair 440s, while Airbus Industrie A.300Bs are on order. Another Spanish airline, Aviaco (q.v.) is a subsidiary of Iberia.

I.C.A.O.: *see* International Civil Aviation Organisation.

Icarus: the son of Daedalus (q.v.) in Greek mythology, Icarus must figure as the first victim of a flying fatality, since he flew too close to the sun, melting the wax in his wings, and fell to his death.

I.C.B.M.: *see* intercontinental ballistic missile.

Icelandair-Flugfelag Islands: Icelandair was formed in 1937 as Flugfelag Akureyrar H.F., operating a Waco seaplane between Reykjavik and Akureyri. The first international services were started in 1945, with a Convair Catalina flying-boat operating to Glasgow and Copenhagen. Civilianised Liberator bombers were used on the international services from 1946, the aircraft being leased from Scottish Aviation, and in 1948 the airline introduced its own aircraft, Douglas DC-4s. Vickers Viscount turboprop airliners were obtained in 1957, followed by Douglas DC-6Bs in 1961 and 1963, since when Fokker F-27 Friendships and Boeing 727s have been introduced.

The Icelandic Government acquired a 13·2 per cent shareholding in the airline in 1950, the year the present title was adopted, and the other major shareholder is the Icelandic Steamship Company, with 39·6 per cent of the capital. Currently thirteen domestic points are served by the airline, which also operates to Scandinavia and Greenland, and to the United Kingdom, with a small fleet of Boeing 727s, Douglas DC-6Bs, and Fokker F-27 Friendships.

Il-2, Ilyushin: one of the first designs by Sergei Ilyushin (q.v.), the Il-2 was also the first military aircraft from this particular Soviet design bureau. A single-engined, low-wing monoplane, the Il-2 was used throughout World War II as a ground-attack aircraft. Early versions were single-seat, but later models included a rear-gunner.

Il-12 'Coach', Ilyushin: Ilyushin's first major civil design, the Il-12, NATO code-named 'Coach', first flew in 1944 and was basically a Viking-equivalent, DC-3 replacement, of low-wing monoplane design. A heavy and clumsy airframe structure restricted the freight payload, but the aircraft was produced in quantity for paratroop, glider-towing, and passenger-carrying duties. Up to twenty-seven passengers could be carried, and two 1,775 hp Shvetsov ASh-82FNV radial engines provided a

maximum speed of 226 mph and a range of up to 1,240 miles. A development was the Il-14 (q.v.).

Il-14 'Crate', Ilyushin: a development of the Il-12 (q.v.) with a much-refined airframe structure, the Ilyushin Il-14, NATO code-named 'Crate', entered production in 1953. Two 1,900 hp Shvetsov ASh-82T radial engines give a maximum speed of 260 mph and a range of up to 1,900 miles, while freight or up to thirty-two passengers can be carried. Developments have included the ultra-lightweight Il-14P of 1955 and the stretched Il-14M of 1956. Il-14Ps were produced in East Germany, while Avia in Czechoslovakia produced the Il-14P and also the Il-14M, the latter being designated the Avia-14.

Il-14 'Coot', Ilyushin: the Ilyushin Il-18, NATO code-named 'Coot', first flew in July 1957, and no less than twenty development aircraft were built before the first production deliveries took place in early 1959. The aircraft is in widespread Soviet Bloc service, and developments include the Il-18D with extended range and the Il-18E with up-rated engines. The standard aircraft uses four 4,250 shp Ivchenko AI-20M turboprops for a cruising speed of 388 mph and a range of up to 2,300 miles while carrying the maximum payload of 29,760 lb of freight or 122 passengers.

Il-28 'Beagle', Ilyushin: Russia's first jet bomber, the Ilyushin Il-28, NATO code-named 'Beagle', first appeared in 1950, since which date several thousand have been built, although the aircraft is being withdrawn from frontline Soviet service. Standard Warsaw Pact equipment for many years, the aircraft has also been supplied to Egypt and to China. Two 5,950 lb thrust Klimov VK-1 turbojets, developed copies of the Rolls-Royce Nene, are mounted in pods on the wing undersurface of this high-wing monoplane and provide a maximum speed of about 580 mph and a range of up to 1,500 miles, while the maximum warload is estimated to be 4,500 lb. A defensive armament of two 20 mm cannon in the nose and two 20 mm cannon in a rear-turret is fitted. The peculiar design feature of the aircraft is its straight wing combined with a swept tailplane. A development is the Il-28U (q.v.) trainer.

Il-28U 'Mascot', Ilyushin: the training development of the Il-28 (q.v.), the Il-28U, NATO code-named 'Mascot', is in extensive service with Soviet Bloc air arms, and differs from the standard aircraft only in having a second pilot's cockpit in front of and below the standard cockpit. Performance is also similar to that of the Il-28.

Il-62 'Classic', Ilyushin: first introduced during the mid-1960s, the Ilyushin Il-62 NATO code-named 'Classic', has since entered service with a number of East European airlines and with the Civil Aviation Administration of China. Four rear-mounted 23,150 lb thrust Soloviev D-30 turbofans provide a maximum cruising speed of 560 mph and a range of up to 5,700 miles for this 212 seat aircraft, which bears a startling resemblance to the British Vickers Super VC.10 (q.v.), even to the point of being the same size as the original V.C.10 project before this was reduced to meet a B.O.A.C. requirement.

I.L.S.: *see* instrument landing system.

Ilya Mourometz, Sikorsky: one of Sikorsky's last designs before the outbreak of the Russian Revolution and his departure for the United States, the Ilya Mourometz (aptly enough meaning 'the giant') was originally intended to be a sixteen-seat airliner, but was instead used as a highly-successful heavy bomber in 1916 and 1917. A total of seventy-three aircraft were built. A biplane, with cabin accommodation for the crew and a defensive armament, the aircraft used four 100 hp Argus engines.

Ilyushin, Sergei Vladimirovich: the head of the Soviet design bureau which carries his name, Sergei Ilyushin was the son of a Russian peasant and entered aviation as a mechanic's helper in the Imperial Russian Air Service in 1915, rapidly rising to become chief mechanic on the Sikorsky Ilya Mourometz (q.v.) heavy bombers by the time of the Russian Revolution in 1917; by this time he had also learned to fly. After the Revolution Ilyushin became chief mechanic at the Zhukovski Air Academy,

where he gained a degree in engineering by studying in his spare time.

Ilyushin's first design was a glider, and this proved to be sufficiently successful for him to be given the post of Chief of the Red Air Fleet Experimental Factory at Moscow, where he worked throughout World War II and produced his first powered designs, including the Il-2 (q.v.) ground-attack aircraft and the Il-12 (q.v.) transport. Since the late 1940s the Ilyushin design bureau has also designed the Il-14 (q.v.), the Il-18 (q.v.), and the Il-62 (q.v.) airliners and the Il-28 (q.v.) jet bomber.

Immelman, Leutnant Max: the first German air ace in World War I, Leutnant Max Immelman flew with Hauptmann Oswald Boelcke's (q.v.) squadron and was the inventor of the 'Immelman Turn' manoeuvre – a climbing half loop with a roll off the top. A total of fifteen victories were accredited to Immelman before he was shot down in June 1916, his Fokker monoplane breaking up in mid-air following combat with a Royal Flying Corps aircraft.

Impala, Aermacchi M.B.326: the South African name for the Aermacchi M.B.326 (q.v.).

Imperial Airways: *see* B.O.A.C.

index, registration: *see* international civil aircraft markings.

Indian Airlines Corporation: the nationalised Indian domestic airline, the Indian Airlines Corporation was formed in 1953 on the state take-over of eight privately-owned airlines operating domestic and regional air services in the sub-continent. A fleet of Boeing 737-200s, Fokker F-27 Friendships, Vickers Viscount 700s, licence-built Hawker Siddeley 748s, and some Douglas DC-3s is operated.

infra-red homing device: one method of anti-aircraft missile guidance, used primarily in air-to-air missiles, the infra-red homing device detects the hottest spot within its range, which should be the jet pipe of an enemy jet aircraft, and, via a programming device which operates the missile's control surfaces, homes on to the target. The American Sidewinder (q.v.) and the British Firestreak (q.v.) are two of the most widely-used first-generation air-to-air missiles using an infra-red homing device, and work only when fired in pursuit at the jet pipe of the opposing aircraft. More recent missiles, such as the American Falcon (q.v.) and the British Red Top (q.v.), will work even if the target aircraft is flying towards the missile.

A persistent problem with infra-red homing devices is their poor performance at low altitude, and this is particularly marked in tropical regions with strong surface heat emissions to confuse the equipment.

instrument landing system: first introduced in 1929, the original concept of an instrument landing system was that it should eventually make blind landings possible, but in fact air transport operations have never used I.L.S. for this purpose, and today autoland (q.v.) makes landings in zero-zero visibility possible. I.L.S. is basically an extension of the airways concept for final approach to the break-off height, where a landing or overshoot decision must be taken by the captain of the aircraft. The aircraft using I.L.S. flies down an invisible narrow radar path, in line with the runway centre line and at the same angle as the glide path, and radar information is projected on flight deck instruments showing the extent of any deviation by the aircraft from the I.L.S. flight path.

Integrated Satellite System: the U.S. spy satellite (q.v.) system.

Inter-American Treaty of Reciprocal Assistance: sometimes known as the Rio Treaty, the Inter-American Treaty of Reciprocal Assistance was signed in 1949 by the United States and the Latin American countries, with the exception of Ecuador and Nicaragua, although both of these countries participate in the Organisation of American States. The terms of the Treaty pledge the signatories to intervene on behalf of any member attacked by an outside force, subject to a decision by a meeting of foreign ministers, and provide means for a peaceful settlement of any internal disputes between signatory nations. Cuba was expelled from membership in 1962, and it may be assumed that Chile is no longer a participant under the terms of the Treaty.

interceptor: basically a development of the fighter, the interceptor is a fast military aircraft with the emphasis on speed and a fast rate of climb rather than on the fighter concept of a highly manoeuvrable aeroplane. Inasmuch as high performance was not matched by outstanding manoeuvrability, the Messerschmitt Bf.109 was one of the first aircraft to display interceptor characteristics, while other World War II aircraft loosely described as interceptors included the Lockheed Lightning and the rocket-powered Messerschmitt Me.163 Komet. However, the true interceptor only arrived with the jet age and the need to tackle very high flying jet bombers, placing reliance on the use of air-to-air missiles rather than on cannon.

The first generation of true interceptors included the British Gloster Javelin and de Havilland Sea Vixen, and the American Convair F-102 Delta Dagger, McDonnell F-101 Voodoo, and Lockheed F-104A Starfighter. Second generation aircraft include the B.A.C. Lightning and Dassault Mirage III, and the latest generation of interceptors covers the so-called 'air superiority fighters' which have Mach 3·0 speed and a high rate of climb – such aircraft include the Mikoyan-Gurevich MiG-23 'Foxbat', the McDonnell Douglas F-15, and the Grumman F-14 Tomcat.

intercontinental ballistic missile: basically, a ballistic missile is an unmanned rocket which flies to its target on a high parabolic trajectory, and in a long-range missile, mainly referred to as an intercontinental ballistic missile (I.C.B.M.), much of the trajectory lies outside of the earth's atmosphere. Guidance of the missile, which is normally a three-stage rocket consisting of a booster, a sustainer, and a re-entry vehicle warhead, is usually only during the descent, and is by an inertial guidance system which is free from interference by electronics countermeasures. I.C.B.M.s are used only as strategic weapons aimed at the enemy's homeland centres of industrial and military importance, and the qualifying minimum range of an I.C.B.M. is 5,000 nautical miles – below this the correct term is intermediate-range ballistic missile (q.v.). Main I.C.B.M.s are the American Titan (q.v.) and Minuteman (q.v.), and the Soviet 'Scarp' (q.v.) and 'Savage' (q.v.).

interdiction bombing: a concept of bombing solely aimed at centres of considerable military, industrial, and communications importance, as opposed to blanket bombing of wide areas, the object of interdiction bombing is to minimise civilian casualties. The R.A.F. during World War II effectively developed the concept, both for humanitarian reasons and to avoid any undue waste of munitions. The high precision bombing required needs to take place amid a reasonable degree of air superiority if aircraft taking the extra time required for precision bombing are not to be shot down. It is also desirable that interdiction bombing should only be used against a sophisticated opponent – interdiction bombing is ineffectual against an underdeveloped opponent concentrating mainly on waging guerrilla warfare, as American experience against North Vietnam has shown.

Interflug–Gesellschaft fur Internationalen Flugverkehr: Interflug is the airline of the German Democratic Republic and was formed in 1954 with the title of Deutsche Lufthansa. Operations started in September 1955, although scheduled international services did not start until February 1956, with a service between East Berlin and Warsaw, followed by the start of domestic services in June 1957. The present title was adopted in September 1963. Currently Interflug operates a domestic route network, in addition to international services to the major centres in Eastern Europe, and to the Middle East and West Africa. As with most airlines in communist states, it undertakes the full range of general aviation activities. The fleet includes Ilyushin Il-18s and Il-62s, Antonov An-24s, and Tupolev Tu-134s and Tu-154s.

intermediate-range ballistic missile: an intermediate-range ballistic missile has a range of at least 1,500 nautical miles, and is generally a strategic weapon. The American Polaris (q.v.) and Poseidon (q.v.) missiles come into this category, as do the Soviet 'Saddler' (q.v.), 'Sasin' (q.v.), 'Skean' (q.v.) and 'Scrooge' (q.v.), and the French S.S.B.S. (q.v.). *See also* ballistic missile *and* intercontinental ballistic missile.

International Air Transport Association: the International Air Transport Association, formed in 1945, is an airline operators' association, the qualifying condition for membership being that an airline must be authorised to operate scheduled services by a government which subscribes to the 1944 Chicago Convention (*see* International Civil Aviation Organisation). Although governments themselves are responsible for the negotiation and operation of international air traffic agreements (q.v.), such things as fares, in-flight facilities for passengers, and inter-airline agreements are the province of I.A.T.A., although government confirmation of I.A.T.A. fare scales is sometimes necessary.

The I.A.T.A. rules and regulations, which cover (amongst other things) such items as seat pitches, meal service, etc., in international commercial aviation, are confirmed by the Annual General Meeting, at which each member airline has one vote. Operations are handled by the Executive Council, and there is an enforcement organisation which checks for breaches of the rules, and a court procedure by which airlines in breach of the rules may be fined.

Almost all international scheduled airlines are I.A.T.A. members, and all I.A.T.A. airlines are covered in this book, with the associate member airlines and those of the Communist Bloc which, although often not I.A.T.A. members, tend to follow I.A.T.A. practices.

international civil aircraft markings: all aircraft have a registration index or a serial number, for civil and military aircraft respectively; a means of identification being even more necessary for aircraft than for motor vehicles since aircraft identification plays an important part in the air traffic control procedure. Some countries, such as the United States and the Soviet Union, use a system of registration numbers prefixed by the international mark (in these cases 'N' and 'CCCP' respectively), and in the case of the United States a registration suffix letter is also included.

With the larger American airlines the suffix indicates the operator's name, Pan American (q.v.) aircraft having registrations suffixed 'PA', and so on. Most countries use an all-letter registration index, prefixed by an international index mark which is followed by a hyphen, e.g. G-BSST, with 'G' as the international mark for the United Kingdom. All international civil index marks are given in this book, but the major ones are:

CCCP-	Soviet Union
CF-	Canada
CS-	Portugal
D-	Federal Germany
EC-	Spain
F-	France
G-	United Kingdom
HA-	Hungary
HB-	Switzerland
I-	Italy
JA-	Japan
LN-	Norway
LQ/LV-	Argentina
N-	United States
OE-	Austria
OH-	Finland
OK-	Czechoslovakia
OO-	Belgium
OY-	Denmark
PH-	Netherlands
PP/PT-	Brazil
SE-	Sweden
SP-	Poland
SX-	Greece
TC-	Turkey
VH-	Australia
VP-Y	Rhodesia
VT-	India
XA/XB/XC-	Mexico
YR-	Rumania
YU-	Yugoslavia
ZK/ZL/ZM-	New Zealand
ZS/ZT/ZU-	South Africa

International Civil Aviation Organisation: originally formed as a result of the 1944 Chicago Convention on Air Transport to implement certain of the decisions taken at that convention, the International Civil Aviation Authority is today a body within the United Nations Organisation. The Chicago Convention concerned itself with a number of matters embodied in some ninety-six articles to which most nations subscribe and which cover such matters as rights to fly (see Five Freedoms of Air Transport), standards for aerial navigation and the provision of route equipment, and ground formalities. I.C.A.O. involvement is generally beneficial to air transport by bringing about a considerable degree of standardisation of procedures and equipment which benefits both the industry and its customers – the confusion which would result from a vast number of differing national standards need not be described.

Intruder, Grumman A-6: see A-6 Intruder, Grumman.

Iran Air–Iran National Airlines Corporation: The Iran National Airlines Corporation was formed in 1962 by a Government decree under which the Iranian Airways Company, or Iranair, acquired the capital and equipment of the ailing Persian Air Services, which dated from 1955. Iranian Airways itself dated from 1944, when it was formed with Trans World Airlines (q.v.) holding 10 per cent of the share capital, and started charter operations the following year, the first scheduled services being introduced in 1946. Athens, Baghdad, Beirut, Cairo, Paris, and Rome were soon added to the network, and in 1949 the airline acquired Eagle Airlines, a small company formed in 1948 with a single de Havilland Dove. Throughout the 1950s a fleet of Douglas DC-3s and DC-4s was operated, with a few Convair 240s, and Vickers Viscount 700 turboprop airliners were acquired in 1959.

After the formation of the Iranian National Airlines Corporation in 1962, the airline was extensively re-organised and became a wholly state-owned concern Technical assistance was provided by Pan American (q.v.) from 1964 for a three year period, and Boeing 727 turbojet airliners were introduced, while the international route network was extended. Currently, the airline operates to major points in the Middle East and Western Europe with a fleet of Boeing 707-386s, 727s, and 737s, and Douglas DC-6s.

Iraqi Airways: the Government of Iraq formed Iraqi Airways in 1945 as a branch of the State Railways, and operations started in 1946 with a service between Baghdad and Basra, using de Havilland Dragon Rapide biplanes. A small route network with services to most of the major Middle East centres had been developed by the end of the first year of operations, and Douglas DC-3s had been introduced. B.O.A.C. (q.v.) provided assistance from 1946 to 1960, a period during which the airline expanded rapidly and introduced Vickers Vikings and, later, Vickers Viscount 700s. Hawker Siddeley Trident 1Es were introduced in 1965, and these and the Viscounts form the present fleet. Services have been introduced to many East European destinations in recent years, reflecting the political climate in Iraq since 1960, although some West European destinations are also served.

I.R.B.M.: *see* intermediate-range ballistic missile.

Iroquois, Bell 204/205: *see* Bell 204/205 Iroquois.

Irvin, Leslie Leroy (1865–1965): the inventor of the modern parachute, Leslie Leroy Irvin, an American, made the first free-fall parachute descent from an aeroplane on 19 April 1919. He later formed the Irving Parachute Company to produce his designs, initially for the U.S. Army, but later for a world-wide range of customers – the difference between the company's name and his being due to an unrectified mistake.

Islander, Britten-Norman BN-2: *see* BN-2 Islander, Britten-Norman.

Israel Aircraft Industries: the largest aerospace concern in the Middle East, Israel Aircraft Industries was formed in 1953 as Bedek Aircraft. The word 'Bedek' means maintenance and repair in Hebrew, and this was the company's function for many years. Today Bedek continues as the maintenance and repair division of I.A.I., while the parent company manufactures components for Bedek, in addition to manufacturing the Israeli-designed Arava (q.v.) light transport and the 1123 Commodore Jet (*see under* Aero Commander) executive aircraft, which was acquired from North American Rockwell in 1967. Guided weapons production included the Gabriel surface-to-surface missile for naval use.

I.S.S.: the initials stand for Integrated Satellite System. For details *see under* spy satellite.

J

J.1 'Blechesel', Junkers: the Junkers J.1 'Blechesel' (rather unkind, it means 'Tin Donkey') was the world's first all-metal aeroplane and was also one of the first designs by Professor Hugo Junkers. Powered by a 120 hp Mercedes engine, the mid-wing, single-seat J.1 monoplane first flew in December 1915, and although only one such aircraft was built, the experience gained was valuable in designing the series of all-metal Junkers transport aircraft which followed. Maximum speed of the J.1 was 105 mph.

J-1 Jastreb, Soko: a single-seat light attack aircraft, the Soko J-1 Jastreb low-wing monoplane is a development of the Soko G2-A (q.v.) Galeb tandem twin-seat trainer, which is also in service with the Jugoslav Air Force. The Jastreb is powered by a single 3,000 lb thrust Rolls-Royce Viper turbojet, giving a maximum speed of 500 mph and a range of up to 850 miles. Three 0·50 in. machine guns are fitted and there are eight wing strongpoints for bombs or rockets.

J21, SAAB: see SAAB-21.

J29 Tunnan, SAAB: see SAAB-29 Tunnan.

J32 Lansen, SAAB: see SAAB-32 Lansen.

J35 Draken, SAAB: see SAAB-35 Draken.

J37 Viggen, SAAB: see SAAB-37 Viggen.

JA-: international civil registration index mark for Japan.

Jaguar, B.A.C./Breguet: development of the Jaguar high-wing monoplane started in 1965 to meet a joint British and French requirement for an advanced jet trainer aircraft with strike capability. The design was largely based on that of the Breguet Br.121 project, airframe design and development being divided between B.A.C.

and Breguet, with design leadership for the latter company. Aero-engine design and development is divided between Rolls-Royce (the design leader) and Turbomeca. Each of the participating countries is taking 200 aircraft, the British models being designated the Jaguar B, for the tandem twin-seat trainer, and the Jaguar S, for the single-seat strike aircraft. The French models are designated the Jaguar A strike aircraft and the Jaguar E trainer. Production plans for a Jaguar M carrier-borne strike aircraft for the Aeronavale have been abandoned because of weight problems.

The first flight of a Jaguar was of a Jaguar E prototype in September 1968; deliveries to the Armée de l'Air started in 1971, and to the Royal Air Force in 1972. Weight problems meant that the R.A.F. had to abandon plans to use the Jaguar as their new advanced jet trainer, and instead the HS 1182 is being developed for this role, while the R.A.F. is only taking a relatively small number of two-seat trainer Jaguars for conversion training, but is increasing its strike aircraft order. The Jaguar A and S use two Rolls-Royce/Turbomeca Adour turbofans of 6,750 lb thrust with reheat for a maximum speed of 1,120 mph and a range of up to 1,900 miles, and a single under-fuselage and four underwing strongpoints can carry up to 10,000 lb of bombs, rockets, and missiles. Two 30 mm cannon are also fitted.

J.A.L.: see Japan Air Lines.

Japan Air Lines–Nihon Koku Kabushiki Kaisha: Japan Air Lines was formed in 1951 when the Supreme Commander Allied Powers gave permission for a Japanese domestic airline to be formed using aircraft and crews supplied by a non-Japanese operator – initially Martin 202s of Northwest Orient Air Lines. The following year, however, J.A.L. was allowed to operate its own aircraft, Douglas DC-4s, and international services, starting with a Tokyo–San Francisco route. Two de Havilland Comet II jet airliners were also ordered, but cancelled after the Comet I disasters. The airline was reformed in 1953, with a 50 per cent government shareholding, since increased to 58 per cent to provide capital for expansion.

A network of domestic services and international services to Asian destinations and across the Pacific was established by the late 1950s, and Douglas DC-7 airliners were introduced. A trans-Polar service to Paris was inaugurated in 1959 using Air France Boeing 707s, pending delivery of Douglas DC-8 jet airliners in 1960. A number of other services to Europe have since been established, including some across the North Atlantic. The airline acquired Japan Domestic Airways in 1971, after operating that airline's route network for a number of years. J.A.L. has a 7·6 per cent interest in the domestic operator, All-Nippon Airways, and a 51 per cent interest in Southwest Air Lines of the Ryukyu Islands.

Today, Japan Air Lines operates an extensive domestic and international network, including services over the North Pole and via Moscow, with a fleet of Boeing 747s, 727s, Douglas DC-8s (including 'Super-Sixty' versions), N.A.M.C. YS-11s, and Beech 18s.

Jastreb, Soko J-1: *see* J-1 Jastreb, Soko.

J.A.T.: *see* Jugoslovenski Aerotransport.

Javelin, Gloster: the world's first production delta-wing jet fighter, the Gloster Javelin first flew in prototype form in November 1951, the first flight of a production model designated the Mk.4 not taking place until July 1954. A number of improved Marks followed, all officially designated as all-weather fighters with the exception of the conversion trainer T.Mk.3. A tandem twin-seat cockpit was standard on all Javelins, and two 12,300 lb thrust Bristol Siddeley Sapphire 203/204 turbojets with reheat provided a maximum speed of 650 mph. Two 30 mm cannon were fitted, and four Firestreak air-to-air guided missiles or packs of unguided rockets could be carried on four underwing strongpoints. Withdrawal from service by the sole user, the Royal Air Force, took place during the late 1960s. A 'T' tailplane was fitted to all Marks of the aircraft.

'Jenny', Curtiss JN-4: *see* JN-4 'Jenny', Curtiss.

jet: the jet engine was the result of independent research in the United Kingdom by Air Commodore Sir Frank Whittle, R.A.F. (q.v.), and in Germany by Dr Hans von Ohain (q.v.) during the 1930s, resulting in the Gloster E.28/39 (q.v.) Whittle and the Heinkel He.178 (the world's first jet aircraft). The first jet combat aircraft were the Messerschmitt Me.262 (q.v.) and the Gloster Meteor (q.v.), while the first jet airliner was the de Havilland Comet I (q.v.).

Jet engines may be divided into several distinct types, of which the original and still the most common is the turbojet (q.v.), in which a series of turbine blades suck air into the engine and compress it prior to ignition, and the escaping gases drive another set of turbine blades to provide the necessary propulsion for the ingestion and compressor blades. A development of the turbojet is the turbofan (q.v.) or by-pass engine, while advanced technology (q.v.) engines use double or triple spools to enable the differing sets of turbine blades to rotate at their individual optimum speeds instead of at a common speed. Lower altitude and lower speed needs for the jet have been satisfied by the turboprop (q.v.), basically a propeller and turbojet combination. Ramjets (q.v.), without any moving parts, are used as reheat (q.v.) or afterburners for supersonic flight, being inserted as a part of the jet pipe behind a turbojet or turbofan engine; and also power some missiles, although it is necessary for rocket boosters to raise the speed to one at which a ramjet can work. *See also* engine.

Jet Provost, B.A.C. 145/167: *see* B.A.C. (145/167) Jet Provost.

JetRanger, Bell 206: *see* Bell 206 JetRanger.

JetStar, Lockheed C-140: *see* C-140 JetStar, Lockheed.

jetstream: a horizontal stream of air moving at a higher speed than the surrounding air, but still moving along the course of the wind; normally found between 15,000 and 40,000 feet up.

Jetstream, Scottish Aviation: the Jetstream twin-turboprop feeder liner and executive low-wing monoplane was designed and developed by Handley Page and entered production with that company after its first flight in May 1967. A private-venture project, the cost of development and of tooling for a high level of production caused serious cash-flow problems for Handley Page, and the company went bankrupt as a result. Some production was continued by a company using the aircraft's name before it was taken over by Scottish Aviation in 1972 in order to meet an R.A.F. order for twenty aircraft as navigational trainers. Two 895 shp Garrett AiResearch TPE.331 turboprops give the aircraft a maximum cruising speed of 300 mph and a range of up to 2,000 miles, while a crew of three and up to eighteen passengers may be carried. Earlier production versions and some of the prototypes used 850 shp Tubromeca Astazou XIV turboprops, but the American engines were introduced in order to meet a U.S.A.F. order, and further hopes of large American orders faded with Handley Page's collapse.

JN-4 'Jenny', Curtiss: one of the earlier landplanes from the Glenn Curtiss company, the JN-4 biplane was a World War I training aircraft produced in both the United States, where it was named the 'Jenny', and in Canada, where it was named the 'Canuck' and had a number of airframe modifications compared with the American aircraft. A 90 hp Curtiss engine powered the twin-seat JN-4, which was used after the war by barnstormers or aerobatic stuntmen, and also to fly the first experimental United States airmail service in May 1918 between Washington, D.C., Philadelphia, and New York City.

Johnson, Air Vice Marshal James Edgar ('Johnny'), C.B., C.B.E., D.S.O., D.F.C., R.A.F.: the highest-scoring R.A.F. pilot during World War II, and also the Allied pilot with credit for the highest number of Luftwaffe aircraft destroyed, 'Johnny' Johnson joined the R.A.F. in 1939 as a Flight Sergeant, rising to the rank of Group Captain by the end of the war, and accounting for no less than thirty-eight Luftwaffe aircraft. Early in his career he flew in a squadron commanded by Group Captain Douglas Bader (q.v.). During the war he won the D.S.O. and two bars, the D.F.C. and bar, and a number of American decora-

tions. After the war he remained with the Royal Air Force until his retirement in 1966 as an Air Vice Marshal, holding also the C.B. and C.B.E.

Ju.52/3MM, Junkers: the Junkers Ju.52/3 transport first flew in 1932, and remained in production throughout the 1930s and World War II, a total of 3,234 aircraft being built in Germany and additional aircraft built by C.A.S.A. in Spain. An all-metal, low-wing monoplane, early versions used three 575 hp B.M.W. radial engines, and later Ju.52/3M aircraft used three 725 hp B.M.W. 132A radials for a maximum speed of 180 mph and a range of up to 550 miles, while eighteen passengers could be carried. The aircraft saw service with a number of airlines before, during, and after World War II, British European Airways being prominent amongst the latter, as well as with the Luftwaffe and a number of other air forces, of which the Spanish and the Swiss have been the last to operate the aircraft.

Ju.87 Stuka, Junkers: perhaps the most famous dive-bomber of all time, or more likely the most infamous, since the Ju.87 was one of the most hated aircraft in history, partly due to its ugly shape – a single-engined, low crank-wing monoplane with twin-seat tandem cockpit – and partly due to the wailing noise made while diving towards its target with the deliberate intention of weakening the resolve of any near-by anti-aircraft gunner. The Ju.87 was used extensively by the Luftwaffe before and during World War II, and has been frequently described as obsolescent on the outbreak of World War II – but it was nevertheless as effective an aircraft as the dive-bomber concept could produce. A fixed undercarriage was fitted, and the aircraft had a mechanism to throw the under-fuselage bomb clear of the propeller, while additional bombs were carried on racks under the wing. A 1,000 hp Junkers Jumo engine provided the power.

Ju.88, Junkers: a mid-wing, twin-engined medium bomber monoplane, the Junkers Ju.88 served with the Luftwaffe throughout World War II on a variety of operations, including bombing, reconnaissance, glider-towing, and so on. After 1942 the aircraft

169

was joined on the production line and supplemented in service by the Ju.188 development.

Ju.90, Junkers: the last Junkers civil design, the Ju.90 entered Lufthansa service in the summer of 1938 and was a very fast low-wing airliner monoplane with four radial engines of 830 hp each giving a maximum speed in excess of 250 mph, while forty passengers could be carried. A wartime military transport and maritime-reconnaissance development was the Ju. 290.

Jugoslovenski Aerotransport-J.A.T.: Jugoslovenski Aerotransport was established in 1946 by the Government of Jugoslavia after a pre-war airline, Aeropout, had ceased operations on the German occupation of the country. Domestic operations started in 1947 with Douglas DC-3s, and by the end of the year services were also being operated to Rumania, Hungary, and Czechoslovakia. Political difficulties and post-war shortages meant the suspension of services for much of 1948 and 1949, while Jugoslavia emphasised its independence of the other East European powers. Throughout the 1950s J.A.T. concentrated on services to Western Europe, with only a few East European routes, and Convair 440 Metropolitans and Douglas DC-6B aircraft were placed in service. A small number of Ilyushin Il-14s were obtained in 1957.

The airline introduced its first jets, Aerospatiale Caravelles, in 1962. Currently, services are operated throughout Europe as well as 'domestically, with a fleet of McDonnell Douglas DC-9s, Aerospatiale Caravelles, and Convair 440s.

jumbo jet: a popular term used for the Boeing 747 airliner, and in view of the aircraft's size the inference is apt.

June Bug, Curtiss: the first design by Glenn Curtiss, the June Bug was flown by its designer on 20 June 1908, making Glenn Curtiss the first American to fly after the Wright brothers. Two weeks later the aircraft won the 'Scientific American' trophy for the first American aircraft to fly over a measured course in public, with a flight of 5,090 feet.

Junkers: Professor Hugo Junkers founded the company which bore his name after designing and building the J-1 (q.v.), which was the world's first all-metal aeroplane. During the inter-war period, Junkers built a succession of highly-successful all-metal airliners, including the F-13 (q.v.), the G-24 (q.v.), and the Ju.52/3 (q.v.), and ending with the Ju.90 (q.v.) of 1938 and its wartime military transport and maritime-reconnaissance development, the Ju. 290. Military aircraft from the company included the famous Ju.87 (q.v.) dive-bomber, and the Ju.88 (q.v.) medium bomber and its development, the Ju.188.

JY-: international civil registration index mark for Jordan.

K

Ka-25 'Hormone', Kamov: a co-axial rotor helicopter for naval use, the Kamov Ka-25, NATO code-named 'Hormone', uses two 900 shp Glushenkov turbines for a maximum speed of 135 mph and a range of 240 miles. A small, short-fuselage helicopter, the Ka-25 has as a distinguishing feature a triple fin. A crane development, the Ka-25K, uses a removable gondola under the cabin to give the operator a clear view aft. Naval versions are fitted with anti-submarine equipment.

Ka-26 'Hoodlum', Kamov: a light utility helicopter using the standard Kamov co-axial rotor, the Kamov Ka-26, NATO code-named 'Hoodlum', uses two 325 hp M-14 radial engines for a maximum speed of 105 mph and a range of 240 miles. A small-fuselage helicopter with a twin-boom tail and a twin-fin, the Ka-26 is frequently operated from ice-breakers and whalers.

Kaman: one of the smaller American helicopter manufacturers, Kaman Aircraft was founded in 1945 by Charles Kaman, and for a number of years produced the HUK-1 and HOK light helicopters. These were followed by the H-43 (q.v.) Husky rescue and fire-fighting helicopter, mainly for the U.S.A.F., and later by the UH-2 (q.v.) Seasprite for the United States Navy, which has operated the type since 1962. A number of new helicopter projects are under evaluation, including the K-700, which would be basically an up-rated Husky with cabin accommodation for twelve, the K-800, a gunship version of the Seasprite, and the Sealite, an anti-submarine and airborne-early-warning shipborne helicopter project.

Kamikaze: a Japanese term meaning 'Divine Wind', it was originally used in aviation for a long-distance Japanese experimental aircraft of the inter-war period, although the notoriety associated with the term comes from its association with the suicide bombers of the late World War II period.

The concept of suicide bombers, targeted on to Allied warships by their pilots was readily acceptable to airmen in the Japanese Army Air Force and Japanese Navy Air Force because of the Shinto philosophy which dictated that death in battle was a sure way to Heaven for the warrior. Initially, the Mitsubishi A6M2 Type O fighter-bombers – the Zero – was used carrying a single 500 lb bomb, but Oka (Cherry Blossom!) piloted bombs were later dropped from heavy bombers. Amongst a number of special aircraft types developed for the J.N.A.F.'s Kamikaze operations were the Aichi D4Y4, the Mitsubishi D5Y1, and the Nakajima Kitsuka – the last, with the Yokosuka Oka, being jet-powered. The J.A.A.F. used the Nakajima Ki.115 and Ki.20.

The first Kamikaze unit was formed in October 1944 by the J.N.A.F., and used Zero fighters; this unit made the first Kamikaze attack on 25 October, when two fighters crashed on to the deck of the escort carrier, U.S.S. St Lo, which sank as a result of the damage inflicted. Altogether the United States Navy lost three escort carriers to suicide attack, out of a total of thirty-four warships lost and 288 damaged. No British carriers were lost as a result of suicide attacks, although several were damaged. The first use of the Oka piloted bomb was on 1 April 1945, when a number of American warships were damaged.

Kamov: the Nikolai Kamov design bureau specialises in helicopters, usually light utility or shipborne machines using the Kamov co-axial rotor (i.e. contra-rotating rotor). One of the earliest designs from this bureau was the Ka-15 'Hen' of 1952, from which was developed the larger Ka-20 'Harp' of the early 1960s, and more recently the Ka-25 'Hormone' (q.v.) and the K-26 'Hoodlum' (q.v.).

'Kangaroo': a fifty foot long Soviet air-to-surface missile used with Tupolev Tu-95 'Bear' bombers, the 'Kangaroo' is of primitive design, but is capable of carrying a nuclear warhead.

Kansan, Beech T-11: *see* Beech 18.

Kawanishi: a Japanese aircraft manufacturer of the inter-war and World War II period, Kawanishi usually specialised in flying-boats and seaplanes. Early aircraft included the ESK1 reconnaissance-seaplane of the early 1930s, followed by the H6K1 Type 97 four-engined flying-boat of the mid-1930s. Wartime aircraft consisted of the Ki.4 and Ki.61 Type 3 fighters, and the H8K2 four-engined flying-boat, which was code-named 'Emily' by the Allies and was reputed to be the best Japanese flying-boat of World War II.

Kawasaki: a Japanese industrial group which entered aviation after the end of World War I, Kawasaki produced a number of aircraft for the Japanese Army Air Force between the wars, including licence-built versions of the Dornier F, known in Japan as the Type 87, the Type 88 and Type 93 bombers, and the Type 95 fighters. Later aircraft included the Ki-45 twin-engined night fighter, known as the 'Dragon Slayer' to the Japanese, but code-named 'Nick' by the Allies.

After World War II Kawasaki returned to aircraft manufacture during the mid-1950s to supply aircraft for Japan's rearmament programme, including licence-built Bell, Boeing-Vertol, and Hughes helicopters, and the Lockheed Neptune maritime-reconnaissance aircraft, which has since been developed into the P-2J, with turboprop engines and an extended fuselage.

KC-130 Hercules, Lockheed: tanker version of the C-130 Hercules (q.v.).

KC-135 Stratotanker, Boeing: tanker version of the C-135 Stratolifter (q.v.).

'Kelt': a rocket-powered development of the 'Kennel' (q.v.) air-to-surface anti-shipping missile, 'Kelt' is frequently carried under the wings of Soviet Tupolev Tu-16 'Badger' bombers and has been in service at least since the late 1960s. Approximately thirty feet in length, the missile uses a liquid-propellant rocket and has a range of approximately 100 miles.

'Kennel': an anti-shipping air-to-surface missile, the Soviet 'Kennel' is carried under the wings of Tupolev Tu-16 'Badger' bombers and in appearance is similar to a scaled-down version of the Mikoyan-Gurevich MiG-15 fighter. Turbojet-powered, the 'Kennel' is about twenty-eight feet in length and has a range of about fifty miles. A radio guidance system is believed to be used.

KH-4, Kawasaki: a four-seat development of the Bell 47 (q.v.) helicopter, which Kawasaki builds under licence in Japan. A future development is the KHT-1 with a rigid three-blade rotor.

King Air, Beechcraft: one of the first light turboprop aircraft in the world, the Beechcraft King Air first flew in January 1964, and has remained in production ever since, although the latest production aircraft, the King Air 100, has a higher performance than the original 90. Two 715 shp United Aircraft PT6A-28 turboprops give the Ego aircraft, which has accommodation for a crew of two, up to eight passengers or 4,800 lb of freight, a maximum speed of 285 mph and a maximum range of 1,540 miles.

Kiowa, Bell 206: *see* Bell 206 Jet Ranger/OH-58 Kiowa.

'Kipper': a Soviet air-to-surface missile of about thirty feet in length, using turbojet propulsion, the 'Kipper' has a range of about 100 miles. Possibly out of service.

Kiran, Hindustan HJT-16: *see* HJT-16 Kiran, Hindustan.

'Kitchen': Soviet air-to-surface anti-shipping missile of about thirty-six feet in length and probably using a liquid-propellant rocket. Little other information is available on the 'Kitchen', which is only seen occasionally, carried by Tupolev Tu-22 'Blinder' bombers.

kite: the most primitive form of heavier-than-air flight, with propulsion being provided by the pull on the tow-line and lift by the incidence of the kite to the wind. The origins of the traditional kite are lost in antiquity, but Lawrence Hargrave (q.v.), an Australian, invented the box-kite in 1893, combining the biplane and tandem-wing concepts and boxing these in at the wing ends, the resulting design providing lift and stability. A number of aircraft were designed using the box-kite concept as a basis, including those by the Voisin brothers (q.v.)

K.L.M. Royal Dutch Airlines–Koninklijke Luchtvaart Maatschappij N.V.: the oldest airline in the world to retain its original identity, K.L.M. Royal Dutch Airlines was formed in 1919 and commenced a service to London from Amsterdam the following year using a D.H.9 single-engined biplane of Aircraft Transport and Travel, making this service the oldest in the world to be operated by the same airline. The airline soon acquired its own aircraft, and in 1929 started its first intercontinental air service between Amsterdam and Djakarta, using a Fokker F-VIIB tri-motor. K.L.M. was also the first European operator of Douglas DC-2 and DC-3 aircraft before the outbreak of World War II, and after the war commenced transatlantic services using Douglas DC-4s, while a fleet of DC-3s was also operated for European services. Services were started in the Netherlands Antilles in 1935, and these continued during World War II, although all other operations ceased.

Currently, some 70 per cent of K.L.M.'s share capital is held by the Dutch Government, leaving the remainder in private hands; while wholly owned subsidiaries of K.L.M. include K.L.M. Aerocarto, an aerial survey company, K.L.M. Noordzee Helicopters, K.L.M. Air Charter, and N.L.M., the Dutch domestic airline. A 25 per cent share in an independent charter airline, Martinair, is also held. K.L.M. operates a fleet of Boeing 747 and McDonnell Douglas DC-8 (including some of the 'Super Sixty' series), DC-9, and DC-10 airliners. K.L.M. is a member of the international KUSS maintenance consortium.

KM-2, Fuji: a four-seat cabin development of the Beech T-34 (q.v.) Mentor trainer, which was built under licence in Japan by Fuji, the KM-2 uses a single 340 hp Lycoming IGSO-480-AIF6 engine for a maximum speed of 230 mph and a range of 570 miles. The KM-2 is a primary training aircraft, but another T-34 development,

with the same fuselage and cabin structure as the KM-2, is the LM-1 and LM-2 Nikko liaison aircraft.

knot: a term of speed indicating nautical miles per hour – it is never a measure of distance.

Koolhoven: originally formed after World War I by Fritz Koolhoven, a Dutchman who had designed some of the Deperdussin aircraft before the war, the Koolhoven firm produced a number of light aircraft designs during the 1920s and early 1930s, some of which were built under licence in the United Kingdom by Desoutters.

K.S.S.U.: originating in 1968 as K.S.S., K.S.S.U. continues as a maintenance grouping for the airlines concerned, which were originally K.L.M., Swissair, and S.A.S. for Boeing 747 maintenance; but it has since been extended to include U.T.A. and the McDonnell Douglas DC-10 airbus. The group is sometimes referred to as KUSS.

KUSS: *see* K.S.S.U.

Kuwait Airways: the Kuwait Airways Corporation dates from 1953, when it was formed as Kuwait National Airways, commencing operations the following year with two Douglas DC-3 aircraft on routes to Iraq, Syria, the Lebanon, Jordan, and Iran. Two Handley Page Hermes aircraft were introduced in 1955, and in 1957, the year in which the Kuwait Airways Company title was adopted, two Douglas DC-4s were obtained. Technical assistance provided by B.O.A.C. from 1958 for a period of five years, during which time Vickers Viscount turboprop airliners were introduced and services to India and Pakistan started.

Until 1962 the airline's share capital was divided equally between the Kuwait Government and private interests, but the airline became wholly state-owned in 1962, when the present title was adopted. Hawker Siddeley Comet 4 airliners were leased in 1962, and some of these aircraft were purchased in 1963. Hawker Siddeley Trident 1Es were operated for a period during the late 1960s, but the present fleet consists entirely of Boeing 707–369C airliners.

L

L-29 Delfin, Aero: the standard Soviet Bloc basic jet trainer, the Czech Aero L-29 Delfin first flew in prototype form in April 1959, using a single Bristol Siddeley Viper turbojet. Production aircraft use a 1,920 lb thrust Czech-designed M-701 turbojet for a maximum speed of 391 mph and a range of 540 miles. A tandem twin-seat aircraft with a mid-wing and a 'T' tailplane, the Delfin can be equipped with bombs, rockets, and cannon for training or counter-insurgency duties.

La Bris, Jean-Marie (1808–72): a French sea captain, Jean-Marie La Bris was one of the earlier pioneers of gliding, making daring attempts at flights in gliders based on the shape of the albatross which were launched from horse-drawn carts. The first such attempt by La Bris was made in 1857, and one of his later attempts is alleged to have lifted off its cradle. A feature of the La Bris gliders, unlike those of Lilienthal (q.v.), was that the pilot sat in the fuselage rather than hanging from it.

Lancaster, Avro: the biggest and best of Great Britain's World War II heavy bombers, the Avro Lancaster was developed in considerable haste after the outbreak of war in 1939, and first flew in 1942. A four-engined, mid-wing monoplane with a twin-fin, the Lancaster was equipped with nose, dorsal, and tail gun-turrets. Four 1,280 hp Rolls-Royce Merlin in-line engines gave a maximum speed in excess of 300 mph, while the aircraft could carry up to a 22,000 lb bomb load with the famous Grand Slam bomb, and did so frequently after this weapon was introduced in March 1945. The aircraft was built in Canada as well as in the United Kingdom.

Perhaps the most famous raid by the Lancaster was that of 617 squadron, R.A.F. on the Moehne and Eder Dams in the Ruhr on 16 May, 1943, but the aircraft was also used for other famous raids, including one against the German battleship *Tirpitz* in a Norwegian Fjord. The aircraft also carried the predecessor of the Grand Slam, the 12,000 lb Tallboy, and was used on many routine bombing missions. A successor, the Lincoln (q.v.), started to enter service in 1945, relegating the Lancaster to maritime-reconnaissance, photographic, and survey work, particularly in Canada where a massive post-war survey operation was conducted by Royal Canadian Air Force Lancasters. A number of civilianised Lancasters served after the war, as Lancastrians, with airlines, including British South American Airways, and on Eagle Airways services during the Berlin Airlift.

Lance: an American surface-to-surface missile built by the Ling-Temco-Vought Aerospace Corporation, the Lance entered U.S. Army service in 1971. A single stage, twenty foot long missile with a range of about thirty miles, the Lance uses a liquid propellant rocket and inertial guidance; either a nuclear or a high explosive warhead can be fitted. It is the likely replacement for the Honest John (q.v.) missile in European service.

L.A.N.-Chile–Linea Aerea Nacional de Chile: the nationalised airline of Chile, L.A.N.-Chile, dates from 1929 when the Government formed an airline, Linea Aeropostal Santiago-Arica, which was placed under air force management and given an initial fleet of eight de Havilland Gipsy Moths. The airline assumed its present name in 1932 when it became civilian-controlled, and continued operations with the varied fleet of Ford Trimotors, Curtiss Condors, and Potez 56s which had largely taken over from the Gipsy Moths. The airline did not start its first international service until 1946, when Douglas DC-3s inaugurated a service to Buenos Aires; during the immediate post-World War II period the fleet consisted of DC-3s and Martin 202s, the first four-engined aircraft, Douglas DC-6Bs, not being introduced until 1955. During the late 1950s a number of new international services were started.

Sud Caravelle jet airliners were obtained

in 1962, and after an extensive reorganisation in 1964, partly with the object of reducing the airline's subsidy, a programme of modernisation was embarked upon during the late 1960s, including the introduction of a fleet of Hawker Siddeley 748 airliners. The current fleet consists of Boeing 707-320B and 727, Aerospatiale Caravelle, Hawker Siddeley 748, Douglas DC-6B and DC-3 airliners and a Cessna 310, while the route network extends throughout South America and to Panama, Miami, and New York. There is also an extensive domestic network covering some thirty points in Chile.

Langley, Samuel Pierpoint (1834–1906): perhaps one of the unluckiest of the pioneers of aviation, the American inventor and astronomer Samuel Pierpoint Langley produced a number of highly successful tandem-wing steam-powered model aircraft during the early 1890s. One model in fact succeeded in flying for a distance of 4,200 feet. All of Langley's aircraft, whether models or full-size, were somewhat inaccurately referred to as 'Aerodromes' by him. A successful quarter-size model with a petrol engine convinced Langley in 1901 of the soundness of his plans, and he proceeded to build a full-sized aircraft using an engine built by two engineers, S. M. Balzer and C. M. Manly. A United States Army contract to build and test the aircraft was awarded but, when attempts were made to catapult Langley's Aerodrome from the roof of a houseboat moored in the Potomac River in October and December 1903, the aircraft fouled the launching gear on each occasion and crashed. This experience discouraged the U.S. Army seriously, and made it extremely difficult for the Wright brothers to sell their ideas to the Army.

Claims made in later years that the Aerodrome could have flown successfully were it not for the mishaps on launching the aircraft, have since been discredited, in spite of the efforts of Glenn Curtiss (q.v.), because examination of the design showed serious structural and control shortcomings and the aircraft was also underpowered.

Lansen, SAAB J32: *see* SAAB-32 Lansen.

Latécoère: a French flying-boat manufacturer prominent during the inter-war period, Latécoère's two most notable products were the 521, which first flew in January 1935, and the 631, on which work started in 1937. Both aircraft suffered a disturbed history. The prototype 521, a large, six-engined, double-deck, high-wing monoplane, was damaged at its moorings in a storm in 1936, but was rebuilt to establish new distance, altitude, and speed records for its class in 1937. The 631, also a large six-engined monoplane, was nearing completion at the time of the German invasion of France in 1940, and was dismantled and hidden in the Forest of Landes during the war, being re-assembled and flown in 1945.

Latham, Hubert (1883–1912): an Englishman, Hubert Latham was one of the pioneers of powered flight, achieving a number of distinctions including that of being the first man to attain an altitude of 500 feet in an aeroplane, and the first to fly an aeroplane in a gale-force wind. Latham's achievements were gained while flying Antionette monoplanes. His attempts to fly the English Channel were marked with a conspicuous lack of success: the first attempt, on 19 July 1909, ended in a ditching in the sea near Calais after engine failure, and the second, on 27 July, two days after Blériot's successful crossing, also ended in an engine failure and ditching the aircraft in the sea near Dover.

Latham established the world's first air speed record over a 100 kilometre distance at the Reims International Aviation Meeting in August 1909, with a speed of 42 mph. A further air speed record of 48·21 mph was also established by Latham in an Antionette on 23 April 1910. Less seriously, Latham is reputed to be the first man to smoke a cigarette while flying.

launching pad: the term used to describe the take-off point for a rocket, usually a sounding rocket or a space vehicle launcher. Missiles are normally fired from a launcher.

Launoy: a French professor, Launoy produced with Bienvenu (q.v.) in 1784 a twin rotor helicopter model, which reputedly influenced Sir George Cayley's (q.v.) model of 1796.

leading edge: the edge of an aerofoil, usually a wing or a fin, which first meets the air during forward flight. *See also* trailing edge.

Lear Jet: Lear Jet Industries was formed during the early 1960s as the Swiss American Aviation Corporation with the intention of building the Lear Jet 23 in Switzerland, but moved to Wichita, Kansas, in 1962, when the title Lear Jet Corporation was adopted, after the founder, William Lear. A subsidiary acquired during the mid-1960s was the Brantly Helicopter Corporation (q.v.), a light helicopter manufacturer; Lear Jet itself has become a subsidiary of the Gates Corporation.

The first Lear Jet product was the Model 23, followed by the stretched Model 24 (q.v.) and the Model 25, all aircraft being low-wing monoplanes with rear-mounted jet engines, and intended for the executive market. A development, the Lear Jet Model 40 for feeder line and corporate use, has been shelved.

Lear Jet Model 24: first flown in February 1966, the Lear Jet Model 24 is a stretched development of the earlier Model 23, which had first flown in October 1963. Two 2,950 lb thrust General Electric CJ610-6 rear-mounted turbojets give the current 24D version of this six-to-eight-passenger aircraft a maximum cruising speed of 534 mph and a maximum range of 2,020 miles.

Lebaudy: the two Lebaudy brothers, Paul and Pierre, were sugar refiners who became interested in the idea of dirigible airships after 1896, although their first ship, built with the assistance of their technical director, Julliot, was not completed until the end of 1902. Painted yellow, the Lebaudy I was soon nick-named 'Jaune', and was 183 feet in length, and powered by a 40 hp Daimler petrol engine which drove two metal propellers. It was in fact the first practical airship, since although Count Ferdinand von Zeppelin had started his experiments in Germany in 1900, he had still to produce a practical airship at this stage.

The Lebaudy I was severely damaged in a landing accident at Chalais in November 1903, although without injury to the crew, and was later rebuilt and flown as the Lebaudy II. A number of other Lebaudy airships were built during the years before 1909, and proved reliable and safe in service. The Lebaudy I had proved itself before its accident by travelling thirty-eight miles in one hour, forty minutes, on 12 November 1903.

Leonardo da Vinci (1452–1519): the famous Italian artist Leonardo da Vinci produced a number of designs for man-powered aeroplanes and helicopters. Although he did not build any models, still less full-sized aircraft, based upon his designs, he is nevertheless given the credit for the first constructive appraisals of flight, although still imitating bird movements by the action of forms of blades or paddles on the air. Leonardo did not confine his technical interests to aviation, for he was also responsible for early tank designs.

Leonov, Lt-Col. Alexei Arkhipovich: Lieutenant-Colonel Alexei Arkhipovich Leonov became the first man to 'walk' in space when he left his Voskhod 2 spacecraft for a total of twenty minutes during its orbits of the Earth on 18–19 March 1965. The Russian spacecraft made a total of seventeen orbits during the flight, which lasted for just over twenty-six hours. Leonov's companion, who remained in the spacecraft, was Colonel Pavel Ivanovich Belyavev.

Let L-410: the first notable product from the Let Narodny Podnik, a new Czechoslovak design bureau, is the L-410, a high-wing S.T.O.L. monoplane with retractable undercarriage, seating for up to twenty passengers, and a crew of two; two 715 shp United Aircraft PT6A-27 turboprops give a maximum cruising speed of 229 mph and a range of up to 705 miles. Four aircraft have been giving service with the regional airline, Slov-Air, for some time, and further developments are expected, although little definite information is available. All-freight configurations are available.

Levasseur, Leon (1863–1922): chief designer and engineer of the Antionette (q.v.) company, Leon Levasseur was one of the major supporters of the monoplane during the early days of aviation, and as a result Antionette aircraft were invariably monoplanes. Levasseur's first designs for Antionette were produced in 1907, but it was not until 1908 that any real promise could be seen. Generally, Levasseur designs employed wing warping, but a few, particularly during the early years, used ailerons.

A number of record-breaking attempts were made with Levasseur's designs, and although the Antionette monoplanes did not acquit themselves well in attempts to fly across the English Channel, a good performance was recorded at the First International Aviation Meeting at Reims in 1909, and a world air speed record was established in 1910 – all of these attempts, whether successful or not, were made with Hubert Latham (q.v.) at the controls.

Li-2 'Cub', Lisunov: licence-built version of the Douglas DC-3 (q.v.).

Liberator, B-24 Consolidated: *see* B-24 Liberator, Consolidated.

Libyan Arab Airlines: formed in 1964 by royal decree as Kingdom of Libya Airlines, Libyan Arab Airlines adopted its present title in 1971. Operations started in 1965, and the airline took over the activities of N.A.A. Libiavia and United Libyan Airlines. Currently a fleet of Aerospatiale Caravelle 6Rs, Boeing 727-200s, and Fokker F-27 Friendships is operated, while the airline also manages a Dassault Mystère 20 and a Lear Jet 24 for the Government of Libya. Services are operated to Western Europe and throughout North Africa, as well as to domestic destinations

licence: licensing of civil aviation generally has the dual purpose of ensuring safety standards and of preventing over-capacity on a particular route or group of routes. Flight-deck crew are licensed, with special licences for pilots, navigators, and flight engineers; maintenance engineers on the ground crew also require licences for the particular aircraft types handled. Scheduled and certain types of charter services also require licences, while in the United Kingdom and many other countries' airlines require an air operator's certificate, or its

equivalent, as a guarantee of competence and reliability, and the aircraft itself will have a certificate of airworthiness, or its equivalent, both on a type and an individual basis – withdrawal of a C.O.A. means that an aircraft is grounded. Typical regulatory bodies include the Civil Aviation Authority in the United Kingdom; and the Civil Aeronautics Board (for route and service control) and the Federal Aviation Authority (for aircraft and aircrew) in the United States.

licence-production: frequently, in order to minimise the foreign exchange costs of buying from aboard, while also avoiding the development costs of a new unilateral design, aircraft will be licence-built in the purchasing country. Generally this applies to military designs rather than civil, although there are some exceptions to this, because air forces will frequently buy aircraft in hundreds, while an airline will, at the most, buy in dozens. Notable licence-production programmes include those in Europe and Japan for the Lockheed F-104G (q.v.) Starfighter, and the Bell, Sikorsky, and Boeing-Vertol helicopters; those in Canada and Australia for the North American F-86 (q.v.) Sabre; and production of the de Havilland Vampire (q.v.) and Venom (q.v.) and the Hawker Hunter (q.v.) in Europe. Generally the market for licence produced aircraft is restricted in some way to prevent any loss of work to the designing company and country.

lift: the end result of the passage of the atmosphere at speed over aerodynamic surfaces, in this case the mainplane, and essentially an upward force.

lifting aerofoil fuselage: an aircraft in which the fuselage is intentionally designed to act as a part of the aerodynamic surface of an aircraft, or which uses a deep-section wing as part of the accommodation, is generally known as a lifting aerofoil fuselage. Few in number, typical aircraft included the Junkers G.38, with accommodation for a dozen of its passengers in cabins in the wing inboard leading-edges, and the French D.B.71 trimotors, in which a twin-boom fuselage layout was adopted, with accommodation in the wings between the booms and the centre fuselage. Both aircraft appeared around 1930.

lifting body: during the late 1960s the American Northrop concern produced a lifting body design for the National Aeronautics and Space Administration. Known as the HL-10, the wingless craft has been used to simulate the control methods necessary for a space shuttle vehicle returning to the Earth's surface. Rocket propulsion is provided, although return to the Earth's surface by the lifting body is by gliding.

Liftmaster, McDonnell Douglas C-118: *see* C-118 Liftmaster, McDonnell Douglas.

Lightning, B.A.C.: the British Aircraft Corporation Lightning interceptor was developed from the English Electric P.1A research aircraft by English Electric during the mid-1950s, and first flew in prototype form in April 1957, using the designation F.Mk.1A. During the late 1950s and early 1960s, deliveries of the F.Mk.1 and the F.Mk.1A to the R.A.F. were put in hand, to be followed by the F.Mk.2, which first flew in July 1961, and incorporated many detail improvements, as well as being able to carry Red Top missiles in addition to the older Firestreak. Further modifications were incorporated in the Mk.3, which first flew in June 1962, with up-rated engines and a square-tipped tail, while the 30 mm cannon of earlier models were deleted. Provision for over-wing ferry tanks first appeared with the F.Mk.3, and these and the other new features were incorporated in the F.Mk.6, the current service version. Operational Lightnings have all been single-seat, but side-by-side twin-seat conversion trainer versions included the F.Mk.4 and the F.Mk.5, respectively variants of the F.Mk.1 and the F.Mk.3. Export multi-role versions for Saudi Arabia and Kuwait were the F.Mk.53 and the T.Mk.55.

Two 16,300 lb thrust Rolls-Royce Avon 302 turbojets with reheat, mounted one on top of the other, power the mid-wing aircraft, giving a maximum speed of some 1,400 mph (Mach 2·3 plus), with probably one of the highest rates of climb of any comparable warplane of its generation, and

a far longer range. Armament consists of two Firestreak or Red Top missiles in the interceptor role, while six 1,000 lb bombs or rockets can be carried on under- and over-wing strongpoints on the multi-role versions.

Lightning, Lockheed P-38: *see* P-38 Lightning, Lockheed.

Lilienthal, Otto (1848–96): the first man in history to design, build, and fly in practical gliders, Otto Lilienthal, a German, made some 2,000 flights before falling to his death while flying a monoplane hang-glider in 1896. Lilienthal's gliders, which he built with some assistance from his brother Gustav, were usually monoplane in configuration, and always hang-gliders from which the pilot hung by his shoulders and controlled the glider by shifting his body. Tentative experiments in 1891 led to a succession of highly successful flights during the years before Lilienthal's death, including some with a series of biplane gliders in 1895. At the time of his death, Lilienthal was working on a powered glider with ornithoptering wing-tips.

The most prominent follower and pupil of Lilienthal was Percy Pilcher (q.v.), a Scotsman, although the Wright brothers are also supposed to have paid some attention to his work before designing their gliders.

Lincoln, Avro: a stretched and extended-range development of the Avro Lancaster (q.v.) heavy bomber, the Avro Lincoln first flew in 1944 and was designed mainly for operations in the Far East against Japan, but World War II ended before the aircraft could be used. The four-engined, mid-wing monoplane, with nose, dorsal, and tail gun turrets, used four 1,760 hp Rolls-Royce Merlin piston engines for a maximum speed of about 350 mph and a range of about 2,000 miles, while an eleven-ton bombload could be carried. After the war the Lincoln remained in R.A.F. service until replaced by the English Electric Canberra jet bomber, from 1951.

Lindbergh, Colonel Charles Augustus: the first man to make a non-stop solo flight across the Atlantic, Colonel Charles Lindbergh made his flight in a single-engined Ryan monoplane, 'Spirit of St Louis', from Long Island, New York, to Paris on 20–21 May 1927.

Ling-Temco-Vought: Ling-Temco-Vought's aerospace associations come from Chance-Vought (q.v.), which Ling-Temco acquired in 1961, while the aerospace company was engaged in production of the F-8 (q.v.) Crusader carrier-borne fighter. The Crusader has since been developed into the A-7 (q.v.) Corsair II strike aircraft. The other L.T.V. interests include guided weapons manufacture and Boeing 747 subcontract work. The major constituent parts of L.T.V. today are the L.T.V. Aerospace Corporation and L.T.V. Electronics Inc.

L.T.V. built and operated the XC-142 experimental V./S.T.O.L. transport, but further work on this four-engined tilt-wing aircraft has been shelved, and production is unlikely.

Liore et Olivier: a French aircraft manufacturer of the inter-war years, Liore et Olivier came into prominence during the late 1920s with the LeO 20 fighters and the LeO 206 four-engined bomber for the French Aviation Militaire. During the 1930s the company produced the LeO 45 bomber for the newly-established Armée de l'Air, and the LeO 257 seaplane for the Service Aeronautique, while Air Orient, a predecessor airline of Air France, received LeO 242 flying-boats. Production ceased with the fall of France in 1940, and was not restarted after the end of World War II.

Lisunov: a Soviet design bureau, although its only aircraft of any note was the Li-2, a licence-built version of the Douglas DC-3, which was supplied to the Soviet Union during the closing years of World War II.

LM-1/LM-2 Nikko, Fuji: a development of the Beech T-34 (q.v.) Mentor, also built under licence in Japan by Fuji, the Fuji LM-1 Nikko liaison aircraft was built with a four-seat cabin fuselage in place of the T-34's tandem twin-seat cockpit. First flight was in June 1955. The LM-1 uses a 225 hp Continental O-470-13A piston-engine, while the later LM-2 uses an up-rated engine. Maximum speed is 210 mph and the range is 520 miles. A training development is the KM-2 (q.v.).

LN-: international civil registration index mark for Norway.

load-factor: the proportion of the available accommodation of an aircraft, an air service, or an airline which is filled, whether it is passenger, freight, or mixed traffic, is the load factor, which is expressed as a percentage of the available capacity.

load ton-miles: the load-ton miles figure is derived from the multiplication of the mileage operated by the capacity used.

Lockheed Aircraft: one of the major U.S. aircraft manufacturers, the Lockheed Aircraft Corporation's first significant aircraft was the Vega of 1927, a single 200 hp engine high-wing monoplane with accommodation for six passengers or mail. The Vega was followed by a development, the Air Express, which established a number of trans-U.S.A. speed records. Another successful aircraft was the Sirius, one of which was used by Charles Lindbergh (q.v.) on a number of record-breaking flights. Lockheed's first twin-engined airliner was the Electra, which appeared in 1930 and was followed by a number of developments, including the Lockheeds 10, 12, and 14, and the Lodestar, for civil use; and the Hudson, a maritime-reconnaissance version of the 14 used by the Royal Air Force during the late 1930s and the early years of World War II.

World War II saw a number of aircraft developed by Lockheed playing an important role, including the P-38 (q.v.) Lightning, which was one of the best long-range escort fighters used by the Allies; the Ventura and the Harpoon, for maritime-reconnaissance duties; and the first Constellation transports. After the war, Lockheed produced the first operational U.S. jet fighter, the F-80 (q.v.) Shooting Star, and its jet trainer development, the T-33, the latter seeing service with many of America's allies. The wartime Constellation was developed into an airliner, with further developments, the Super Constellation and the Starliner, appearing during the early 1950s. The company's specialisation in maritime-reconnaissance aircraft produced the Neptune at this time for several NATO and other Western air forces and air arms, and the aircraft was also built under licence in Japan.

A turboprop airliner introduced during the latter half of the 1950s, the Electra II, was not as successful as the company might have expected, although a military transport, the C-130 Hercules, more than compensated by becoming the most successful peacetime military transport aircraft to date. A supersonic interceptor, the F-104 Starfighter, was produced not only in the United States but in Europe and Japan as well. The 1960s were marked by the production of a successor to the Neptune, the P-3 (q.v.) Orion, a turboprop maritime-reconnaissance aircraft. A heavy jet transport for the U.S.A.F. was built, carrying the designation C-141 (q.v.) Starlifter, along with an executive jet, the C-140 (q.v.) JetStar, and the SR-71 (q.v.) high altitude reconnaissance aircraft – basically a successor to the earlier Lockheed U-2 of the late 1950s.

A number of problems struck Lockheed more or less at once during the early 1970s. Having been unsuccessful during the late 1960s in a design competition for a supersonic jet airliner, the concern produced an airbus, the L-1011 (q.v.) TriStar, with which both Lockheed and the engine manufacturer encountered difficulties; these came on top of problems with the AH-56 Cheyenne combat helicopter, which introduced a number of radical new features, and the C-5A Galaxy, the largest aircraft in the world. Fortunately, all the problems proved to be solvable, but not without some uncomfortable moments for Lockheed. Currently, however, Lockheed has probably more advanced technology in its products than any other U.S. airframe manufacturer. The latest product is the S-3A (q.v.) Viking carrier-borne anti-submarine aircraft for the U.S. Navy.

Lodestar, Lockheed: the Lockheed Lodestar was the ultimate development of the Lockheed 10, 12, and 14 series of twin-engined airliners, and was in fact sometimes referred to as the Lockheed 18 Lodestar. After a first flight in September 1939, most

of the Lodestar production was allocated to the U.S.A.A.F., U.S. Navy and the R.A.F., using the U.S.A.A.F. designation of C-56. A few aircraft did go to civil operators, however. The Lodestar used two Pratt and Whitney radial engines.

long-haul: usually, an air journey or an aircraft designed with a range of more than 2,500 miles.

Longhorn, Maurice Farman: the Maurice Farman Longhorn biplane, so-called because of the extravagant forward sweep of the undercarriage skids to prevent the aircraft from toppling forward on landing, was a notably easy aircraft to fly, although the performance was unspectacular, and was used as the standard training aircraft by the British Army before the outbreak of World War I. A single 70 hp Renault engine drove a pusher propeller.

L.O.T.-Polskie Linie Lotnicze: the Polish State airline, L.O.T., was formed in 1929 by the Government of Poland to take over two private-enterprise airlines, Aerolot and Aero, which dated from the early 1920s and operated domestic services, which L.O.T. continued. The new airline introduced its first international services during 1930, operating to the neighbouring East European states and to Greece. Steady progress was made throughout the rest of the decade, and by 1939 the airline was operating to many European capitals with a fleet of Lockheed Electras and Douglas DC-2s.

Operations were suspended during the German occupation of Poland in World War II, although many of the aircraft had been flown to England when the German invasion started. The airline was reformed in March 1945, using ex-Polish Air Force Lisunov Li-2s, and operations were quickly put in hand with Soviet assistance. The route network was expanded during the late 1940s and throughout the 1950s, while Ilyushin Il-12 and Il-14 airliners were acquired, with some Convair 240s, and these aircraft have since been followed by Ilyushin Il-18s and Il-62s, and Antonov An-24s, with some Tupolev Tu-134s.

The current fleet consists of Ilyushin Il-62, Il-18, and Il-14, Tupolev Tu-134, and Antonov An-24 airliners, while services are operated throughout Europe and to the Middle East.

Louvrié, Charles de: one of the first men to foresee the advent of the jet engine, the Frenchman, Charles de Louvrié, produced the first practical design for a jet-powered aircraft in 1865. The aircraft was to have an engine which used vaporised petroleum as a fuel. No attempt was made to build the aircraft, even in model form.

Lovell, Captain James Arthur, U.S.N.: one of the crew of Apollo VIII (q.v.), which was the first spacecraft to break clear of the Earth's gravitational hold and the first to orbit the moon, after being launched by a Saturn 5 rocket on 21 December 1968. At the time of the Apollo VIII mission, Lovell held the rank of Commander. He probably holds the record both for the highest number of space missions and for the greatest length of time spent in space by any one man, having also been on the Gemini 7 (4–18 December 1965) and Gemini 12 (11–15 November 1966) spaceflights, and on the Apollo XIII flight (11–17 April 1970) which did not complete its mission due to the failure of the service module.

low-wing: a mainplane configuration with the wing or mainplane running through or under the fuselage.

LQ-: international civil registration index mark for Argentina.

L.T.V.: *see* Ling-Temco-Vought.

Lufbery, Major Raoul: the first American pilot to gain fame and distinction during World War I, Raoul Lufbery was a mechanic before joining and eventually becoming the leading fighter pilot in the Lafayette Squadron – an all-American unit formed in the French Aviation Militaire in 1915 before the United States entered the war. Lufbery flew a Spad XIII while with the Lafayette Squadron. He joined the United States Army when America entered the war, and was put in command of the 94th Squadron shortly afterwards. Raoul Lufbery fell to his death on 19 May 1918 from his Nieuport 28 after it caught fire following a battle with a German Albatros.

At the time of his death he had seventeen confirmed victories, making him the third-ranking U.S. pilot of the war.

Lufthansa–Deutsche Lufthansa A.G.: Lufthansa's history dates from the formation of an airline of the same name in 1926 on the merger of Deutsche Aero-Lloyd and Junkers Luftverkehr, the latter being a subsidiary of the aircraft manufacturer. These airlines in turn owed their existence to mergers of smaller airlines, the oldest of which dated from 1919. The new airline was owned by regional airlines, the central and the provincial governments, and private interests, and inherited a fleet of more than a hundred (often very small) aircraft and an extensive European and domestic route network.

The airline immediately turned its sights on to services across the Atlantic and to the Far East. The Atlantic operations were largely left to the Condor Syndicate, an associate of Lufthansa which also established a number of South American airlines between the wars. Sea-air services were operated on both the North and the South Atlantic, the former consisting of catapulting a small seaplane with mail from a steamer as it approached the American or European coast, while the latter involved a Dornier Wal flying-boat service between Germany and the Canary Islands, catching the outward bound steamer, and meeting the incoming vessel. A Dornier X twelve-engined flying-boat was used on a flight to South America in 1932, and sometimes Zeppelins replaced the steamer on the South American route. Other trial flights by flying-boat were made in 1934 and 1936, with a small fleet of ships fitted to act as refuelling and servicing bases. Dornier Do-18 diesel-engined flying-boats and Blohm and Voss Ha-139 flying-boats were also used occasionally.

Services in Europe were not neglected, amongst the improvements being a chain of radio beacons for night-mail flight within Germany, while Junkers Ju.52/3M tri-motors and, later, Focke-Wulf Fw.200 Condor and Junker Ju.90 four-engined airliners were placed in service. Most operations ceased with the outbreak of war in 1939, and many South American operations were taken over by the governments concerned.

The airline was re-established in 1954, and operations were re-started in 1955 with a fleet of Lockheed Super Constellations, Convair 340s, and Douglas DC-3s, with aircrew provided initially by Trans World Airlines, Eastern Airlines, and British European Airways. Operations reached all six continents during the mid-1960s, a polar route to Tokyo having been started in 1964, and transatlantic services some time before that. Vickers Viscount turboprop airliners were introduced in 1958, followed by Boeing 707 jets in 1960. In 1972 the airline became the first to operate an all-freight version of the Boeing 747.

Currently, Lufthansa operates a fleet of Boeing 747s, 737s, 727s, and 707s, and McDonnell Douglas DC-10s. The airline is a member of the ATLAS (q.v.) maintenance consortium.

Lunardi, Vincenzo: an Italian diplomat, Vincenzo Lunardi made the first flight in a hydrogen balloon in Great Britain on 15 September 1784, journeying from Moorfields in the City of London to North Mimms in Hertfordshire, where he jettisoned his ballast and a pet cat before travelling on to Ware. He later arranged for the first balloon ascent by a woman in Great Britain, on 29 June 1785. It had been intended that he should accompany the woman, a Mrs Sage, with two other companions, but because of a weight problem Lunardi and another member of the crew had to forgo their chance of a flight on this occasion.

LV-: international civil registration index mark for Argentina.

L.V.G.: the German Luft-Verkehrs Gesellschaft started its existence before the outbreak of World War I, building biplanes of conspicuously clean lines which were later developed into reconnaissance and bomber aircraft during the war. The most famous aircraft from L.V.G. included the C.III and C.V reconnaissance-bombers with Benz engines, and the company was also one of those engaged in building the Rumpler-Etrich Taube (q.v.) for the German Military Air Service in 1914.

LX-: international civil registration index mark for Luxembourg.

Lynx, Westland W.G.13: *see* W.G.13 Lynx, Westland.

LZ-: international civil registration index mark for Bulgaria.

M

M-4, Douglas: a biplane designed for airmail services, the single-seat, single-engined Douglas M-4 first flew in July 1926, and was used extensively by the United States Army Air Corps on many of the pioneer airmail services in the United States.

Macchi: *see* Aermacchi.

McCudden, Major James Byford, R.F.C.: after joining the Royal Engineers as a bugler in 1910, James McCudden, an Irishman, transferred to the Royal Flying Corps in 1913 as an air mechanic, but became an observer and had reached the rank of Sergeant by 1915, being selected for flying training the following year. McCudden later rose to the rank of Major, and had fifty-seven confirmed, and many unconfirmed, victories flying F.E.2., D.H. 2., and S.E.5 aircraft, by the time of his death in a flying accident in July 1918 when the engine of his aircraft failed.

McDonnell Aircraft: founded by James McDonnell, a former engineer with Martin (q.v.), in 1939, McDonnell Aircraft was initially engaged in experimental work, including a high-altitude fighter with two engines, the XP-67, and in developing the United States Navy's first carrier-borne jet fighter, the FH-1 Phantom, on which work started in 1942, although the aircraft did not fly until after the end of World War II. A development of the Phantom was the F2H-2 Banshee, also for the U.S. Navy, while an interest in missile development started with the RTV-2 Gargoyle experimental ground-to-air missile and the KDH-1 Katydid radio-controlled target drone. A number of experiments with helicopters also took place during the late 1940s.

A number of aircraft followed the Phantom and the Banshee, including the XF-88 Demon and the F-101 Voodoo interceptor for the U.S.A.F. and R.C.A.F. Real prosperity came to the company with the F-4 (q.v.) Phantom II for the United States Navy and United States Marine Corps initially, although since then the aircraft has been adopted by the U.S.A.F. and many of the more important Western air arms. The Phantom II put McDonnell into the position of being able to acquire Douglas Aircraft (q.v.) in 1967, to form McDonnell Douglas (q.v.).

McDonnell Douglas: formed in 1967 from a merger of Douglas Aircraft and McDonnell, the new McDonnell Douglas took over an existing workload which consisted of the McDonnell F-4 (q.v.) Phantom II, a number of guided weapons, and the Douglas A-4 (q.v.) Skyhawk, DC-8 (q.v.), and DC-9 (q.v.) airliners. Since the takeover the F-15 (q.v.) air superiority fighter and the DC-10 (q.v.) air bus have been introduced, although design work on both of these projects was really pre-merger. The Breguet 941S S.T.O.L. transport has been operated experimentally in the United States as the McDonnell Douglas 188E, and the concern holds a licence for production of the Hawker Siddeley Harrier in the United States. McDonnell Douglas is also a NASA contractor.

Mace: a U.S. tactical cruise missile, the Mace is deployed by the U.S.A.F. as a surface-to-surface weapon for long-range battlefield support, and can carry a nuclear warhead. Inertial guidance is used to avoid interference by electronics countermeasures in the Mace B, although the earlier Mace A used a radar-mapping system. Only the Mace B remains, and in limited numbers. A turbojet powerplant is used, but rocket boosters are used for launching. Most Mace missiles have been replaced by the Pershing (q.v.).

Mach: an indication of airspeed relative to the speed of sound, which is Mach 1·0, regardless of altitude.

M.A.D.: *see* magnetic anomaly detector.

Magister, Potez CM.170: *see* CM.170 Magister, Potez.

magnetic anomaly detector: the M.A.D.

'stinger' protruding from the tail is a feature of many modern maritime-reconnaissance aircraft. Basically, the magnetic anomaly detector detects submerged submarines by the slight change in the earth's magnetic field caused by the submarine's presence in the area. The range of M.A.D. is extremely short.

'Mail', Beriev Be-12: *see* Be-12 'Mail', Beriev.

Malan, Group Captain Adolph G. ('Sailor'), D.S.O., D.F.C. R.A.F.: a South African flying with the Royal Air Force during World War II, Group Captain Adolph 'Sailor' Malan was one of the R.A.F.'s top three fighter pilots during the war, with a total of thirty-five confirmed victories. He was awarded the D.S.O. and the D.F.C., with a bar to each, for his achievements. After the war he returned to South Africa to farm.

Malaysian Airlines System: Malaysian Airlines System officially commenced operations in 1973 on the cessation of operations by Malaysia-Singapore Airways, although Malaysian Airlines in fact has existed since 1971. The history of the airline dates from the formation in 1947 of Malayan Airways, which became Malaysian Airways in 1963, on the formation of the Federation of Malaysia, and Malaysia-Singapore Airlines in 1966, when Singapore left the Federation. During the period of existence as Malaysia-Singapore Airlines, the Governments of Malaysia and of Singapore each held a 42·75 per cent interest, the remainder being held by B.O.A.C., Qantas, and others.

A fleet of Boeing 737-200s, Fokker F-27 Friendships, and Britten-Norman Islanders is operated on domestic services and services to Far East destinations. Traffic arrangements with Mercury Singapore Airlines cover the long-haul operations of the former Malaysian-Singapore Airlines.

Malev-Magyar Legikozlekedesi Vallalat: formed in March 1946 as the Hungarian-Soviet Airlines Company, Malev did not adopt its present title until 1954, when full control of the airline passed to Hungary. Initially the airline had operated only domestic services with a fleet of Lisunov Li-2s, which were later supplemented by Ilyushin Il-12s, but by the time of the Soviet handover of the airline to Hungary a small international network was being operated. Today Malev operates throughout Europe, the Middle East, and North Africa, with a fleet of Tupolev Tu-134s and Ilyushin Il-18s.

Mallard, Grumman: the first post-war Grumman civil amphibian, the Mallard retained the high-wing monoplane layout introduced before the outbreak of World War II, and used two radial engines.

'Mallow', Beriev Be-10: *see* Be-10 'Mallow', Beriev.

Malta Airlines: the collective title for the Malta Airways Company and its associate, Air Malta, Malta Airlines does not operate any aircraft of its own at present, but with B.E.A. (q.v.) operates jointly-licensed services to Italy and the British Isles. This arrangement has existed since 1946, when the airline was formed, although plans to operate the airline's own aircraft, with Pakistan International Airlines providing assistance, exist for the future.

M.A.N.: M.A.N. Turbo G.m.b.H. is the main German aero-engine manufacturer, holding Rolls-Royce and General Electric licences for production of the power-plants for a number of European collaborative aircraft projects. The company's origins derive in part from the formation of B.M.W. (q.v.) before the start of World War I.

'Mangrove', Yakovlev Yak-25: *see under* Yak-25 'Flashlight', Yakovlev.

Mannock, Major Edward 'Mickey', V.C., D.S.O., M.C., R.F.C.: Britain's greatest World War I fighter pilot, Major Edward 'Mickey' Mannock re-joined his Territorial Army unit after the start of World War I following his repatriation from Turkey because of poor health. Initially he served with the Royal Engineers, eventually being accepted for flying duties in spite of an astigmatism in his left eye. Mannock joined his first R.F.C. squadron in April 1917, flying Nieuport Scouts, and during the next year he rose from the rank of Lieutenant to that of Major, and was put in command of No. 74 Squadron, flying S.E.5a's. He was killed on 26 July 1918, being shot down by infantry fire. Although the confirmed victories credited to him amounted to no less than seventy-three, it is generally considered that even this high score fails to give the true picture, since many aircraft which should have been added to his total were added to those of other pilots at his insistence. Awarded the D.S.O. and two bars and the M.C. and bar during his flying service, he was awarded the Victoria Cross posthumously.

Marauder, Martin B-26: see B-26 Marauder, Martin.

maritime reconnaissance: the use of aircraft for maritime-reconnaissance duties, that is for the survey of the ocean for enemy warships, and submarines in particular, dates from World War I. Maritime-reconnaissance aircraft should not be confused with the use of ships' aircraft for 'spotter' duties, since in its generally-accepted sense a maritime-reconnaissance aircraft is also able to engage the enemy; the obvious reason being that immediate action is often vital if the enemy ship or submarine is not to escape. Another object of maritime-reconnaissance in peace and war is rescue, and today many such aircraft also have to keep a lookout for oil pollution.

Amongst the earliest maritime-reconnaissance machines were the airships of the Royal Naval Air Service during World War I, although from 1915 torpedo-bomber operations were mounted, and at the Battle of Jutland, in May 1916, the German Fleet's cruisers and destroyers were kept under observation by British aircraft. Amongst the aircraft used on these duties during the war were Short biplanes and the Italian Aermacchi M.8s. After the war the tendency for flying-boats to be used for maritime-reconnaissance duties soon became apparent, with such aircraft as the Seaplane Experimental Station F.5, the Supermarine Southampton, the Short Singapore, and the Blackburn Iris, while Germany produced the Dornier Wal (q.v.) for civil and naval use. Amongst the few landplanes built specifically for maritime-reconnaissance duties between the wars was the French Bloch M.B.210.

The aircraft of World War II started to appear during the 1930s, some of the more impressive types coming from Germany, where the Heinkel He.59 seaplane was followed during the late 1930s by the four-engined Focke-Wulf Fw.200 (q.v.) Condor and the Junkers Ju.90 (q.v.) landplanes, the Heinkel He.115 twin-engined seaplane, and the Dornier Do.24 (q.v.) trimotor seaplane. New American aircraft at the time included the Consolidated PBY-5 Catalina (q.v.), which first flew in 1935, while two years earlier the British Supermarine Walrus (q.v.) had made its appearance. The Short Sunderland first flew in 1937. Other aircraft included the Martin PBM-1 Mariner flying-boat, the Blohm und Voss BV.138 twin-boom flying-boat, and the Japanese Navy's Kawanishi H8K2 four-engined flying-boat. Landplanes also included the Lockheed Hudson, which was later replaced by the Ventura and the Harpoon from the same stable. Vickers Wellingtons (q.v.) were used by the R.A.F., while both the Americans and the British used versions of the Consolidated B-24 (q.v.) Liberator, and the United States Navy had the Liberator specially developed into a new aircraft, the Privateer.

After World War II the importance of maritime reconnaissance increased, rather than diminished, due to the Soviet Union's vast naval ambitions and large fleet of both conventional and nuclear submarines. The Allies soon gained, and have since maintained, a considerable superiority in maritime-reconnaissance and anti-submarine techniques, although the Soviet Union has not relinquished such measures herself, and the Beriev (q.v.) design bureau in particular has specialised in such aircraft, while Tupolev designs are also often used.

The first of the post-war generation of maritime-reconnaissance aircraft was the American Lockheed P-2 (q.v.) Neptune, which first flew in 1945, and was in production throughout the late 1940s and into the 1950s, with licence-built turboprop developments only recently being produced in Japan by Kawasaki. The Neptune was soon joined by the Martin Marlin flying-boat and by the Avro Shackleton (q.v.). A generation of carrier-borne aircraft included the Grumman S-2 (q.v.) Tracker, the Fairey Gannet (q.v.), and the Breguet Br.1050 (q.v.) Alize. During the late 1950s the Canadair CP-107 (q.v.) Argus replaced the Avro Lancaster (q.v.), which had been relegated to maritime-reconnaissance duties after the war. A Lockheed aircraft, the P-3 (q.v.) Orion, became the first production turboprop maritime-reconnaissance aircraft, while during the late 1960s the Franco-German Breguet Br.1150 (q.v.) Atlantique also used turboprop engines, as did the Shin Meiwa PS-1 (q.v.) flying-boat put into production in Japan. The first turbojet aircraft in this field was the British Hawker Siddeley Nimrod (q.v.) of the late 1960s. A new carrier-borne aircraft from Lockheed, the S-3 Viking, uses two turbofan powerplants.

Modern maritime-reconnaissance aircraft are armed with depth charges and air-to-surface missiles, and are equipped with autolycus (q.v.), magnetic anomaly detector (q.v.), and surface scanning radar equipment, all of which can be supplemented by the use of sonar buoys, which are parachuted into the sea to transmit back their findings to the aircraft.

markers: an alternative name for radio beacons providing information from which an aircraft's position may be determined.

markings: the identification letters, numerals, and symbols on an aircraft. Civil aircraft have an international civil aircraft marking (q.v.), followed by the registration index which may sometimes, particularly in the United States, be suffixed by a letter indicating the operator's identity. Military aircraft markings, which evolved early in World War I, carry the national colours in wing, fuselage, and tail markings, accompanied by squadron identification letters and a serial number. Normally the national colours on the wings and fuselage are in the form of roundels, but in some cases rectangles or triangles are used. Tail markings are usually in the form of flashes or stripes, although these are by no means always used, one exception to the use of national tail markings normally being Royal Navy aircraft.

Martel: an air-to-surface missile developed jointly by the United Kingdom and France, the Hawker Siddeley Dynamics-Engins Matra Martel (Missile Anti-Radar and Television) was produced during the late 1960s and early 1970s to equip Hawker Siddeley Buccaneers and Nimrods, B.A.C.-Breguet Jaguars, Dassault Mirages, and Breguet Atlantique aircraft. A relatively small missile of only twelve feet or so in length, the Martel uses a solid-fuel rocket. The British versions use television guidance from the attacking aircraft which continues after the aircraft has turned away from the target, while the French version homes on the target's radar emissions. A high explosive warhead is used, and Martel has a range of about forty miles.

Martin: the former Glenn L. Martin concern, now a division of the Martin Marietta Corporation, is no longer involved in aircraft manufacture, concentrating instead on weaponry and space research. Nevertheless, during a long history dating from the first Glenn Martin aeroplane flights in 1909, the company has been associated with a number of interesting aircraft, including a number of biplane designs built between 1909 and the outbreak of World War I in Europe.

Although Martin (in common with many other American aircraft manufacturers) did not have overmuch success in getting combat aircraft into U.S. Army service during America's involvement in World War I, one of the Martin aircraft, the MB-1 bomber biplane with two 400 hp Liberty engines, became standard U.S.A.A.C. equipment for several years after the war ended. The MB-1 was followed by a development, the NBS-1, with up-rated engines, and also available in commercial transport versions. During the 1920s Martin also built the MO-1 single-engined observation monoplane for the U.S. Navy, and the MS-1, a small float-plane with folding wings for stowage aboard submarines.

The late 1920s saw the company developing a biplane dive-bomber for the U.S. Navy, while during the early 1930s the P2M trimotor flying-boat with three Wright 575 hp Cyclone radial engines was produced for the Navy and commercial users. A considerable variety was evident in Martin's product range throughout the 1930s, for while the four-engined M-130 'China Clipper' flying-boat was helping Pan American to establish new air services, the B-10 twin-engined, all-metal, mid-wing bomber monoplane was establishing speed records for its class, and starting a trend in bomber design that lasted through the rest of the 1930s and the 1940s. One development of the B-10 concept was the B-26 (q.v.) Marauder of World War II, which was also a twin-engined mid-wing bomber, for the U.S.A.A.F. and R.A.F. Another Martin aircraft of the war period was the PBM-3 Mariner flying-boat for the U.S.N.

After World War II Martin made a rare attempt at the civil landplane market with the 404, a twin-engined forty-passenger airliner for U.S. domestic airlines. A new flying-boat, the P5M Marlin, was built for the U.S. Navy, while experiments with jet aircraft were put in hand, although production did not start until Martin obtained a licence to build the English Electric Canberra (q.v.) jet bomber as the Martin B-57 (q.v.). Following the end of Canberra production, Martin effectively left the aircraft design and production field.

Martin, Sir James, C.B.E.: an Ulsterman, Sir James Martin was invited by the Ministry of Aircraft Production in 1944 to develop a means of assisted escape for fighter pilots. Even before the advent of the jet age, increasing fighter speeds were making escape by the traditional 'over-the-side' method both difficult and hazardous, and the early jet aircraft, such as the Gloster Meteor, emphasised this need for a means of throwing a pilot well clear of his aircraft. It soon became clear that the best method of achieving this end would be by forced ejection of the seat by explosive charge, with the occupant of the seat falling from it after ejection and parachuting to earth in the normal way.

Experiments started in 1945, and the immediate problem which arose was that of the considerable 'g' force which built up due to the sudden acceleration of the seat; but this was cured by using a two-stage explosive charge, with a minor explosion

to start the seat moving, and a second explosion to accelerate the seat and clear the aircraft. Dummy tests were made from the second seat in an old Boulton Paul Defiant during the summer of 1945, while later experiments used a Gloster Meteor jet fighter The first live shot using an ejector seat was from a Gloster Meteor by one of Martin Baker Aircraft's fitters, Mr Bernard Lynch, on 24 July 1945; the aircraft was flying at a speed of 320 mph and a height of 8,000 feet at the time. Soon, a number of British aircraft, including the Gloster Meteor, the Supermarine Attacker, the Blackburn Firebrand, the Westland Wyvern, the English Electric Canberra, and the de Havilland Venom, were fitted with the Martin Baker Mk.1 ejector seat developed from the experimental units.

An automatic seat, the Mk.2, was developed during the late 1940s for use in those cases where the pilot lost consciousness after initiating the ejection procedure – the main feature being a barostat which released the parachute at a reasonable height (premature opening could mean a pilot suffocating or freezing to death). *See also* ejector seat.

Marut, Hindustan HF-24: *see* HF-24 Marut, Hindustan.

'Mascot', Ilyushin Il-28U: *see* Il-28U 'Mascot', Ilyushin.

Matra: Engins Matra is the major French aircraft armaments manufacturer, commanding a worldwide market for rockets and rocket pods, and a less extensive, but still important, market for guided weapons produced either unilaterally or in co-operation with other European firms. One of the major products at the present time is the Martel (q.v.).

'Max', Yakovlev Yak-18: *see* Yak-18 'Max', Yakovlev.

Maxim, Sir Hiram Stevens (1840–1916): American-born, but a naturalised Briton, Sir Hiram Maxim was already famous as the inventor of the Maxim machine gun by the time he turned his attention to aviation in 1889. One of the first pioneers to test his aerofoils in a wind tunnel, Maxim also devoted some considerable effort to the development of a steam engine with a high power-to-weight ratio – not without success. A test rig was built in 1894, consisting of a large unmanned biplane powered by two 180 hp steam engines driving two pusher propellers – the whole contraption running along a length of tramway track with further retaining rails to prevent it from becoming airborne. On a test run in July 1894 the test rig lifted off the track before fouling the retaining rails and being brought to a standstill. Later it made a number of public exhibition runs for charities.

An attempt to build an aeroplane designed by Maxim in 1910 was a failure – it was unable to lift itself off the ground.

Mayo, Short: not in fact an aircraft, but two aircraft, the Short Mayo composite aircraft consisted of a Short S.21 Maia flying-boat and a Short S.20 Mercury seaplane – both aircraft being four-engined high-wing monoplanes. The S.21 flying-boat carried the smaller seaplane aloft on top of its fuselage, releasing it to complete the journey with mail and newspapers: on a transatlantic flight to Montreal in July 1936 the Mercury was released in mid-air west of Ireland, and the total flying time amounted to just over twenty hours. Experiments continued until 1939, including a Dundee to South Africa flight, but stopped with the outbreak of World War II. The Mercury was scrapped in 1941, and the Maia was destroyed in a German air raid on Poole.

M.B.151, Bloch: the Bloch M.B.151, a single-engined, single-seat fighter of low-wing monoplane construction, entered service with the Armée de l'Air in 1938, some aircraft also being supplied to Greece and to Rumania before the outbreak of World War II the following year. One of the few modern French fighter aircraft at the start of the war, its production ceased on the German invasion of France in 1940. A 1,080 hp Gnome Rhône engine gave the aircraft a maximum speed of the order of 320 mph.

M.B.200, Bloch: a high-wing bomber monoplane with two engines, the Bloch M.B.200 was the company's first military aircraft, and entered service with the Armée de l'Air in 1934, forming the backbone of the pre-World War II French bomber force. A development, the MB. 210 low-wing bomber, was used by the French Navy as a land-based maritime-reconnaissance aircraft.

M.B.326 (Impala), Aermacchi: possibly the most successful Italian military aircraft, the Aermacchi M.B.326 jet trainer first flew in prototype form in December 1957, using a single 1,750 lb thrust Bristol Siddeley Viper 8 turbojet. Production aircraft use the up-rated 2,500 lb thrust Viper II, and the first of these was delivered to the Italian Air Force in October 1960. Variants of the basic aircraft have included the M.B.326B for Tunisia and the M.B. 326F for Ghana, while the aircraft has also been built under licence in Australia as the M.B.326H, and in South Africa as the M.B.326K Impala armed jet trainer and counter-insurgency aircraft. A tandem twin-seat aircraft with a low-wing, the M.B.326 has a maximum speed of 500 mph and a range of 690 miles.

M.B.B.: *see* Messerschmitt - Bölkow-Blohm.

M.B.F.R.: Mutual and Balance Force Reductions, or M.B.F.R., is a Soviet concept for equal reductions in military strength in Europe by the Warsaw Pact and NATO countries. Talks on M.B.F.R. are taking place during 1973, coinciding with the other Soviet proposal, the European Security Conference, and further talks on Strategic Arms Limitation (SALT, q.v.). The weakness in NATO acceptance of the idea of M.B.F.R. is that Warsaw Pact forces outnumber those of NATO by 960,000 men to 580,000 men, while the equipment differential is even more marked. Therefore only a proportional cut would be acceptable, but this is not the Soviet idea – hence the definition of M.B.F.R. by Western military commanders as 'More Bonuses For Reds'.

Me.109, Messerschmitt: *see* Bf.109, Messerschmitt.

Me.110, Messerschmitt: a long-range escort

fighter and bomber interceptor, the Messerschmitt Me.110 also saw extensive use as a fighter-bomber and glider-tug throughout most of World War II. Two 1,395 hp Daimler Benz engines powered this low-wing, twin-fin monoplane.

Me.163 Komet, Messerschmitt: the world's first and only rocket-powered interceptor, the Messerschmitt Me.163 Komet first flew in 1944, and was soon in use against Allied bombers. A single Walter liquid-fuelled rocket powered the aircraft, which had a single-seat cockpit and a swept wing. Maximum speed was 590 mph, and endurance was limited to eight minutes, although this was sometimes extended by towing the aircraft into position and by intermittent gliding. Take-off was normally by use of a catapult, and the Me.163 had a landing skid instead of a wheeled undercarriage.

Me.262, Messerschmitt: the world's first operational jet aircraft, the Me.262 fighter first flew in 1942, but did not enter service until 1944, when it was used as a light jet bomber on Hitler's orders. Two Junkers Jumo turbojet engines powered the aircraft, which was single-seat and of low-wing construction. An interesting feature was the use of a tailwheel, instead of the tricycle undercarriage more usually associated with jet aircraft, on some versions.

Me.323 Gigant, Messerschmitt: aptly named, the Messerschmitt Me.323 Gigant (Giant) was a powered version of the Me.321 glider, which could carry 130 fully-equipped troops and required three Me.110 glider tugs to become airborne. Six 925 hp engines powered the Me.323, which in appearance showed its glider ancestry clearly. A high wing with extensive external strut-bracing was employed.

medium-haul: air services or an aircraft capable of a range of between 1,000 and 2,500 miles.

Mentor, Beech T-34: *see* T-34 Mentor, Beech.

Mercedes Benz: today known as a manufacturer of motor vehicles, before Germany's defeat at the end of World War II Mercedes Benz was also one of the leading aero engine manufacturers, with a history dating from pre-1900 experiments with airships and the activities of the former Mercedes and Daimler Benz concerns. Amongst the aircraft which used Mercedes Benz engines were the Messerschmitt Bf. 109 (q.v.) of World War II, and before this the Junkers J.1 (q.v.).

Mercure, Dassault: a short- and medium-haul second generation jet airliner by Avions Marcel Dassault, the Mercure is effectively the French aircraft industry's replacement for the Aerospatiale Caravelle (q.v.). The first flight of the Mercure prototype took place in May 1971, and the aircraft has entered service with the domestic airline, Air Inter. Two 22,000 lb thrust Pratt and Whitney JT8D turbofans give a maximum cruising speed of 590 mph and a range of up to 1,200 miles, while between 124 and 155 passengers can be carried.

Mercury Singapore Airlines: the Singapore successor to Malaysia-Singapore Airlines, Mercury Singapore Airlines (M.S.A.) commenced operations in 1973 to Europe and the Far East, in effect taking over Malaysia-Singapore's long-haul routes and continuing to maintain traffic links with Malaysia on those routes, with a fleet of Boeing 707-320s, 737-100s, and 737-200s, and Fokker F-27 Friendships. *See also* Malaysian Airlines.

Mer-Sol Balistique Stratégique: the French submarine-launched ballistic missile, the Mer-Sol Balistique Stratégique is produced by Aerospatiale, the standard M-1 version being already in service, and an improved M-2 version scheduled to enter service in 1974. France has five missile-carrying submarines, each with sixteen missiles and nuclear propulsion, but in all cases the whole package is inferior to those of the United States and Royal Navies in every way. At present the M.S.B.S. carries an atomic warhead, but a hydrogen warhead will be available in 1976 with a new missile, the M-20, while a further development will be the M-4, which will require extensive modifications to the submarine force because of its increased diameter. The M.S.B.S. M-1 is a two-stage rocket of thirty-four feet in length and five feet in diameter, using gimballed nozzles and roll rockets for control. The warhead is only 500 kilotons, and range is a mere 1,450 miles.

Messerschmitt: the German Messerschmitt concern came into prominence during the late 1930s with the Bf. 109 (q.v.) fighter and Me. 110 (q.v.) fighter-bomber, both of which were extensively developed. Other notable aircraft included the Me. 321 glider and its powered development, the Me. 323 (q.v.) Gigant, the Me. 163 (q.v.) Komet rocket-powered interceptor, and the Me. 262 (q.v.) jet fighter. The company closed down in 1945, but was restarted in the mid-1950s, merging with Bölkow to form Messerschmitt-Bölkow-Blohm (q.v.) in 1964.

Messerschmitt, Professor Wilhelm: originally a designer of gliders and light aircraft, Professor Wilhelm (Willy) Messerschmitt became one of the leading aircraft designers in Germany before and during World War II, providing his country with a range of aircraft which were the equal of any then available in the world. Messerschmittt's first notable design was the Bf. 10 (q.v.) fighter, while others which were also of importance included the Me. 163 (q.v.) Komet, the first rocket-powered aircraft, and the Me. 262 (q.v.), the first operational jet fighter. After World War II ended, Messerschmitt worked in Argentina for a period before returning to Germany to restart aircraft manufacture.

Messerschmitt-Bölkow-Blohm: formed from the merger of Bölkow and Messerschmitt (q.v.) in 1964, Messerschmitt-Bölkow-Blohm has developed the Bo.208 junior light aircraft, which is now built in Sweden, and the Bo.105 helicopter, as well as participating in a number of European collaborative projects, such as the C-160 (q.v.) Transall, the ELDO space programme, the A.300B (q.v.) airbus, and the Panavia 200 (q.v.) Panther. Subcontract work on Dornier's licence-production of Bell UH-1D Iroquois helicopters has also been undertaken.

Meteor, Gloster: the world's second operational jet fighter, and the first and only Allied jet aircraft to reach operational status during World War II, the Gloster Meteor first flew in March 1943, and entered service with the R.A.F. in July 1944. A single-seat, low-wing monoplane, the Meteor used two Whittle turbojets initially, but later aircraft employed various versions of the Rolls-Royce Derwent. A number of developments of the basic aircraft were introduced during its production life of ten years, including tandem twin-seat training, night fighter, and photographic reconnaissance aircraft. After being retired from first-line duties, many aircraft were converted to target-tug and communication duties. After the war the aircraft established a number of air speed records, with later aircraft showing a considerable improvement in performance against the early versions – speed alone increased from 480 mph to 576 mph for standard aircraft, with special versions managing well in excess of 600 mph. The Meteor has seen extensive worldwide service with a good number of air forces and air arms. A typical version with two 3,700 lb thrust Rolls-Royce Derwent turbojets would be capable of a maximum speed of 575 mph and a range of 800 miles. Many Meteors were in fact built by Armstrong-Whitworth.

meteorological briefing: more usually referred to as a 'met' briefing, the meteorological briefing is the result of the pre-flight inquiry made by a pilot into the weather conditions which can be expected along the course of his flight, over a target area in the case of a military pilot, and at any diversionary airfields.

meteorology: the science of weather study and forecasting.

Metroliner, Fairchild: originally developed by Swearingen, production of the Metroliner has been taken over by Fairchild-Hiller since the early 1970s. A twin-engined, pressurised feeder-liner, the Metroliner uses two 904 shp Garrett AiResearch TPE331-303 turboprops for a maximum cruising speed of 305 mph and a range of up to 1,700 miles, while accommodation exists for up to twenty passengers.

Metropolitan, Convair: the name usually given to the Convair 240, 340, and 440 series of airliners; *see under* Convair.

Mi-4 'Hound', Mil: the second major Soviet helicopter to enter production, the Mil Mi-4, NATO code-named 'Hound', appeared during the early 1950s. Although the fuselage bears a strong resemblance to that of the Sikorsky S-55, the Mi-4 has tail doors fitted. A single 1,700 hp ASh-82V piston-engine gives a maximum cruising speed of 130 mph and a range of 250 miles, while up to fourteen passengers can be carried in addition to a crew of two. The Mi-4 is in widespread civil and military use in the Soviet Bloc. A successor is the Mil Mi-8 (q.v.) 'Hip'.

Mi-6 'Hook', Mil: until the advent of the Mi-12 (q.v.) the largest helicopter in production and service, the Mil Mi-6, NATO code-named 'Hook', first appeared in 1957. Two 5,500 shp Soloviev D-25 shaft turbines drive a single main rotor and provide a maximum speed of 217 mph and a range of 650 miles, while accommodation exists for a crew of four and up to 120 passengers. Tail doors and a rear-loading ramp are fitted, while a distinguishing feature is a stub wing to provide stability. A crane version, the Mil Mi-10 (q.v.) (Harke), is also available.

Mi-8 'Hip', Mil: the Mil Mi-8, NATO code-named 'Hip', is essentially a Mi-4 replacement helicopter, with accommodation for up to twenty-eight passengers, using two 1,500 shp Isotov TB-2-117 turbines to drive a single main rotor and provide a maximum speed of 150 mph and a range of 400 miles

Mi-10 'Harke', Mil: a crane development of the Mil Mi-6 (q.v.), the Mil Mi-10, NATO code-named 'Harke', is in the main identical to the basic aircraft to below the level of the cabin windows, except for the omission of the stub wing. The shallow fuselage provides accommodation for up to twenty-eight passengers, and a fifteen-ton load can be carried; with still higher weights for the latest version with an up-rated Aoloviev D-25 turbine. The Mi-10 first appeared in 1961.

Mi-12 'Homer', Mil: replacing the Mil Mi-6 as the world's largest helicopter, the Soviet Mil Mi-12, NATO code-named 'Homer', is a compound helicopter with two large intermeshing rotors at the wingtips. Four 6,500 shp Soloviev turbines provide for a maximum speed of 160 mph and a payload of up to 88,636 lb, while it is believed that the wings, with reverse taper (i.e. the tips are broader than the roots), contribute a significant amount of lift while cruising.

Middle East Airlines Airliban: one of the major airlines in the Middle East, Middle East Airlines dates from 1945, when the airline was formed with a fleet of three de Havilland Dragon Rapide biplanes, later supplemented by Douglas DC-3s. Pan American (q.v.) acquired a 36 per cent interest in M.E.A. in 1949, allowing further expansion of the fleet and the route network, while an association with B.O.A.C. in 1955, after the Pan Am agreement expired and the British airline acquired a 48·5 per cent shareholding, enabled M.E.A. to buy Vickers Viscount turboprop airliners. The first jets, de Havilland Comet 4s, were operated on charter from B.O.A.C. in 1960, before the airline acquired its own Comet 4Cs in 1961.

A period of complete independence for M.E.A. came in 1961, and lasted until 1963, when Air France acquired a 30 per cent interest in the airline as a result of M.E.A. merging with Air Liban, an airline with services throughout the Middle East and West Africa. A further merger occurred in 1969, with Lebanese International Airways. Today, M.E.A. operates a fleet of Boeing 707-320s and 720-120s, and Convair CV-990As, and operates an extensive route network throughout the Middle East and to Africa, Asia, and Europe. A serious blow to the airline occurred in December 1968, when an Israeli attack on Beirut Airport destroyed six of its aircraft, but the airline has since recovered from this.

'Midget', Mikoyan-Gurevich MiG-15UTI: *see* MiG-15UTI 'Midget', Mikoyan-Gurevich.

mid-wing: a form of monoplane configuration, the mid-wing aircraft, as the term suggests, has the wing running through the middle of the fuselage, either between decks, or between the cabin and the bomb bay, fuel tanks, or baggage holds of the aircraft. Mid-wing aircraft are fewer in number than low-wing or high-wing types since most designers avoid this type of layout if possible, because in many cases the wing structure would cause an obstruction or restriction in the accommodation inside the fuselage.

MiG-3, Mikoyan-Gurevich: the second joint design by Mikoyan and Gurevich after the start of the Mikoyan-Gurevich design bureau, the MiG-3 was a piston-engined fighter and one of the few World War II aircraft built by the Soviet Union to have a standard of performance comparable with Western types. The MiG-3 first appeared in 1940, and had a maximum speed of 400 mph.

MiG-9, Mikoyan-Gurevich: the first Soviet turbojet fighter, the Mikoyan-Gurevich MiG-9 appeared after the end of World War II and utilised a German-designed airframe and engines captured by Russian troops during the closing days of the war.

MiG-15 'Fagot', Mikoyan-Gurevich: developed with the assistance of captured German research documents and designs, and with an engine developed from Rolls-Royce Nene and Derwent turbojets sold to the Soviet Union, this Mikoyan-Gurevich single-engined, single-seat jet fighter was amongst the first such aircraft to employ swept wings. The aircraft first flew in December 1947, and later formed the backbone of the Communist air power during the Korean War, where it proved itself to be inferior to British and American designs, even to the point of being shot down by piston-engined aircraft. Maximum speed from the 5,950 lb thrust Klimov VK-1 turbojet of later versions was about 600 mph, with a range of about 600 miles. The aircraft was also built in Czechoslovakia and Poland. A few remain in service, mainly in the less advanced African states. The MiG-15UTI (q.v.) as a twin-seat training development.

MiG-15UTI 'Midget', Mikoyan-Gurevich: the MiG-15UTI is a tandem twin-seat development of the MiG-15 (q.v.), using a Klimov RD-45FA turbojet. The aircraft is still in widespread use as an advanced jet trainer.

MiG-17 'Fresco', Mikoyan-Gurevich: a development of the MiG-15 (q.v.), the Mikoyan-Gurevich MiG-17, NATO code-named 'Fresco', entered production in 1953 in the Soviet Union, as well as later being built in Poland, Czechoslovakia, and China (as the Shenyang F-4). A number of versions were built, the basic aircraft being known as the 'Fresco A' in the West; the 'Fresco B' had detail airframe modifications, the 'Fresco C' had afterburning, and the 'Fresco D' had elementary all-weather radar equipment fitted. The performance of the 'Fresco C' and 'Fresco D' versions amounts to a maximum speed of 700 mph and a range of 800 miles from a single 6,990 lb thrust Klimov VK-1A turbojet with afterburning, while cannon can be fitted and either rockets or two 550 lb bombs carried on under-wing strongpoints.

MiG-19 'Farmer', Mikoyan-Gurevich: a departure from the normal Mikoyan-Gurevich policy of single-engined aircraft, the MiG-19, NATO code-named 'Farmer', retained the similarity of line which marked the aircraft as a development of the MiG-15 (q.v.) and MiG-17 (q.v.), but employed two engines. Development in A, B, C, and D versions was similar to that of the earlier aircraft, although all MiG-19s used reheat. The two 8,820 lb thrust Klimov VK-5 turbojets with reheat give the single seat aircraft, of which many remain in service on fighter-bomber duties, a maximum speed of 850 mph and a range of 1,200 miles. Rockets, missiles, and bombs can be fitted to the underwing strongpoints to supplement the basic armament of 30 mm cannon. The MiG-19 was built in China as the Shenyang F-6.

MiG-21 'Fishbed', Mikoyan-Gurevich: a return to the single-engined concept, but with a delta-shaped main lane, the Mikoyan-Gurevich MiG-21, NATO code-named 'Fishbed', first appeared in 1956 and is a Mach 2·0 interceptor of markedly short range compared with Western aircraft. A number of versions have been built, of which the most common are those known in the West as the 'Fishbed C', and the all-weather radar-equipped 'Fishbed D' – the basic aircraft's all-weather capability is limited. The MiG-21 has been built in

Czechoslovakia as well as in the Soviet Union, and production continues in India and China (Shenyang F-8). A conversion trainer version is the MiG-21UTI (q.v.). The single 13,000 lb TRD Mk.R37F turbojet with reheat gives the single-seat aircraft a maximum speed of Mach 2·0 (1,320 mph) and a range of 750 miles, while two 'Atoll' air-to-air missiles can be carried and some aircraft are still equipped with 30 mm cannon. A successor is the MiG-23 (q.v.).

MiG-21UTI 'Mongol', Mikoyan-Gurevich: a variation on the 'Fishbed C' theme, the Mikoyan-Gurevich MiG-21UTI conversion trainer, NATO code-named 'Mongol', uses the same powerplant as the MiG-21 (q.v.), but has a tandem twin-seat cockpit.

MiG-23 'Foxbat', Mikoyan-Gurevich: a world record holder on a number of different counts, the Mikoyan-Gurevich MiG-23 'Foxbat' is the MiG-21 (q.v.) successor in the Soviet Union. In design completely different from its predecessors, the MiG-23 has a high wing, twin fins, and two 33,000 lb thrust turbofans, with reheat which give a maximum speed of the order of Mach 3·0. A missile armament is assumed, but little detail is known in the West, although the aircraft first appeared in 1967.

Mikoyan, Artem: the head of the Soviet design bureau which carries his name and that of his partner, Mikhail Gurevich, Artem Mikoyan was a graduate of Moscow Technical High School and entered aviation by way of transport aircraft design for another of the Soviet design bureaux. The first project on which he and Gurevich were actually in charge of design was the MiG-1 fighter of the late 1930s, although the first to be built in any real numbers was the MiG-3 (q.v.) of 1941.

The main characteristic of Mikoyan designs has been the thought given to ease of construction, a feature based on the limitations of Soviet industry, particularly during the early days of the Mikoyan-Gurevich design bureau. During the period immediately after the end of World War II in Europe, Mikoyan was quickly able to put captured German plans into production resulting in the MiG-9 (q.v.) jet fighter, and

later used the results of German swept-wing research, with Rolls-Royce Nene turbojets bought from Britain, to build the MiG-15 (q.v.), which was one of the first swept-wing jet aircraft.

Although Russian aircraft design has lagged behind that of the West until recently, the latest product from the Mikoyan-Gurevich design bureau, the MiG-23 (q.v.) 'Foxbat', is a potent rival to Western interceptor and fighter aircraft.

Mil: very much a product of the post-World War II evolvement of the helicopter, the Russian Mil design bureau has been headed since its inception by Mikhail L. Mil. Although the bulk of production has been concentrated on the smaller Mi-2, Mi-4 (q.v.), and Mi-8 (q.v.) designs, the bureau has the reputation of building very large helicopters which consistently set payload records for this type of aircraft; the first of these was the Mi-6 (q.v.), but while this still led the world, its commanding position was taken by the giant Mi-12 (q.v.). Current production models are the Mi-8, the Mi-6 and its crane development, the Mi-10 (q.v.), and the Mi-12. Most machines are available in civil and military forms, but few Mil designs are ever shipborne.

Milan, Dassault: a version of the Dassault Mirage 5 (q.v.).

Miles Aircraft: a British manufacturer of light aircraft, including training, observation, and communications types, during the period immediately before, during, and after World War II, Miles Aircraft became a part of Handley Page (q.v.) in 1948. Amongst the aircraft produced by Miles during its existence were the Magister and Master trainers, and the Aerovan, a light transport not unlike the present Short Skyvan in appearance, although piston-engined and smaller.

Minuteman: in service in two versions, the American Minuteman intercontinental ballistic missile is a three-stage rocket capable of carrying a nuclear warhead for more than 6,000 miles in the Minuteman 2 version, and more than 7,000 miles in the Minuteman 3; this can also carry a warhead comprised of three multiple independently-targeted re-entry vehicles. Guidance is essentially inertial. Main contractor on the missile is Boeing (q.v.), and the Minuteman is fired from superhardened concrete silos in the United States.

Mirage III, Dassault: the most successful French military aircraft since the heyday of French military aviation before 1918, the Dassault Mirage III delta-wing supersonic fighter first flew in prototype form in November 1956, followed by pre-production Mirage III-As, before the first production Mirage III-B tandem twin-seat conversion trainers and III-C single-seat fighters first flew, in late 1959 and October 1960 respectively. The aircraft has since sold to a number of countries in Europe, the Middle East, and Latin America, and to Australia and South Africa, French production being supplemented by licence production in Switzerland and Australia. III-D and III-E fighter-bomber and III-R reconnaissance versions have also been developed and placed in service. The Mirage III-E uses a single 13,760 lb thrust SNECMA Atar 09C turbojet with reheat, sometimes with rocket assistance, for a maximum speed of 1,400 mph and a range of up to 1,500 miles. Two 30 mm cannon are fitted, and rockets, missiles, or two 1,000 lb bombs on two underwing strongpoints can be carried.

Mirage IV-A, Dassault: the Dassault Mirage III (q.v.) was scaled up to meet an Armée de l'Air requirement for a supersonic bomber capable of delivering nuclear weapons, resulting in the Mirage IV, which first flew in June 1959 in prototype form. The first production machine flew in December 1963. Two 15,400 lb thrust SNECMA Atar 09K turbojets with reheat provide for a maximum speed of 1,450 mph and a range of 2,000 miles. Sixteen 1,000 lb bombs, four Martel missiles, or a nuclear bomb can be carried on underwing strongpoints or inside a fuselage recess.

Mirage 5, Dassault: a simplified development of the Mirage III (q.v.), the Dassault Mirage 5 is essentially an aircraft for the developing countries and is stripped of sophisticated electronics and the rocket assistance which is an option with the Mirage III series. The first flight of a Mirage 5 was in 1967, and the aircraft has found no shortage of customers. A further development is the Milan, basically a Mirage 5 with retractable foreplanes to reduce landing approach speed. A single 14,110 lb thrust SNECMA Atar 9C turbojet with reheat gives the Mirage 5 a maximum speed of 1,385 mph and a range of up to 1,500 miles. Two 30 mm cannon are fitted and up to 8,000 lb of bombs, rockets, or missiles can be carried on underwing and under-fuselage strongpoints.

Mirage F.1, Dassault: an interceptor development of the Mirage series, the Dassault Mirage F.1 breaks away from the delta wing of earlier models. and has a separate mainplane and tailplane, with a high wing. First flight was in December 1966. A single 14,000 lb thrust SNECMA Atar 9K turbojet with reheat gives a maximum speed of 1,385 mph and a range of 1,500 miles, while air-to-air missiles are normally carried and cannon usually fitted. The aircraft is in Armée de l'Air service. Another version of the Mirage series, the F.2 fighter, has still to enter production.

Mirage G.8, Dassault: the production development of the Mirage G variable-geometry fighter-bomber which first flew in November 1967, the Dassault Mirage G.8 is a tandem twin-seat aircraft, with two SNECMA Atar 9K turbojets of 14,000 lb thrust with reheat, which first flew in May 1971. Maximum speed is Mach 2·5 and the range is about 2,000 miles. No orders have been received yet and only two aircraft have been built.

M.I.R.V.: *see* multiple independently-targeted re-entry vehicle.

missiles: missiles fall into several categories, and are dealt with in this book under these specific headings: air-to-air missile (q.v.), air-to-surface missile (q.v.), surface-to-air missile (q.v.), and surface-to-surface missile (q.v.); as well as under individual type names.

Mitchell, North American B-25: *see* B-25 Mitchell, North American.

Mitchell, Reginald J. (1895–1937): one of the greatest British aircraft designers, Reginald Mitchell was responsible for designing for his employer, Supermarine (q.v.), the series of Schneider Trophy seaplanes which not only won the Trophy outright for Great Britain, but also led to the development of the Supermarine Spitfire (q.v.) fighter. The first Mitchell design for Supermarine was the S.4 seaplane of 1925, which established a world speed record for its class of 226 mph, and was followed by the S.5, S.6, and S.6B (q.q.v.), which won the Trophy in 1927, 1929, and 1931. The S.6B led directly to the Spitfire, which was Mitchell's last design before he died at the early age of forty-two.

Mitchell, Major-General William: in the middle of a brilliant career in the United States Army, William Mitchell persuaded Congress to allocate funds for military aircraft in 1915. He later became second-in-command of the U.S. Army Air Corps and was responsible for the formation of a fleet of Martin MB-2 heavy bombers. After successful bombing exercises against redundant battleships, he declared that these proved the overwhelming superiority of air power – a statement which brought him into conflict with his superiors and to an eventual court martial. After his death, he was reinstated with the rank of Major-General.

Mitsubishi: in common with the other major aircraft manufacturers in Japan between the wars, Mitsubishi was an industrial group which entered aviation during the early 1920s, and initially undertook licence production of European designs. One of the first Japanese-designed aircraft built by the company was the Type 10 carrier-borne fighter for the Japanese Navy Air Force in the mid-1920s, and this was followed by licence-built Blackburn and Junkers designs in the late 1920s and early 1930s. A number of other fighters for both the Japanese Navy Air Force and the Japanese Army Air Force preceded the famous Mitsubishi design, the A6M2 Type 0 carrier-borne fighter, the 'Zero', built at the outbreak of World War II. Mitsubishi also produced the Ki 67 Type 3 bomber for the Japanese Army Air Force during

the war, and the D5Y1 Kamikaze suicide aircraft for the Navy.

After the end of the war, aircraft manufacture was stopped until 1952 and Japan's limited re-armament. Initially, most aircraft were licence-built once again, including Sikorsky helicopters, but more recently the Mu-2 (q.v.) communications and light transport aircraft has been in production. An advanced jet trainer project, the T.2, is under development.

mobility: a strategic concept based upon the ability to reinforce troops in any particular theatre of war at short notice, as well as moving away from fixed bases as far as possible, mobility entails the provision of a large number of transport aircraft (usually heavy freighters for equipment and jet transports for troops), and amphibious warfare vessels such as commando carriers and assault ships. Equipment has to be air-portable, meaning that it should be stripped of any unnecessary weight and fit into the available space on board the transport aircraft in service. Tactical or battlefield mobility is more or less simply based on an adequate supply of helicopters, particularly medium and heavy lift types.

Mohawk, Grumman OV-1: see OV-1 Mohawk, Grumman.

'Mongol', Mikoyan-Gurevich MiG-21UTI: see MiG-21UTI 'Mongol', Mikoyan-Gurevich.

monocoque: the term monocoque means a hollow structure without internal bracing, and in the sense in which it is used in aviation, is usually applied to a fuselage which is hollow and has only the minimum of stiffening. All modern aircraft are of monocoque construction, but the first aircraft in history to be built in such a way was the appropriately named Monocoque Deperdussin (q.v.) of 1912. The design did not become common until well after the end of World War I.

Monocoque Deperdussin: the first monocoque monoplane, the Monocoque Deperdussin of 1912 had a wooden fuselage, and no external bracing for the wings. A high-wing aircraft with a single seat and a single 160 hp Gnome engine giving a maximum speed of 126 mph, it was able to establish a number of speed records. Oddly, wing warping was used, although in many other respects the aircraft was well ahead of its time.

monoplane: an aircraft with only one set of wings in the mainplane is a monoplane – the concept being based on the idea that the wing is continuous through, under or over the fuselage. All aircraft in production today are monoplanes with the sole exception of the Antonov An-2 (q.v.). Early monoplanes included those of Blériot (q.v.) and the Antionette (q.v.) concern, plus a number of individual types such as the Monocoque Deperdussin (q.v.). The concept fell into disrepute at the start of World War I, and monoplanes were not built, during the war or for some time afterwards with a few exceptions, including the products of the Junkers factory. The advantage of the monoplane is its low drag and reduced weight characteristic.

Montgolfier, Joseph (1740–1810) and Étienne (1745–99): the two Montgolfier brothers were papermakers living at Annonay, near Lyons. They had noticed that the smoke of a fire carried charred paper and ash upwards, and, after experiments with paper bags, built a large linen and paper sphere which made an ascent over a large fire in the middle of the market at Annonay on 5 June 1783. Further experiments followed before the two brothers travelled to Paris, where they sent a cock, a duck, and a sheep on the first successful aerial voyage in history. Tethered experiments followed, carrying men and women aloft, before the first man-carrying flight in history was made on 21 November 1783, when two aristocrats, Pilâtre de Rozier (q.v.) and the Marquis d'Arlandes, travelled for five miles over Paris. The balloon used in the man-carrying flight had a charcoal and straw fire in a brazier suspended under it to maintain height.

The hot-air balloon invented by the Montgolfier brothers was named the Montgolfière (q.v.), after its inventors.

Montgolfière: the fashion of the late eighteenth century, notably in France, was to name inventions after the inventor, and the Montgolfière, or hot-air balloon, was so named after the brothers Montgolfier (q.v.) of Annonay, near Lyons. In spite of the hydrogen balloon, or Charlière (q.v.), being superior in so many ways and arriving on the scene within a few days of the Montgolfière, the hot-air balloon remained popular throughout the nineteenth century because of its lower costs, and has now made something of a comeback for publicity and sporting use, with blowlamps replacing the charcoal and straw fires in braziers of the early hot-air balloons.

Mooney Aircraft: the Mooney Aircraft Corporation first came into prominence during the post-World War II period with a range of light aircraft, all of which have been single-engined, although the company has been responsible for such innovations as the first single-engined pressurised aircraft – the former top-of-the-range model, the Mk.22 Mustang. More recently, the company has become a part of Butler Aviation International, trading under the name of Aerostar, and there has been considerable streamlining of the product range. The remaining models are basically developments of the former middle-range Lycoming-engined Mooney M-20 series, renamed the Aerostar Ranger, Chaparral, and Executive.

'Moose', Yakovlev Yak-11: *see* Yak-11 'Moose', Yakovlev.

Morane-Saulnier: formed during the years before World War I, the French Morane Saulnier concern was at first a proponent of the monoplane, with all-moving tail surfaces a characteristic of the marque. A Morane-Saulnier *'monoplan de chasse'* with a 100 hp Rhône rotary engine was, in 1915, the first aeroplane to be fitted with a machine gun firing straight ahead through the propeller disc, the propeller itself being protected by steel deflector blades. Before this, however, the famous Morane-Saulnier Parasol reconnaissance aircraft had evolved into one of the first fighters.

During the inter-war period Morane-Saulnier continued to produce military aircraft for the then French Aviation Militaire, including the M.S.225C fighter of the late

1920s, and the M.S.406 (q.v.) fighter of the late 1930s – one of the better French aircraft in service with the Armée de l'Air in 1939. Production ceased with the fall of France, but restarted after the end of World War II, with the M.S.733 Alcyon basic trainer and liaison aircraft, and during the 1950s the M.S.760 Paris communications aircraft and basic jet trainer for the Armée de l'Air and the Aeronavale. A number of light aircraft, including the Rallye-Club, were put into production during the late 1950s, but production passed to Aerospatiale when Morane-Saulnier went out of business.

Morava: a nationalised Czechoslovak manufacturer. The main product of Morava is the L-200, a light-twin with four seats which uses 210 hp M337 piston-engines and is in service with many East European airlines as an air taxi.

Mosquito, de Havilland D.H.98: *see* D.H.98 Mosquito, de Havilland.

'Moss', Tupolev Tu-114: *see* Tu-114 'Moss', Tupolev.

Moth, de Havilland D.H.60: *see* D.H.60 Moth, de Havilland.

motor: the term 'motor' is sometimes used to refer to a piston engine for an aircraft, although the context is usually one such as 'trimotor', 'engine' being the normal expression.

'Moujik', Sukhoi Su-7UTI: *see* Su-7UTI 'Moujik', Sukhoi.

Moy, Thomas: an Englishman, Thomas Moy built and tested a large model tandem-wing monoplane in 1875. There were no special features in the design, which used a 3 hp steam engine to drive two propellers, and was named the 'Aerial Steamer'. The model did, however, manage to lift itself six inches off the ground under its own power – the first actually to do so without a down-ramp run.

Mozhaiski, Alexander Feodorovitch (1825–1890): the man credited by the Soviet Union with building the first aircraft to fly, Alexander Feodorovitch Mozhaiski in effect only repeated du Temple's (q.v.) effort of 1874 when, in 1884, he built an aircraft which succeeded in travelling through the air for about eighty feet after a down-ramp run. A British-built steam engine drove three propellers in the design, but the force which counted was obviously that of the down-ramp run, since sustained flight was out of the question.

M.R.C.A.: *see* multi-role combat aircraft.

M.S.406, Morane-Saulnier: perhaps the best French fighter in service at the outbreak of World War II, by which time about a thousand were in Armée de l'Air service, the Morane-Saulnier M.S.406 used a single 860 hp piston engine. The aircraft conformed to the then standard fighter shape, with low wing, hump back, and single-seat cockpit.

M.S.B.S.: *see* Mer-Sol Balistique Stratégique.

Mu-2B, Mitsubishi: a seven-seat short take-off utility transport and communications aircraft, the prototype Mitsubishi Mu-2 first flew in September 1963, the production Mu-2B making its first flight in March 1965. Two AiResearch TPE-331 turboprops of 575 hp each provide for a maximum speed of 310 mph and a range of 1,240 miles. The aircraft is in service with the Japanese Ground and Air Self-Defence Forces.

multiplane: the multiplane concept is unusual, since it means an aircraft with several sets of wings, i.e. more than three, for which the term is triplane. One of the most notable multiplane designs was the model built by Horatio Phillips (q.v.).

multiple independently-targeted re-entry vehicle: working on the principle that a number of small nuclear devices are more effective than a single large device because the destructive power does not increase in proportion to the increase in size, the M.I.R.V., or multiple independently-targeted re-entry vehicle was evolved by the United States. Basically, a M.I.R.V. ballistic missile differs from the standard I.C.B.M. only in that the last stage contains a number of warheads instead of one, and each warhead is independently directed to a different target. The American Minuteman (q.v.) 3 can carry a three M.I.R.V.

warhead, while the Poseidon (q.v.) carries ten re-entry vehicles. The Soviet Union lags behind the United States in this field at present, and for this reason the Strategic Arms Limitation Talks (SALT, q.v.) allowed a launcher numerical superiority to the U.S.S.R., although as SALT does not prevent any up-dating of missiles, the Soviet Bloc may not only catch up on but overtake the United States in terms of the number of warheads available.

multi-role combat aircraft: among the problems which face any air force or air arm is that of the limited number of duties which any one aircraft may be expected to perform, coupled with the spares and support problems of operating a number of different aircraft types, and it was partly to solve these problems that the concept of the multi-role combat aircraft was evolved. Few aircraft are really ideal for tackling fighter, interceptor, ground-attack and bomber duties, but the Luftwaffe has attempted this with its Lockheed F-104G (q.v.) Starfighters, and the new Panavia 200 Panther for the R.A.F., Luftwaffe, and Italian Air Force is the aircraft for which the phrase was actually invented. Whether it will work in practice remains to be seen.

Musketeer, Beechcraft: the prototype Beechcraft Musketeer first flew in October 1961, and production started the following year. Basically, the aircraft is a two-to-four-seat private-owner type with a low wing and a single Lycoming piston engine. A number of models are now available, including the Custom four- or six-seat version with a 180 hp Lycoming 0-360-A4G engine; the Super, also with up to six seats, and using a 200 hp Lycoming 10-360-A2B engine; and the Sport, with two or four seats and a 150 hp Lycoming 0-360-E2. piston engine giving a maximum speed of 130 mph and a range of 800 miles.

Mustang, North American F-51: see F-51 Mustang, North American.

Mya-4 'Bison', Myasischev: the only design of any importance by V. M. Myasi-chev known in the West, the Mya-4, NATO code-named 'Bison', first flew in 1954 and has since equipped the U.S.S.R.'s long-range bomber and reconnaissance forces, being basically an equivalent aircraft to the Boeing B-52 (q.v.). Four 19,180 lb thrust Mikulin AM-3D turbojets (mounted in the wings, unlike the B-52s' underwing nacelles) give a maximum speed of 560 mph and a range of about 7000 miles, while a bomb load probably well in excess of 50,000 lb can be carried.

Myasischev, V. M.: a Soviet design bureau head, the only project of any importance by V. M. Myasischev known in the West today is the Mya-4 (q.v.) 'Bison' heavy bomber.

Mystère, Dassault: the first flight of the Mystère aircraft series was during the early 1950s. The latest and most popular version was the Dassault Mystère IVA which first flew in September 1952, and was subsequently put into production for the Armée de l'Air, the Indian Air Force, and the Israel Defence Force/Air Force. A low-wing aircraft with a swept leading edge and a single-seat cockpit, the Mystère IVA uses a single 7,716 lb thrust Hispano-Suiza Verden 350 turbojet for a maximum speed of 695 mph and a range of 600 miles, while a warload of up to two 1,000 lb bombs, rockets, or napalm tanks could be carried on two underwing strongpoints, in addition to fuselage-mounted 30 mm cannon.

Mystère 20/30, Dassault: sometimes known as the Fan Jet Falcon, the Dassault Mystère 20 executive and communications jet first flew in May 1963, and the latest version uses two rear-mounted General Electric CF-700 turbojets of 4,800 lb thrust for a maximum cruising speed of 500 mph and a range of up to 2,000 miles, while up to fourteen passengers may be carried. A development is the Mystère 30, which has been scaled-up to become a forty-two-passenger airliner using two 5,500 lb thrust Avco-Lycoming ALF-502D turbojets for a maximum cruising speed of 512 mph and a range of 2,000 miles.

N

N-: international civil registration index mark for the United States of America.

nacelle: today generally applied to the engine housing, or engine nacelle, the term nacelle correctly refers to any small construction on an aircraft, even including the fuselage, or crew nacelle, on aircraft with a tailplane mounted on outriggers.

Nagasaki: the second city to be devastated by an atom bomb (q.v.), Nagasaki was bombed on 9 August 1945, and this led directly to Japan's surrender in World War II. Some 38,000 persons were killed and 21,000 injured, out of a population of 173,000.

Nakajima: one of the three major industrial groups to enter aircraft production in post-World War I Japan, Nakajima initially devoted its efforts to licence production of British and French aircraft, before building Japanese designs such as the Nakajima Type 5 advanced trainer, the A2N1 carrier-borne fighter and, later, the Type 91 fighter monoplane. During the late 1930s Nakajima produced for the Japanese Navy Air Force the B4Y1 Type 96 carrier-borne fighter and the B5N1 Type 97 carrier-borne strike aircraft, while the Japanese Army Air Force used the Nakajima Ki-27 Type 97 fighter and the Ki-34 Type 97 transport. Wartime aircraft from Nakajima included the J1N1 Type 2 fighter and reconnaissance aircraft for the J.N.A.F., the Ki-44 Type 2 and Ki-84 Type 4 fighters, and the Ki-49 Type 0 bomber for the J.A.A.F. The Kitsuka turbojet-powered and the Ki-115 and Ki-201 Kamikaze suicide aircraft also came from this concern. All production activities ceased with the end of World War II.

N.A.M.C.: *see* Nihon Aeroplane Manufacturing Company.

narrow beam radar: narrow beam radar is used to track attacking aircraft once they have entered the area under surveillance, and this is done by use of a radar beam which oscillates around the target making evasion difficult. Once the radar is locked on to the aircraft, it can be used for guidance by surface-to-air missiles, and other anti-aircraft measures can be related to the radar beam.

NASA: *see* National Aeronautics and Space Administration.

National Aeronautics and Space Administration: the National Aeronautics and Space Administration is best known for its control and operation of the United States space programme, but this state-backed organisation is also responsible for promoting aeronautical research generally, and in recent years has sponsored supercritical wing research and experiments with short, vertical, and quiet take-off airliners.

National Airlines: originating in 1934 as a regional airline in the South-East of the United States, National Airlines now operates extensively throughout the United States, particularly through the Southern States and up the Eastern seaboard, with a fleet of Boeing 747s and 727s, and McDonnell Douglas DC-10s and DC-8s. The most recent development by the airline has been the transatlantic service from Miami to London, operated since 1970.

NATO: the North Atlantic Treaty Organisation, NATO, dates from 1949, and is a multilateral military alliance which has a membership including Belgium, Canada, Denmark, France, West Germany, Greece, Iceland, Italy, Luxembourg, the Netherlands, Norway, Portugal, Turkey, the United Kingdom, and the United States. Greece and Turkey joined in 1952, followed by West Germany in 1955. France is no longer a fully active member, and Canada has been showing a lukewarm attitude to the Alliance for some years now.

Control of NATO is through the North Atlantic Council, which is now based in Brussels, although originally situated in Paris. Subordinate to the Council there are two major Commands, ACLANT (q.v.) for the North Atlantic and SHAPE (q.v.)

for Europe, and under these a number of subsidiary commands and the small ACCHAN for the English Channel area.

nautical mile: a nautical mile, as opposed to a statute mile, is 6,080 feet in length, or slightly more than nine furlongs. Unless speed, range, and distance are specifically stated to be in nautical miles, it must be assumed that statute miles are being used.

Navajo, Piper PA-31: one of the larger aircraft in the Piper range, the PA-31 Navajo first flew in prototype form in September 1964, with deliveries starting in August 1967. Between six and nine seats can be fitted, according to layout, and the two 300 hp Lycoming 10-540-M piston engines give this low-wing monoplane a maximum cruising speed of 213 mph and a range of up to 1,360 miles.

navigation lights: aircraft are fitted with navigation lights on the wing tips and, as with shipping, the colours are green for starboard and red for port. The object of the lights is to give other aircraft an indication of an aircraft's heading.

navigator: alternatively known as the observer, the navigator relieves the pilot of navigation and, in military aircraft, certain other duties such as bomb-aiming or missile control. Civil navigators are sometimes junior pilots, but in any case a navigator's licence is required. U.S. military aircraft usually tend to have two pilots instead of a pilot and a navigator.

Naviplane, Bertin: a twin turboprop-powered passenger and freight hovercraft designed and developed in France by Bertin, the Naviplane is fully amphibious. Few orders have been received for the machine at present.

NC, Curtiss: the Curtiss NC flying-boats first appeared in October 1918, with the first flight of the NC-1. Four of these aircraft, NC-1 to NC-4, made the first attempt on a transatlantic flight in stages between 8 and 31 May 1919, although only one aircraft, NC-4, completed the crossing. A biplane, the NC series used four 400 hp Liberty engines.

N.D.A.C.: the Northern Defence Affairs Committee, N.D.A.C., is a policy committee within the North Atlantic Treaty Organisation (NATO) q.v., and operates through meetings of Ministers of Defence. France, Iceland, and Luxembourg do not participate in the Committee.

Neiva: a Brazilian firm dating from 1950, Neiva initially started by producing gliders, one of the company's first aeroplanes entering production in 1956 following the acquisition of the Paulista CAP-4 high-wing monoplane, which was re-introduced by Neiva as the L-6 and L-7 observation and primary training aircraft for the Brazilian armed forces. An all-metal high-wing monoplane which followed the L-6 series was the C-42 Regente, a four-seat aircraft which first flew in 1961, and this was followed by the T-25 Universal basic trainer for the Brazilian Air Force, which was first flown in 1966.

Neptune, Lockheed P-2: see P-2 Neptune, Lockheed.

neutron kill: in addition to the blast effect, the detonation of nuclear devices against attacking intercontinental ballistic missiles produces a neutron kill effect from the strong radiation emitted, and this neutralises the warhead, or warheads, of the attacking missile. Most A.B.M. devices depend on neutron kill for their effectiveness.

New York Airways Inc.: originally formed in 1949, New York Airways started scheduled helicopter operations in 1952, linking the New York airports with one another and with the centre of New York. Boeing-Vertol 107s were used for several years, but in 1970 the current fleet of four Sikorsky S-61Ls was introduced.

New Zealand National Airways Corporation: formed in 1945, the New Zealand domestic airline, New Zealand National Airways, did not start operations until 1947. Merged into the airline, which was state-owned, were three private-enterprise airlines, Air Travel, Cook Strait Airways, and Union Airways, which had encountered difficulties in maintaining operations due to wartime shortages. Today New Zealand National operates a relatively dense domestic route network serving some twenty-five destinations, and using Boeing 737-200, Vickers Viscount 800, and Fokker F-27 Friendship airliners.

Nieuport, Edouard (1875–1911): Edouard Nieuport, a Frenchman, started aircraft production with a streamlined development of the Blériot monoplanes, and followed this with a number of reconnaissance and experimental types and, during World War I, fighter aircraft. In 1912 a Nieuport was the first aircraft to loop the loop, and in 1916 the Nieuport Nighthawk biplane was one of the first Allied aircraft to be fitted with a propeller-synchronised machine gun. *See also* Nieuport-Macchi.

Nieuport-Macchi: a joint venture which started during World War I, Nieuport-Macchi produced a series of flying-boat fighters, of which the largest was the M.8 biplane, capable of carrying a small bomb load and, in spite of its classification, sometimes used for maritime-reconnaissance.

Nigeria Airways: Nigeria Airways dates from 1958, when it was formed as the West African Airways Corporation (Nigeria) to take over the services operated by the former West African Airways Corporation, an airline formed after World War II with B.O.A.C. (q.v.) assistance to operate services within, between, and to and from the British colonial territories in West Africa. The present title was adopted in 1971. A wholly state-owned airline, Nigeria Airways operates domestically, and to major points in Europe and the Middle East, as well as to New York, with a fleet which includes Boeing 707s and 737-200s, and Fokker F-27 Friendships and one F-28 Fellowship.

Nightingale, McDonnell Douglas DC-9: *see under* DC-9, McDonnell Douglas.

Nihon Aeroplane Manufacturing Company: formed in 1959 by the Japanese Government, with the support of a number of private companies, the Nihon Aeroplane Manufacturing Company's first major project was the N.A.M.C. YS-11, a turbo-prop short-haul airliner and transport aircraft, which has since been followed by the

C-1A, a turbofan-powered S.T.O.L. military transport for the J.A.S.D.F.

Nike-Hercules: a surface-to-air guided missile, the Nike-Hercules was developed to intercept high-altitude bombers, and is now largely obsolescent due to the tendency for bombing operations to be carried out at low level. A few missiles do remain in service with the U.S.A.F., however and these have a range of seventy-five miles, and use a solid-fuel rocket motor and radio-command guidance.

Nimrod, Hawker Siddeley: a maritime-reconnaissance development of the Hawker Siddeley Comet 4 (q.v.) airliner, the Nimrod is the world's first jet-powered maritime-reconnaissance aircraft. Differences between the Nimrod and the Comet include a modified tailplane with a M.A.D. 'stinger' and altered fin, a weapons bay in a double-bubble fuselage effect, and the substitution of Rolls-Royce Spey turbofans for the Avon turbojets of the Comet. First flight of a prototype, developed from a Comet, was in May 1967, while the first production aircraft flew in June 1968, and deliveries to the R.A.F. started in late 1969. The four 11,500 lb thrust Rolls-Royce Spey Mk.250 turbofans give a maximum speed of about 600 mph and a range of about 4,000 miles to the aircraft, which can carry a full load of anti-submarine equipment and weapons.

9H-: international civil registration index mark for Malta.

9J-: international civil registration index mark for Zambia.

9K-: international civil registration index mark for Kuwait.

9L-: international civil registration index mark for Sierra Leone.

9M-: international civil registration index mark for Malaysia.

9N-: international civil registration index mark for Nepal.

9Q-: international civil registration index mark for Ghana.

9U-: international civil registration index mark for Burundi.

9V-: international civil registration index mark for Singapore.

9XR-: international civil registration index mark for Ruanda.

9Y-: international civil registration index mark for Trinidad and Tobago.

NORAD: the North American Air Defence Command (NORAD) is a joint United States and Canadian organisation which includes radar and communications networks, interceptor squadrons, and surface-to-air missile units.

Noratlas, Nord 2501: *see* Nord 2501 Noratlas.

Nord Aviation: dating from 1936, when it was formed as a nationalised concern, Nord Aviation has now been merged with Sud Aviation to form Aerospatiale (q.v.). Since the end of World War II, Nord has produced the Nord 2501 (q.v.) Noratlas transport and the Nord 262 light transport, and has been the French partner in the Franco-German C-160 Transall military transport project. In addition there have been a number of smaller aircraft, sub-contract work on other projects, and guided weapons and sounding rocket development and manufacture.

Nord 262: originally a Max Holste design, the prototype Nord 262 first flew in December 1962, although before this an unpressurised and piston-engined aircraft based on the Max Holste design had made several flights. Deliveries of the 262 started in 1964. Two 1,065 shp Turbomeca Bastan VIC turboprops give this high-wing mono-plane a maximum speed of 235 mph and a range of 700 miles, while up to thirty passengers, or cargo, can be carried.

Nord 2501 Noratlas: first flown in prototype form in September 1949, the Nord 2501 Noratlas, a high-wing, twin-boom, military transport, remains in service in Portugal and Israel, and has been built in Germany as well as France. Two 2,040 hp SNECMA-built Bristol Hercules 738 radial engines give a maximum speed of 272 mph and a range of 1,550 miles with a 4½ ton payload.

North American Air Defence Command: *see* NORAD.

North American Rockwell: North American Rockwell is the result of a merger between North American Aviation, which dates from 1934, and Rockwell Standard, whose aerospace interests included guided weapons production and the Aero Commander (q.v.) range of light and executive aircraft. Many of the most notable North American designs were produced during World War II, including the B-25 (q.v.) Mitchell medium bomber, the F-51 (q.v.) Mustang fighter, and the T-6 (q.v.) Texan basic trainer. After the war these were followed by the T-28 (q.v.) Trojan advanced-trainer and the F-86 (q.v.) Sabre jet fighter, which proved to be highly successful in the Korean War, and was built under licence in Australia, Canada, Japan, and Italy.

During the 1950s North American produced a supersonic development of the Sabre, the F-100 Super Sabre (q.v.), along with the T-2 (q.v.) Buckeye jet trainer and T-39 (q.v.) Sabreliner navigational trainer, and followed these during the 1960s with the A-5 (q.v.) Vigilante carrier-borne nuclear bomber and reconnaissance aircraft. A number of experimental aircraft were also built after the war, including the X-15 (q.v.) and the XB-70A Valkyrie high-speed research aircraft. More recently, the OV-10A (q.v.) Bronco COIN and general-purpose aircraft has been put into production, while development work continues on the B-1 (q.v.) bomber for the U.S.A.F. and on the N.A.S.A. space shuttle, for which North American is the main contractor.

The latest North American project is a V./S.T.O.L. fighter for the U.S. Navy's sea control (*see under* Aircraft Carriers, United States) ships, and this aircraft is scheduled to fly in 1974, using a Pratt and Whitney F401 engine modified for vectored thrust. A Mach 2·0 performance is specified.

North Atlantic Treaty Organisation: *see* NATO

North Pole: the North Pole, in common with the South Pole, has had an attraction for explorers and aviators alike. The attempt to fly over the North Pole in a balloon, by Andree and his companions in 1897, met with disaster, but the American Lieutenant-Commander Richard Evelyn Byrd flew over the North Pole from Spitzbergen on 9 May 1926 in a Fokker Trimotor. He repeated this performance over the South Pole (q.v.) three years later. Regular commercial flights over the North Pole were started in 1954 by S.A.S. (q.v.), using Douglas DC-6 airliners.

Northrop: the American Northrop Corporation's first significant aircraft was the Alpha, a six-passenger, single-engined monoplane airliner of the late 1920s, and this was followed in 1933 by the Gamma mailplane, and in 1935 by another development in the same series, the A-17, used on a variety of duties by the British, American, and Canadian armed forces during the late 1930s and early 1940s. Northrop's most important aircraft of World War II was, however, the P-61 Black Widow night fighter or, in its reconnaissance version, the F-15 Reporter.

After the war Northrop started its famous series of experiments with tailless aircraft, while also working on the F-89 Scorpion fighter. The late 1950s and early 1960s saw the emergence of the highly successful F-5A/B (q.v.) lightweight and low-cost jet fighter and the related T-38 (q.v.) Talon advanced jet trainer. A new development of the F-5 series is the F-5E. Lifting body (q.v.) research has also been undertaken for NASA during the late 1960s and early 1970s, using the Northrop HL-10 and HL-14 lifting bodies. The Northrop A-9A has been a contender for the U.S.A.F.'s ground-attack A-X programme competition.

Northwest Orient Airlines–Northwest Airlines Inc.: dating from 1926, when it was formed as Northwest Airways, the airline's present title of Northwest Airlines Incorporated was taken in 1934, and its aircraft and timetables carry the fleet name of Northwest Orient Airways. After World War II the airline expanded from its regional services to start operating trans-United States services and the first international services, to the Far East. Currently it is operating a route network which includes coast-to-coast services in the northern part of the United States, services across the Pacific and north to Canada and Alaska, and a service to Florida. It operates a fleet which includes Boeing 747s, 707s, 720s, and 727s, and McDonnell Douglas DC-10 airbuses.

notam: a notice of information to airmen, issued by the appropriate air traffic control or navigational authorities.

nuclear weapons: *see* atom bomb and hydrogen bomb.

Nungesser, Captain Charles: a rancher in South America, Charles Nungesser returned to France on the outbreak of World War I in 1914, and joined a cavalry regiment. He learned to fly in 1915 and, after a spell in a bomber squadron, became a fighter pilot, eventually becoming France's third most successful pilot in the war, with a total of forty-three confirmed victories. After the war he became a stunt pilot, but disappeared during a transatlantic flight in May 1927.

O

O-: U.S. Army and U.S.A.F. designation for observation and tactical reconnaissance aircraft.

O-1 Bird Dog, Cessna: Cessna developed the 0-1 Bird Dog light aircraft for liaison and reconnaissance duties after winning an U.S. Army design competition in 1950. The aircraft has since seen extensive service with the U.S. Army, U.S.M.C., and U.S.A.F., as well as with NATO, SEATO, and Latin American air arms, and has been built under licence in Japan by the Fuji concern. A single 213 hp Continental 0-470-11 piston engine powers the three-seat aircraft and provides a maximum speed of 115 mph and a range of 530 miles.

O-2 Super Skymaster, Cessna: a military development of the Cessna 337 (q.v.) Super Skymaster light aircraft, the Cessna 0-2 is unique in having a twin-boom layout, with one engine in front of the cabin driving a tractor propeller in the conventional manner, while the second engine is behind the cabin and drives a pusher propeller: the idea behind this arrangement being to give the simplicity of handling of a single-engined aircraft to a light twin. Two 210 hp Continental TS10-360-A piston engines give a maximum speed of 240 mph and a range of up to 1,500 miles, while up to six persons can be accommodated. A French-built version by Reims Aviation (q.v.), known as the Milirole, is equipped with strongpoints for weapons and can be used for counter-insurgency operations, as well as on the forward air control, liaison, and reconnaissance duties of the U.S.A.F.'s version.

OB-: international civil registration index mark for Peru.

OD-: international civil registration index mark for the Lebanon.

OE-: international civil registration index mark for Austria.

OH-: international civil registration index mark for Finland.

OH-6A Cayuse, Hughes: see Hughes 500.

OH-43D Husky, Kaman: see H-43 Husky, Kaman.

OH-58 Kiowa, Bell: see Bell 206 Jet Ranger.

Ohain, Dr Hans von: one of the pioneers of the turbojet engine, and a rival of Sir Frank Whittle (q.v.), Dr Hans von Ohain's work in Germany during the 1930s led to the development of a centrifugal flow turbojet which was used in the Heinkel He.178 (q.v.), which made a number of successful flights, starting in August 1939. Ohain's work also led to the wartime development of the Messerschmitt Me. 262 (q.v.) and the Arado Ar. 234 (q.v.), although during the early part of his experiments he had largely been backed by Ernst Heinkel.

OK-: international civil registration index mark for Czechoslovakia.

Olympic Airways: formed in 1957 as the successor to T.A.E. National Greek Airlines, which had been formed in 1951 by the merger of T.A.E., Hellos, and Aero Metaforai Ellados, Olympic Airways is the property of the millionaire shipowner, Aristotle Onassis. Douglas DC-3s predominated in the Olympic fleet at first, but these were soon supplemented by Douglas DC-6Bs and de Havilland Comet 4B jet airliners. Currently the airline operates a fleet of Boeing 707-384s, 720Bs, and 727s, with N.A.M.C. YS-11 and Short Skyvan aircraft, plus various light aircraft. Trials have been carried out with two Yakovlev Yak-40 airliners – the first European airline outside the Communist Bloc to try them – and this aircraft may appear in the fleet in the future. The route network includes domestic destinations, Europe, the Middle East, Australia, and the United States.

1½-Strutter, Sopwith: so-called because of the arrangement of the wing bracing struts, the Sopwith 1½-Strutter biplane was the first British reconnaissance fighter to be fitted with a rear-gun ring for the observer and fixed forward-firing guns for the pilot. A 130 hp Clerget rotary engine gave the aircraft a maximum speed of 100 mph. Entry into service was in 1916.

One-Eleven, British Aircraft Corporation: originating from a Hunting Percival design (*see* Percival Aircraft) prior to the formation of the British Aircraft Corporation, the One-Eleven design was enlarged and developed by B.A.C. before construction of the prototype began in April 1961. The first flight of the prototype was in August 1963, and the first production aircraft flew in December of the same year, with deliveries to airlines starting early in 1965. The original production version was the One-Eleven Series 200. This was followed by the slightly larger Series 300 and Series 400 aircraft, and eventually by the stretched Series 500, originally designed specifically for British European Airways, but also ordered by other airlines, and the rough-field Series 475. In terms of the value of sales, the One-Eleven is the most successful British airliner ever. Two 11,400 lb thrust Rolls-Royce Spey Mk.511 turbofans mounted in the rear of the aircraft give the Series 400 a maximum cruising speed of 548 mph and a range of up to 1,240 miles in the airliner version – although the corporate version has a New York-to-Paris capability with ten passengers – while between eighty and one hundred passengers may be carried. The Series 500 with uprated engines can carry up to 120 passengers.

OO-: international civil registration index mark for Belgium.

orbital bomb: an orbital bomb is a nuclear warhead which can be placed in a low orbit until required. The use of such weapons is prohibited by the Outer Space Treaty (q.v.).

orbiting solar observatory: the orbiting solar observatory is basically the old idea of a space station, manned and in permanent orbit of the Earth, its main objects including observation of the solar system free from the distortions of the atmosphere, and observation of the Earth's surface for military, agricultural and meteorological reasons. The United States and the Soviet

Union are both pressing ahead with this type of development. The United States first used an Apollo launcher, but the Soviet Union is also intending to use the space station as a jumping-off point for journeys to the moon and to other planets.

organic airpower: organic airpower is the modern term for the air elements which are attached to and a part of surface forces: notably an army air corps or a naval air arm.

Orion, Lockheed: the first Lockheed aircraft to be fitted with a fully retractable undercarriage, the single-engined Orion low-wing monoplane first flew in 1930, and was noted as being an exceptionally clean design for the period. Maximum speed was about 225 mph, while the cruising speed was about 25 mph slower. Six passengers could be carried, and the aircraft enjoyed a considerable success not only in the United States, but also in Europe with airlines such as Swissair.

Orion, Lockheed P-3: *see* P-3 Orion, Lockheed.

ornithopter: the term ornithopter is restricted to aircraft propelled by flapping wings, and there have been no real flights using this method, which was discredited long before the Wright brothers made their first flight. The power sources for ornithopter flight were many: some attempts used manpower, but others used clockwork, steam engines, gas, internal combustion engines, and even gun cartridges.

Osprey, Hawker: a development of the Hawker Hart (q.v.) series, the Osprey was a carrier-borne fleet spotter and reconnaissance aircraft, although a number of floatplanes were also built and operated from the Royal Navy's battleships and cruisers. The Osprey retained the Hart powerplant, biplane layout, and tandem twin-seating.

O.S.T.: *see* Outer Space Treaty.

Otter, de Havilland Canada DHC-3: *see* DHC-3 Otter, de Havilland Canada.

Ouragan, Dassault M.D.450: one of the first French-designed turbojet aircraft, the Dassault M.D.450 Ouragan first flew in July 1949, and entered Armée de l'Air service shortly afterwards, many of the aircraft passing secondhand to India and Israel. A low-wing fighter-bomber, the Ouragan used a single Hispano-Suiza-built Rolls-Royce Nene turbojet of 5,070 lb thrust for a maximum speed of 584 mph, and could carry up to 2,200 lb of bombs and rockets in addition to the four 20 mm cannon fitted as standard.

Outer Space Treaty: the Treaty on the Exploration and Use of Outer Space, or Outer Space Treaty, was signed by the United Kingdom, the United States, and the Soviet Union in 1967. With a few exceptions, such as Communist China, the rest of the world followed. The Treaty prohibits the deployment and use of weapons in outer space, including such weapons as orbital bombs, but does not exclude the use of weapons which make only a part-orbit, such as ballistic missiles.

outrigger: as with a canoe, an aircraft uses outriggers to support features such as the tailplane in any aircraft in which these are not a part of the actual fuselage. An alternative name for an outrigger on an aircraft is boom, as in 'twin-boom configuration'.

OV-1 Mohawk, Grumman: primarily an observation and battlefield surveillance aircraft, since proposals for armed versions have not been taken up at the present time, the Grumman OV-1 Mohawk has been in production for the United States Army since 1959 in three main versions: the photographic OV-1A, the side-looking-radar equipped OV-1B, and the OV-1C with infra-red mapping equipment. A high-wing, side-by-side twin-seat aircraft with a triple fin, the OV-1 uses two 1,005 shp Lycoming T53-L-3 turboprops for a maximum speed of 317 mph and a range of 1,680 miles.

OV-10A Bronco, North American: the first aircraft to be specifically designed for counter-insurgency duties, the North American OV-10A Bronco first flew in prototype form in July 1965. A twin-boom, high-wing aircraft with tandem twin-seating, two 715 shp Garrett AiResearch T76 turboprops give it a maximum speed of 305 mph and a range of up to 1,400 miles, while 3,200 lb of ordnance can be carried on underwing, under-fuselage and fuselage-side strongpoints. Luftwaffe OV-10A's have a single General Electric turbojet to boost maximum speed to 350 mph for target-towing duties. In service with the U.S. Army.

overshoot: the term 'overshoot' usually refers either to an aborted landing with return to normal flight, or to a landing run ending beyond the length of the runway or landing strip.

Overstrand, Boulton-Paul: a development of the Boulton-Paul Sidestrand medium bomber, the Overstrand retained the same layout and biplane configuration when it first appeared in 1934, but with the addition of an enclosed cockpit for the pilot and a power-operated gun-turret in the nose.

OY-: international civil registration index mark for Denmark.

P

P-: U.S.A.A.C. and U.S.A.A.F. designation for pursuit, or fighter, aircraft up to 1947, when the F- (q.v.) fighter designation was adopted.

P-: United States Navy designation for maritime-reconnaissance aircraft.

P-2 Neptune, Lockheed: a maritime-reconnaissance aircraft designed for the United States Navy during the closing stages of World War II, the Lockheed P-2 Neptune first flew in May 1945, and remained in production for more than twenty years. A number of versions were produced, and the aircraft was also built under licence in Japan for the Japanese Maritime Self-Defence Force by Kawasaki, which has also developed a stretched and turboprop-powered version, the P-2J. The standard aircraft, many of which are still in service, uses two 3,500 hp Wright R-3350-32W piston engines and two 3,400 lb thrust Westinghouse J34 turbojets for a maximum speed of 356 mph and a range of up to 3,685 miles. It can carry up to 8000 lb of bombs, mines, depth charges, or torpedoes. The P-2J uses two 2,850 shp General Electric T64-IHI-10 turboprops and two 3,085 lb thrust turbojets.

P-3 Orion, Lockheed: the successor to the P-2 Neptune (q.v.), the Lockheed P-3 Orion first flew in prototype form as a development of the Electra airliner in August 1958, a further prototype flying in November 1959, and the first production aircraft in April 1961. Deliveries to the United States Navy of the P-3A started in August 1962. The U.S. Navy has also received the P-3B, with up-rated engines, and the P-3C, with augmented avionics. The P-3B is the version used by the Royal Australian Air Force, the Royal New Zealand Air Force, and the Royal Nor-

wegian Air Force, and uses four 4,910 shp Allison T56-A-14 turboprops for a maximum speed of 476 mph and a range of up to 4,800 miles, while a full range of weapons can be carried in the bomb-bay and on underwing strongpoints.

P-26, Boeing: the U.S.A.A.C.'s first all-metal fighter, the Boeing P-26 low-wing monoplane first appeared in 1932, and was powered by a single 550 hp Pratt and Whitney R-1340-21 engine, giving a maximum speed of 235 mph. A number of speed and altitude records were established by the aircraft, which had a single-seat open cockpit and a non-retractable undercarriage.

P-38 Lightning, Lockheed: one of the best fighter aircraft of the war, the Lockheed Lightning first appeared in 1941 and was intended for long-range escort duties for bombers over Germany and for the Pacific theatre of operations; in the latter area it accounted for more Japanese aircraft than any other single Allied aircraft during World War II. More than 9,000 Lightnings were built. These mid-wing, twin-boom monoplanes used either 1,150 hp or 1,475 hp Allison engines, and were used by the U.S.A.A.F. and the R.A.F.

P-39 Airacobra, Bell: one of the mainstays of the United States Army Air Force's fighter power on America's entry into World War II, the Bell P-39 Airacobra was notable for a number of unusual features, including the positioning of the 1,100 hp Allison engine behind the cockpit, although the propeller remained in the usual position in front of the cockpit, with a cannon fitted into the middle of the propeller boss. The aircraft was one of the first to have a tricycle undercarriage. A low-wing monoplane designed as a fighter, the P-39 was most noted as a ground-attack and anti-tank aircraft.

P.149, Piaggio: originally developed as a four-seat touring and training version of the earlier Piaggio P.148, the P.149 was first flown in prototype form in June 1953, but did not enter production until ordered by the Luftwaffe. Deliveries started in 1957, production being undertaken by both

P.166, Piaggio

Piaggio and Focke-Wulf. Some ex-Luft-waffe trainers have been sold to Nigeria, and another P.149 operator is the Austrian Air Force. A single 270 hp Lycoming GO-480 piston engine gives a maximum speed of 192 mph and a range of 680 miles.

P.166, Piaggio: a light transport and com-munications landplane developed from the earlier Piaggio P.136 amphibian, the P.166 and the military P.166M have been in production for some years for a number of civil and military users, including the Italian and South African Air Forces. An unusual feature of this high-wing mono-plane is the use of pusher propellers. Two 380 hp Piaggio-Lycoming IGSO-540-A1C piston engines give a maximum speed of 227 mph and a range of 1,200 miles, while up to ten passengers can be carried.

P.1127 Kestrel, Hawker Siddeley: *see under* Harrier, Hawker Siddeley.

Pacific Security Treaty: more usually known as ANZUS (q.v.).

Pakistan International Airlines: the Pakis-tan International Airlines Corporation dates from 1955, when it was formed by the Pakistan Government to take over the operations of Orient Airways, which dated from 1951 and operated services across India between the two halves of Pakistan. Initially, P.I.A. operated a fleet of Douglas DC-3 Dakotas and Lockheed Super Con-stellations, but turboprop Vickers Viscount airliners were introduced in 1959, and soon followed by Fokker F-27 Friendships. Jet operations with a leased Pan American Boeing 707 started in 1960, a service to New York being inaugurated in 1961. P.I.A. later obtained its own jet aircraft, including a fleet of Hawker Siddeley Tri-dent IEs, which were operated during the late 1960s before being sold to Communist China. For many years, an extensive Si-korsky S-61N helicopter network was operated in East Pakistan, but fixed-wing aircraft took over in 1966. The airline was the first non-Communist airline of any importance to operate to Canton and Shanghai.

The size of P.I.A.'s domestic route net-work was cut back drastically when East Pakistan broke away after the war of December 1971 between Pakistan and India, and some cuts in international ser-vices have also had to be made as a result. Currently, the airline operates to a number of European and Middle Eastern destina-tions, and to London and New York, as well as within Pakistan itself, with a fleet of Boeing 707 and 720 airliners, Fokker F-27 Friendships, and de Havilland Canada DHC-6 Twin Otters.

Pan American World Airways Inc.–Pan Am: originally formed in 1927 by Juan Trippe, chairman until 1968, as Pan American Airways, with a fleet of two Fokker F-VII tri-motors operating a service between Florida and Cuba, Pan American World Airways is today one of the world's major airlines. The airline expanded rapidly during its early years, with Sikorsky amphibians operating services to South America by 1928; and in 1929 Pan American–Grace Airways (PANAGRA) was formed with the W. R. Grace Com-pany to operate services to South America. Pan American sold its 50 per cent interest in PANAGRA to Braniff (q.v.) in 1966. The mid and late 1930s saw Pan American operating trial services across the Pacific in 1935 with a Martin M-130 flying-boat named China Clipper, and the North Atlantic in 1939 with a Boeing 314 flying-boat named Dixie Clipper. The North Atlantic services were operated with Imperial Airways.

Although operations were restricted by World War II, Pan American was able to acquire American Overseas Airlines in 1950, to introduce the world's first 'Round-the-World' service in 1947, using a Lock-heed Constellation, and in 1955 to place the first order for American jet airliners, Boeing 707s, which entered service on the North Atlantic in late 1958. The airline was also the first to operate the Boeing 747 'jumbo' jet airliner.

Pan Am currently operates an extensive international network, the only significant blanks being in Latin America, but the airline does not operate domestic services within the United States. A German domestic network to West Berlin is opera-ted, as is a Caribbean network, and the airline has an interest in Avianca (q.v.) and Ariana (q.v.), while also acting as

U.S. agents for the Dassault Mystère business jet, and owning the Intercontinental Hotel chain. A substantial part of the airline's work includes military charters. The current fleet includes Boeing 747, 707, 727, and 720 airliners.

Panavia: Panavia Aircraft is a German-registered company in which the British Aircraft Corporation (q.v.) and Messerschmitt-Bölkow-Blohm (q.v.) are the major shareholders with 42½ per cent each, Aeritalia (q.v.) being the minority shareholder with 15 per cent. The company is responsible for the design, development, and production of the Panavia 200 (q.v.) Panther.

Panavia 200 Panther: the Anglo-German-Italian multi-role combat aircraft (M.R.C.A.), the Panavia 200 Panther is a Starfighter, Buccaneer and Phantom (in the bombing role) replacement due to enter service during the mid- and late-1970s. Production is being shared amongst the shareholders of Panavia (q.v.), design-leadership on the air frame being vested in Messerschmitt-Bölkow-Blohm. The RB.199 triple-spool turbofans are being produced by the Turbo-Union consortium, in which Rolls-Royce is the dominant member and has engine design-leadership. A twin tandem-seat high-wing monoplane with variable-geometry wings, the Panther will have a maximum speed of Mach 2·0 and an operational radius of up to 1,000 miles. The aircraft is designed for quick convertibility between the reconnaissance, interceptor, interdictor, maritime attack and close support roles. The R.A.F. will receive 385, the Luftwaffe 400, and the Aeronautica Militaire 100 aircraft.

Panther, Panavia 200: *see* Panavia 200 Panther.

P.A.R.: *see* perimeter acquisition radar.

parachute: although parachutes are not really necessary for balloonists, because the balloon itself will often act as a parachute, the parachute and the balloon came into existence at about the same time. The first successful parachute descent was by a Frenchman, Sebastian Lenormand, who jumped from a tower in Montpellier in 1783. The canopy of his parachute had bracing. The parachute moved a step forward in 1797, when another Frenchman, André Jacques Garnerin (q.v.), made a descent of 3,000 feet from a balloon; in this case the parachutist was suspended below the canopy in a car, not unlike that of a balloon.

The first parachute descent from an aeroplane was made in the United States in March 1912 by Captain Albert Berry, but the credit of developing the modern parachute must go to another American, Leslie Leroy Irvin (q.v.), who made a descent in April 1919 from an aircraft at Dayton, Ohio. In later years the Irving Parachute Company became one of the leaders in the development and production of parachutes. By the start of World War II the use of paratroops (q.v.) had been pioneered.

Today the parachute is an important feature still in the escape of aircrew from military aircraft, since the parachute is essential after ejection. There is increasing interest in the parachute for sport in its own right. Supplies can still be dropped by parachute; indeed many modern armies have at least a part of their equipment designed for parachute delivery. Paratroops however, are perhaps a little less important than they were, with the advent of the helicopter for assault purposes.

Parasol, Morane-Saulnier: the Morane-Saulnier Type L, or Parasol, monoplane first flew in 1913. During World War I the aircraft was one of the first French reconnaissance types, and later versions were armed. A single 80 hp Gnome or Rhône rotary engine powered the Parasol, which had a high wing and tandem twin seats.

paratroops: airborne troops dropped by parachute are known as paratroops, and are to be found in most modern armies. The concept of paratroops was largely evolved by Germany and the Soviet Union during the 1930s, and Germany in particular made extensive use of paratroops during the period immediately before the outbreak of World War II, and during the war, the largest paratroop assault being that on the island of Crete in Operation Mercury on 21 May 1941, with some 10,000 paratroops involved, excluding troops landed by glider or by transport aircraft.

Since the war the paratrooper has become less important, due to the advent of ever faster and heavier helicopters, which have the advantage of not leaving the men hanging in the air as easy targets for defenders, and also of requiring relatively little specialised training of troops.

Parnall: a small British aircraft manufacturer, Parnall existed between the two World Wars, when the company produced a number of interesting aircraft, including the Pixie, which won the British Light Aircraft Trials of 1923, and a larger development with a more powerful engine and folding wings, the Pixie III. Most interesting of all was the Peto of 1927, a biplane fitted with floats and folding wings which had the distinction of being the world's first truly submersible aircraft. The Peto was operated from the Royal Navy's 'M' Class submarines, which were fitted with a watertight hangar in front of the conning tower. The experiment was not particularly successful.

payload: the maximum revenue-earning weight which can be carried by an aircraft.

PC-6 Porter/Turbo-Porter, Pilatus: a S.T.O.L. transport in production in Switzerland, the Pilatus PC-6 is also built under licence in the United States by Fairchild Hiller. A high-wing, single-engined monoplane, the PC-6 series can carry up to ten persons. A single 340 hp Lycoming GSO-480-B1A6 or 350 hp IGO-540-A1A piston engine powers the PC-6, while the PC-6A Turbo-Porter uses a 523 shp Turbomeca Astazou IIE/G turboprop, and the PC-6B the version built in the U.S.A., uses a 550 shp United Aircraft PT6A-6 turboprop. The PC-6A has a maximum speed of 175 mph and a range of 620 miles.

PD-808, Piaggio-Douglas: although basically a Douglas design, the design work on the Piaggio-Douglas PD-808 was completed in Italy by Piaggio, which developed the aircraft and put it into production. Intended as an executive aircraft and navigational trainer, the PD-808 first flew in prototype form in August 1964, and has

since entered service with the Aeronautica Militaire. Two rear-mounted 3,350 lb thrust Bristol Siddeley Viper 526 turbojets provide for a maximum speed of 546 mph and a range of up to 1,600 miles, while between four and eight passengers can be carried in addition to the crew of two.

Pe-2, Petlyakov: at a time when Soviet aircraft design standards lagged far behind those of the West, the Petlyakov Pe-2 light bomber was noted for being an exceptional design. Also used on reconnaissance and ground-attack duties, the twin-engined Pe-2 remained in Soviet service throughout World War II.

Pe-8, Petlyakov: originating as a Tupolev (q.v.) design, the TB-7, the alternative Petlyakov designation of Pe-8 was chosen after Tupolev was prevented from finishing his work on this aircraft before being imprisoned during one of the Stalin Régime's political purges during the 1930s. The Pe-8 was the only four-engined heavy bomber to be produced by the Soviet Union during World War II. A mid-wing monoplane, the Pe-8 could carry up to 8,800 lb of bombs.

Pénaud, Alphonse (1850–80): prevented from joining the French Navy because of a hip disease, Alphonse Pénaud concentrated on building model flying machines, first coming into prominence in 1870. A leading advocate of twisted rubber propulsion, Pénaud's first significant design was for a model helicopter in 1870, while in 1871 he designed and built his 'plenophore', a rubber-motor monoplane which has the distinction of being the first inherently stable aeroplane. A strong supporter of Sir George Cayley's teachings, Pénaud himself did much work on stability, and proved the soundness of placing a mainplane in front of a stabilising tailplane.

Pénaud's most famous design, prepared with the help of his assistant, Grauchet, appeared in 1876, although it was never built. A monoplane amphibian, far ahead of its time, the Pénaud aeroplane would have had two tractor propellers, mainplane and tailplane, the latter with a vertical fin and rudder, a joystick for vertical and lateral control, and a cockpit with a glass canopy.

Unfortunately, Pénaud committed suicide before the aircraft could be built.

penetration aids: the term 'penetration aids' refers to devices which assist an intruding missile or aircraft to reach its target without detection, and include the use of chaff to counter radar by producing false images, and the production of false images by using balloons or decoys. Other methods include jamming of radar by electronics countermeasures (q.v.) or radiation, or saturation by the use of multiple re-entry vehicles. A last-ditch measure is the detonation of a warhead once interception is inevitable.

Percival Aircraft: a British aircraft manufacturer, Percival Aircraft was acquired by the Hunting Group in 1954, becoming Hunting Percival until 1957, when the title Hunting Aircraft was adopted. In 1960 it was one of the companies amalgamated to produce the British Aircraft Corporation. The more notable products include the Provost basic trainer, and its jet development, the Jet Provost, which continues in service and production today as the B.A.C. 145/167 (q.v.) basic jet trainer and counter-insurgency aircraft. Communications, light transport, and navigational training aircraft were the Prince and Sea Prince twin-engined high-wing monoplanes. The B.A.C. One-Eleven (q.v.) jet airliner was developed from a Hunting Percival design.

performance and weather minima: performance and weather minima generally relates to the worst conditions and lowest performance in which operations can be permitted (performance is dependent both on weather and on the altitude of an airfield).

perimeter acquisition radar: a co-ordinated unit comprising radar, computer, and communications equipment, perimeter acquisition radar is intended mainly as an anti-missile system, and amongst the applications can be included the United States Safeguard anti-ballistic missile system. A feature of P.A.R. is the use of phased array radar (q.v.) which enables the equipment to be fixed and, by using numerous radar beams, to track more than one intruder at once, with the computer assisting in predicting the likely targets. I.C.B.M.s can be picked up by P.A.R. at about 1,000 miles from their target.

Pershing: the Pershing surface-to-surface tactical missile has been in service for a number of years, although the Pershing 1A, with a number of detail improvements, did not enter service until late 1969 with the United States and Federal German Armies. A Martin Marietta product, the Pershing 1A is about thirty-five feet long and has a range of about 450 miles, using a two-stage solid-fuel rocket, with an inertial guidance system and a nuclear warhead.

Petlyakov, Vladimir: assistant to Andrei Tupolev (q.v.), Vladimir Petlyakov enjoyed a brief period of fame while his superior was imprisoned during the Stalin régime's purges. A Tupolev design, the TB-7, was taken over by Petlyakov and appeared as the Pe-8 (q.v.), Russia's only four-engined heavy bomber during World War II. In addition, Petlyakov produced his own Pe-2 (q.v.) light bomber design.

PH-: international civil registration index mark for the Netherlands.

Phantom I, McDonnell: the first operational jet fighter for the United States Navy's aircraft carriers, the McDonnell FH-1 Phantom (not to be confused with the much later F-4 Phantom II) first flew in prototype form in January 1945, before making its first deck landing on the U.S.S. *Franklin D. Roosevelt* in July 1946.

Phantom II, McDonnell Douglas F-4: *see* F-4 Phantom II, McDonnell Douglas.

phased-array radar: phased-array radar differs from conventional radar in that, instead of using a single beam which can be moved mechanically, a very large number of beams is provided by fixed equipment, enabling several targets to be tracked at once. The larger U.S. aircraft carriers and the guided missile cruiser U.S.S. *Long Beach* use phased-array radar. *See also* perimeter acquisition radar.

Philippine Airlines–P.A.L.: Philippine Airlines started operations early in 1946, and for many years operated an extensive international network in addition to its domestic services. The international routes had to be dropped in 1954, leaving the airline as a purely domestic operator for the next eight years. International services were re-established during the 1960s, however, initially to the Far East and the United States, and expanded until at present the airline operates to many points in Europe as well. Mainly a private-enterprise undertaking, with a 25 per cent state shareholding, P.A.L. today operates a fleet of McDonnell Douglas DC-8, B.A.C. One-Eleven, and Hawker Siddeley 748 airliners.

Phillips, Horatio (1845-1924): an Englishman, Horatio Phillips is credited with the discovery that an aerofoil will provide lift if the upper surface is deeply curved and the lower surface is shallowly curved – which is one of the fundamental principles of the science of aerodynamics. Phillips patented his findings in 1884, 1890, and 1891, and in 1893 built a large model with a venetian blind 'multiplane' wing configuration, which was powered by a small steam engine driving a tractor propeller; the Phillips model left the ground during trials on a circular test track, although the machine remained tethered. A further model made a number of hops in 1907.

Phoenix: the Hughes AIM-54A Phoenix air-to-air guided missile is supposed to be one of the most advanced weapons of its kind in operational service. Originally intended for the General Dynamics F-111B, the Phoenix is now being fitted to the Grumman F-14A Tomcat air superiority fighter of the United States Navy. Thirteen feet in length and with a range of more than sixty miles, the Phoenix uses a solid fuel rocket and radar homing, with a high explosive warhead and a proximity fuse.

PI-: international civil registration index mark for the Philippine Republic.

Piaggio/Piaggio-Douglas: the Italian firm of Piaggio has been engaged in aircraft production for a number of years, although it was not until 1964 that a separate company, Industrie Aeronautiche e Meccaniche Rinaldo Piaggio, S.p.A., was formed for aircraft manufacture. In recent years aircraft production has included the P.14

and P.149 (q.v.) single-engined light trainers, the P.136 amphibian, and the P.166 (q.v.) light transport and communications aircraft. The Piaggio-Douglas PD-808 (q.v.) is an executive jet and navigational trainer, the design of which originated with Douglas but was completed and developed by Piaggio.

Piasecki Helicopters: formed in 1946 as the Piasecki Helicopter Corporation, Piasecki produced a number of tandem-rotor military helicopter designs before the Vertol title was adopted in 1956. The company was acquired by Boeing (q.v.) in 1960, forming Boeing-Vertol (q.v.).

Pilâtre de Rozier, François (1757-85): a Frenchman, François Pilâtre de Rozier made a tethered ascent in a Montgolfier balloon on 15 October 1783, before making an ascent on 21 November with the Marquis d'Arlandes and journeying across Paris. Less happily, Pilâtre de Rozier was one of the first two men to be killed in a balloon accident when, with Jules Romain, he attempted to cross the English Channel on 15 July 1785, using a balloon with a combination of hydrogen and hot air for lift, which caught fire.

Pilatus: dating from 1939, the Swiss company Pilatus is a subsidiary of the Oerlikon Company and has produced a number of training aircraft for the Swiss Air Force, including the P-2 and P-3, while also manufacturing components for the Mirage III-S fighters introduced by Switzerland during the late 1960s. The most recent product has been the PC-6 (q.v.) Porter and Turbo-Porter S.T.O.L. light transport and utility aircraft, which has also been built under licence in the United States by Fairchild Hiller.

Pilcher, Percy (1867-99): a Scotsman, Percy Pilcher was a pupil of Otto Lilienthal (q.v.), the German gliding pioneer, and a notable contributor to the development of the glider in his own right. He built his first hang-glider, called the Bat, in 1895, and eventually managed to fly it after incorporating modifications recommended by Lilienthal, including a tailplane. Three more gliders followed, the last, the Hawk, incorporating a sprung undercarriage and being winched into the air, thus reducing dependence on the wind being in a favourable direction. A design for an aeroplane had been completed and another glider built, but not tested, when Pilcher fell to his death following structural failure of the Hawk in 1899.

Pioneer, Scottish Aviation: originating as an observation aircraft design, the Scottish Aviation Pioneer first flew in 1947, and was subsequently re-engined and modified before entering production as a S.T.O.L. light transport and utility aircraft, the first flight of a production model taking place in June 1953. A five-seat, high-wing monoplane, the Pioneer used a single 540 hp Alvis Leonides 503/7 radial engine for a maximum speed of 145 mph and a range of 650 miles. A much larger twin-engined development was the Twin Pioneer (q.v.).

Pioneer 10: launched on 3 March 1972 by an Atlas-Centaur rocket, the Pioneer 10 unmanned spacecraft attained the highest escape velocity – 32,400 mph – so far at the start of its journey to Jupiter, the object of the mission, and possibly far beyond. The planet Jupiter is the largest in the solar system, and Pioneer 10 is due to pass the planet during the winter of 1973/4 and to take readings and pass information back to Earth. The spacecraft may well pass out of the solar system, and a plaque has been fitted in case of interception by intelligent life, giving information on the origins of Pioneer 10 and also showing representations of human figures.

pioneers: the most prominent pioneers in the various different fields of aeronautics are mentioned elsewhere in this book, and include the inventors of the balloon, the Montgolfier brothers (q.v.) and J. A. C. Charles (q.v.); the pioneers of the airship, including the Tissandier brothers (q.v.), the Lebaudy brothers (q.v.), and Zeppelin (q.v.); the fathers of aerodynamics, Sir George Cayley (q.v.), Horatio Phillips (q.v.), and Alphonse Pénaud (q.v.); the gliding pioneers, including La Bris (q.v.), Lilienthal (q.v.), and Pilcher (q.v.); as well as the pioneers of heavier-than-air powered flight, Ader (q.v.), du Temple (q.v.), Moy

(q.v.), Mozhaiski (q.v.), Langley (q.v.), and the Wright Brothers (q.v.). Many of these gained fame in more than one aspect of the subject. The work of Garnerin (q.v.) on parachutes, Congreve (q.v.) on rockets, and Martin (q.v.) on ejector seats is also worth mention, as are the early pilots, aerial trail-blazers, and spacemen or astronauts.

Piper Aircraft: developed in 1937 by William Piper out of the former Taylor Aircraft Company, which dated from 1931, Piper initially concentrated on building the Cub high-wing monoplane, waiting until after the end of World War II to supplement this with another high-wing monoplane, the Tri-Pacer. A low-wing aircraft, the PA-6 Skysedan, was developed during the early post-war years, but did not enter production. Piper's first low-wing production model was the PA-235 Apache (q.v.) during the mid-1950s, and this was followed in 1956 by the single-engined PA-24 Comanche (q.v.), and in 1959 by the PA-25 Pawnee agricultural aircraft and the low-cost Tri-Pacer replacement, the PA-28 Cherokee (q.v.). The Apache later developed into the Aztec (q.v.), while the Cherokee appeared first in a stretched version, the Cherokee Six, and then with a retractable undercarriage as the Cherokee Arrow, before also being built as a twin-engined aircraft, the PA-34 Seneca (q.v.). Before this another single-engined aircraft, the Comanche, had also had a twin development put into production, the PA-30 Twin Comanche (q.v.) of 1963.

Piper concentrates for the most part on the lighter end of the market (meaning aircraft of below 12,500 lb weight). A larger Piper first appeared in 1964 in the PA-31 Inca, which entered production as the PA-31 Navajo (q.v.), although this and the even larger PA-35 Pocona, which can seat up to seventeen passengers and has still to enter production, are still strictly speaking light aircraft. Today, Piper is one of the top three manufacturers of light aircraft in the world, and with the exception of the PA-18 Super Cub, builds an all low-wing range of aircraft.

piston engine: essentially all non-turbine engines are piston engines, including radial engines, rotary engines, and 'V' engines, but the term is generally used to describe modern in-line or horizontally opposed engines.

Pitcairn: a small American aircraft manufacturer, Pitcairn produced a number of designs during the late 1920s and early 1930s, including the Mailwing single-engined biplane for, as the name clearly implies, airmail services in 1927, and two years later the Super Mailwing, with uprated engine. During the 1930s the company conducted a number of gyroplane experiments, including the PCA-2 tandem twin-seat aircraft, which used engine assistance to start the rotor moving, and the PA-36 of 1934, which was a cabin gyroplane with a variable pitch rotor.

pitch: the angle of the blades of a propeller or a rotor to the surrounding air. For variable pitch *see* propeller.

pitch, seat: the distance between seat backs in an airliner, which has a direct bearing on leg-room. Seat pitch on scheduled services is regulated by the International Air Transport Association on international routes and by national authorities on domestic services, although charter aircraft are subject to less exacting requirements.

PJ-: international civil registration index mark for the Netherlands Antilles.

PK-: international civil registration index mark for Indonesia.

Po-2, Polikarpov: a single-engined Soviet biplane, the tandem twin-seat Polikarpov Po-2 first appeared in 1928, and during the years which followed undertook a wide variety of duties, including basic training, target towing, liaison, observation and communications, as well as ground-attack and night bombing during the early years of World War II. One Po-2 night bomber unit was 'manned' largely by women.

Polaris: a product of the Lockheed Missiles and Space Corporation, the Polaris was the world's first submarine-launched intercontinental ballistic missile when it first appeared in 1959. Sixteen Polaris missiles are carried on U.S.N. and Royal Navy missile submarines. The earlier A-1 and A-2 versions have now been phased out,

but the A-3 remains in service. Thirty-one feet in length, with a range of 2,500 miles (1,500 miles for the A-2 version), Polaris is a two-stage rocket with an inertial guidance system, and is launched by a gas-steam injection system. The A-3 version has a triple independently-targeted re-entry vehicle dispenser for a warhead, and in the British versions the re-entry vehicles and warheads are British-built. On thirty-one out of the forty-one U.S. missile submarines, Polaris is being replaced by Poseidon (q.v.).

Polikarpov: a Soviet design bureau prominent throughout the 1930s and during the early 1940s, Polikarpov failed to produce any really outstanding designs, although the Po-2 (q.v.) biplane undertook a wide variety of duties, and the I-16 fighter monoplane was in service at the outbreak of World War II.

Porter, Pilatus: *see* PC-6 Porter/Turbo-Porter, Pilatus.

Poseidon: the successor to the Polaris (q.v.) on about thirty-one of the United States Navy's missile carrying submarines, the Lockheed UGM-73 Poseidon C-3 intercontinental ballistic missile was first launched on test in August 1968. Poseidon has a range of 2,500 miles, and contains a dispenser for between ten and fourteen multiple independently-targeted re-entry vehicles, each with a nuclear warhead. Thirty-four feet long, using an inertial guidance system of considerable accuracy, Poseidon is launched by a gas-steam injection system.

Potez: a French aircraft manufacturer dating from World War I, Potez was nationalised in 1937 and ceased aircraft production on the fall of France during World War II. Aircraft production did not restart until 1953, and in 1958 Potez acquired Air Fouga, the manufacturer of the CM.170 (q.v.) Magister jet trainer for the French armed forces. A number of light aircraft were built during the late 1950s and 1960s, when Potez was absorbed into Sud Aviation, now Aerospatiale (q.v.).

Pou-de-Ciel: the Pou-de-Ciel, or Sky Louse, sesquiplane was first produced in

1933 by Henri Mignet as a low-cost aircraft for the amateur to build at home. Grounded in the United Kingdom before World War II following a spate of fatal accidents, restrictions on the aircraft were lifted during the post-war years. Engine sizes vary between 17 hp and 35 hp, but aircraft are single-seat only.

powered flight: the definition of powered flight is that it must be sustained by the aircraft's engine after the momentum of the take-off run has disappeared, and that the flight must be controlled. The first such flights were those of the Wright brothers (q.v.) in 1903, but before this there had been a number of powered hops, notably by du Temple (q.v.) in 1874 and by Mozhaiski (q.v.) in 1884, in which the power that counted was that of the take-off run. It is generally accepted that a flight through the air of at least a quarter of a mile is necessary if sustained powered flight is to be proved; the distance over the ground is not particularly relevant because of the effects of head and tail winds.

PP-: international civil registration index mark for Brazil.

Pratt and Whitney Engines: one of the three major aero-engine manufacturers in the world, the American firm of Pratt and Whitney builds a wide range of military and civil engines, and has in fact done so for many years. Major civil powerplants include the JT9D advanced technology turbofan for the Boeing 747 (q.v.), the JT8D for the Boeing 707 (q.v.) and the McDonnell Douglas DC-8 (q.v.), and the small JT12A, one of the applications for which is the North American T-39 (q.v.) Sabreliner. The JFTD12 turbine is used on the Sikorsky S-64 (q.v.) helicopter. Military engines include the F100 for the McDonnell Douglas F-15 (q.v.) and the F401 for the Grumman F-14 (q.v.) Tomcat air superiority fighter, the TF30 for the General Dynamics F-111 (q.v.) and the J52 for the McDonnell Douglas A-4 (q.v.) Skyhawk and the Grumman A-6 (q.v.) Intruder.

precision approach radar: the term 'precision approach radar' refers to radar which provides precise information for guiding an aircraft during the final stages of the landing approach.

pre-emptive attack: a pre-emptive attack is generally regarded as a defensive thrust at enemy forces before they actually open hostilities, with the qualification that the enemy forces must genuinely be so positioned that their intentions can reasonably be assumed to be hostile. The most successful instance of a pre-emptive attack in recent years has been that of the Israeli armed forces against those of the Arab nations at the outset of the Six Days Arab-Israeli War of June 1967.

pressurisation: all modern airliners (with the exception of some third level types), and the larger executive aircraft have cabin pressurisation permitting flight at high altitudes without discomfort for the passengers. The cabin pressure is seldom that of sea level, but is nevertheless still set at a lower altitude than that outside the aircraft. The first pressurised airliner was the Boeing 307 (q.v.) Stratoliner, which first flew in 1938.

Princess, Saunders-Roe: the largest flying-boat ever built, the Saunders-Roe Princess ten-engined flying-boat was intended for B.O.A.C.'s South American services, although later the intention became that of providing a large troop transport. In the event, although three aircraft were built and flown successfully, it never entered civil or military service, mainly due to the 140-ton aircraft being too far in advance of the traffic needs of the day, but also due to the then rapidly declining position of the flying-boat in the face of advances with landplanes.

Privateer, Consolidated: a development of the Consolidated B-24 (q.v.) Liberator heavy bomber, the PB4Y-2 Privateer was a maritime-reconnaissance aircraft for the United States Navy. Four 1,200 hp Pratt and Whitney radial engines powered the Privateer, which was fitted with radar and radio equipment to assist with its duties.

procurement policy: often considered to be inadequate, a procurement policy involves a deliberate method of evaluating and selecting equipment for armed forces, often with other policy decisions over duties or minimising imports in mind. Most governments have an organisation, such as the British Ministry of Defence's Procurement Executive, charged with this duty.

propeller: the propeller, or airscrew as it is often called, is an aerodynamic surface providing propulsion rather than lift. A considerable degree of sophistication is applied to propeller design, and the most efficient working of propeller and engine combined comes with the constant speed propeller, depending upon variable pitch, changing the angle of the blades to the surrounding air with fine pitch for take-off and coarse pitch for cruising, not unlike a motor vehicle changing gear. The first variable pitch propellers were developed in England in 1924 by Hele-Shaw and Beacham, although the development did not become commonplace until the latter half of the 1930s.

Other developments include contra-rotating propellers, with two propellers on the same axis rotating in opposite directions, frequently driven by two engines. Propellers may be either the more common tractor propeller or the less usual (particularly at the present time) pusher propeller, usually mounted behind the wing trailing edge or the fuselage.

It is important not to confuse the propeller with the rotor blades (q.v.) of a helicopter.

propulsion: various systems of propulsion have been used in man's efforts to fly. Man-power proved to be a particularly ineffectual, not to mention exhausting, method of propulsion, not least because of the poor power-to-weight ratio of the propulsive unit. Electric motors and steam engines have also proved to be too heavy and to use an excessively heavy fuel source in each case, while the internal combustion engine has been a comparative newcomer, albeit a successful one. Internal combustion engines have consisted either of petrol engines in either rotary, radial, in-line, or 'V' form, with some experiments with diesel engines, or various forms of turbine engine, ram-jets, or rocket motors using either liquid or solid fuel.

prototype: a pre-production version of an aircraft, built for testing and evaluation, is a prototype. It may differ considerably from the production versions, perhaps even to the extent of having 'pre-production' versions between the prototype and the production aircraft.

Provost, Percival: a side-by-side twin-seat training aircraft built by Percival for the R.A.F. and a number of other air forces and air arms during the 1950s, the Provost used a single 550 hp Alvis Leonides 25 piston engine for a maximum speed of 200 mph. A turbojet development was known as the Jet Provost, and remains in production today as the B.A.C. 145/167 series.

PS-1, Shin Meiwa: the only flying-boat in production during the 1970s, the Shin Meiwa PS-1 has been developed for the Japanese Maritime Self-Defence Force for maritime-reconnaissance duties. The first flight was in October 1967, in prototype form. Four 2,850 shp I.H.I.-built General Electric T64-IHI-10 turboprops power the aircraft, giving a maximum speed of 340 mph and a range of up to 3,000 miles. A very short take-off capability and low landing speed are part of the aircraft's performance characteristics.

PT-: international civil registration index mark for Brazil.

Pullman, Ford-Stout: the first American-designed all-metal airliner, the Ford-Stout Pullman high-wing monoplane appeared in 1924 and was powered by a single 420 hp Liberty engine. Accommodation was available for six passengers.

Puma, Aerospatiale S.A.330: *see* S.A.330 Puma, Aerospatiale.

Pup, Beagle: a two-to-four-seat monoplane for the private owner, the Beagle Pup entered production during the mid-1960s, but went out of production on Beagle's (q.v.) collapse, although a development, the Bulldog military trainer, was taken over by Scottish Aviation. A fully aerobatic low-wing aircraft, the Pup, many of which remain in service, is powered by a 100-hp Rolls-Royce Continental piston engine, although up-rated versions were planned.

Pup, Sopwith: one of the leading aircraft types in use during World War I, the Sopwith Pup fighter biplane first entered service with the Royal Flying Corps in 1915, and was usually a single-seat aircraft with a forward-firing machine gun. A single 110 hp Le Rhône rotary engine powered the aircraft, which was generally considered to be an outstanding machine.

pusher propeller: a pusher propeller is one mounted aft of the engine, meaning that with a wing-mounted engine the propeller will be behind the trailing edge. Aircraft in production at present using pusher propellers include the Cessna 337 (q.v.) Super Skymaster and the Piaggio P.166 (q.v.).

PV-2 Harpoon, Lockheed: a development of the earlier Lockheed PV-1 Ventura, itself a military development of the Lodestar (q.v.) airliner of pre-World War II days, the PV-2 Harpoon was an extended range maritime-reconnaissance aircraft which entered service with the United States Navy before the end of World War II. The twin-engined, mid-wing monoplane and twin-fin layout of the earlier aircraft was retained, but performance and defensive capability were increased.

PZ-: international civil registration index mark for Surinam.

Q

Currently the airline operates throughout the Far East, Middle East, North America and the Caribbean, and to Europe, using a fleet of Boeing 707s and 747s.

Q.T.O.L.: *see* quiet take-off and landing.

Quail: a United States missile, the Quail is a diversionary missile (*see* penetration aids) which can be fired from a B-52 bomber to produce a radar echo similar to that of the B-52, thus misleading defending radar measures.

Qantas Airways Ltd: the origins of the Australian state airline, Qantas Airways, lie in the formation of the Queensland and Northern Territory Aerial Service in 1920, with an Avro 504K and a B.E.2E for pleasure flights and charters throughout Queensland. Subsidised scheduled services started between two Queensland towns in 1922, and during the rest of the decade the airline continued to develop with the introduction of de Havilland D.H.50s, the formation of two flying schools in 1926, and the start of the Flying Doctor Service in 1928, using a specially converted D.H.50.

The title Qantas Empire Airways Ltd was taken in 1934, and in 1935 the airline's D.H.86s took over the Darwin–Singapore section of the Imperial Airways route from London. During the late 1930s Short Empire 'C' Class flying-boats were introduced, and these formed the nucleus of a Royal Australian Air Force flying-boat squadron on the outbreak of World War II in 1939. The links with Imperial Airways continued with that airline's successor, the British Overseas Airways Corporation, during World War II, a Perth–Colombo service being a part of the horseshoe Australia–South Africa route operated by B.O.A.C. during the war.

After the war a variety of aircraft was used, including ex-wartime bomber and transport aircraft and new Short Sandringham flying-boats. Lockheed Super Constellations were introduced in 1954, and in 1958 inaugurated the airline's round-the-world service. The first non-American airline to introduce Boeing 707 jet airliners in 1959, Qantas initially used a special smaller version, the 707-138B, before later replacing these with standard aircraft. Turboprop Lockheed Electras were also introduced in 1959.

The present title was adopted in 1967.

Quebecair: originating as Rimouski Aviation in 1946, the present title of Quebecair was taken in 1953 after a merger with another airline, Gulf Aviation. Two other airlines, Matane Air Service and Royalair, have been acquired since, and today Quebecair operates an intensive network of services, many on a third-level basis, throughout Eastern Canada, with a fleet which includes B.A.C. One-Eleven, Fokker F-27, Douglas DC-3, de Havilland Canada DHC-2 Beaver and DHC-3 Otter aircraft, Cessna light aircraft, and Bell and Hughes helicopters.

Queen Air, Beechcraft: the Beechcraft Queen Air first flew in prototype form in August 1958, and entered production as the Model 65, at that time the company's largest aircraft, before being followed by the A65 with swept fin. An up-rated version was designated the A80, while a model with the powerplants of the A65 and the wingspan of the A80 was designated the Model 70. The most recent version is the pressurised Model 89, which remains in production alongside the A65, Model 70, and B80, while commuter versions have been discontinued. The B80 uses two 380 hp Lycoming IGSO-540-A1D piston engines for a maximum speed of 224 mph and a range of up to 1,600 miles, providing accommodation for up to eleven people, including the pilot.

quiet take-off and landing: the concept of quiet take-off and landing is relatively new, originating with the new generation of advanced technology engines (q.v.) which permit a considerable reduction in noise levels compared with conventional turbojet and turbofan engines. Many of the major manufacturers are developing these new engine types, while the airframe manufacturers also have a number of new designs intended to utilise advanced technology engines. Apart from adopting the new engines with their twin or triple spools, methods of reducing aircraft noise include fitting variable pitch fan blades to existing engine designs (on which work is being undertaken by the British firm of Dowty Rotol), and changing the engine position, sometimes to above the wing, as on the V.F.W.-Fokker 614, or even back into the wing, as on a number of the older British designs, including the de Havilland Comet jet airliners and the V bombers.

R

R2C-1, Curtiss: the Curtiss R2C-1 biplane established two successive world air speed records on 4 November 1923, while being flown by Lieutenant Alford J. Williams, U.S.N. The lower (official) record was of 266 mph, while a later flight produced an unofficial record of 270 mph. An unusual feature of the aircraft, which used a liquid-cooled engine, was the provision of radiators in the upper wing.

R-34: a British airship, the R-34 made the first airship crossing of the North Atlantic, flying from the United Kingdom to the United States between 2 and 6 July 1919, and returning between 9 and 13 July. A crew of twenty men was carried, under the command of Squadron Leader G. H. Scott, R.A.F., and at the time the R-34 was the world's largest airship. The R-34 was destroyed on 24 August 1921, with the loss of the lives of forty of the forty-five persons aboard at the time, when the airship broke up in mid-air.

R-100: undoubtedly the best airship built in Britain, the R-100 was a private-enterprise venture, and made a highly successful return flight across the North Atlantic in 1930.

R-101: Britain's last airship until the 'Europa' of 1972, the R-101 was built by the British Government as a rival to the highly-successful R-100 (q.v.) in 1930, and started a flight to Egypt and India in October. Unfortunately the flight soon ended in disaster, with the R-101 breaking up in mid-air at Beauvais, France, with the loss of forty-seven of the fifty-four passengers, including the Secretary of State for Air and the Director-General of Civil Aviation.

RA-: U.S.N. designation for reconnaissance versions or developments of attack (*see* A-) aircraft types, such as the North American RA-5 Vigilante (*see* A-5).

radar: radar is a means of finding the range and the direction of an object by transmitting ultra-high-frequency radio waves which reflect back to the source, and as such is used extensively in air traffic control and air defence (including early warning systems), and for navigation of aircraft, for which advantages include the ability to operate an aircraft in poor visibility and the ability to operate more safely at high speeds than would be possible by depending solely on visual observations. Radar homing devices are also used to guide surface-to-air missiles (*see* semi-active radar homing).

radar beam riding: a system of missile guidance, radar beam riding is used for many missiles in the surface-to-air role. The intruding aircraft is picked up in a radar beam, which locks on to the aircraft while the missile is launched and follows the radar beam to intercept with the aircraft.

radial engine: an internal combustion engine, the radial engine has stationary cylinders arranged around a common crankshaft. The radial engine came into prominence during the 1930s and 1940s, when its high power-to-weight ratio could be put to good use, but the first appearance of a radial engine was Manley's engine for one of Samuel Langley's (q.v.) models in 1901. A drawback of the radial engine is its high drag characteristic.

radio beacon: a form of air lane marker, the radio beacon transmits a signal which enables the position of an aircraft to be determined.

radio-command guidance: the term 'radio-command guidance' refers to control of guided missiles by radio or wire transmitted steering corrections from an operator who will probably also have been the launcher of the missile.

radome: the radome is the cover on radar or radar antennae, and is often the only part of the equipment visible.

R.A.F.: *see* Royal Aircraft Factory.

ramjet: a ramjet differs from a turbojet or turbofan in having no moving parts. The basis of the ramjet concept is that of a tube with burners around the inside, and it is therefore dependent on there already being a steady stream of air through the ramjet in order that it may function; in other words, a ramjet cannot operate from rest, but it is suitable for augmenting thrust, as in the case of providing reheat (q.v.) on a turbojet engine, or as the main power source when rocket boosters can be provided for take-off, as with certain surface-to-air missiles, including the Bloodhound (q.v.) and the Thunderbird (q.v.).

range: although the range of a missile is purely a matter of the maximum distance which it can cover, that of an aircraft is a more complex and indefinite factor, since to a very considerable extent payload can be traded off in return for an increased fuel load and an extended range, within the capacity of the fuel tanks, although even this can be increased by fitting underwing or drop tanks to military aircraft. Alterations to cruising speed can also affect range considerably. In this book, the range of a military aircraft is usually the ferry range (i.e. the range without armament and with additional tanks), while for civil aircraft the range with standard payload is given. The extent of variation possible can be judged from the example of aircraft in the B.A.C. One-Eleven and McDonnell Douglas DC-9-10 category, capable of carrying more than eighty passengers over 1,000 miles or, in the corporate version, a dozen passengers 3,500 miles. The fact that maximum landing weight is normally lower than maximum take-off weight in an aircraft limits the extent to which range can be reduced in return for increased payload; this is also one reason why an aircraft making an early emergency landing dumps most of its fuel load first, although safety in an emergency is another factor.

Military reconnaissance aircraft, including maritime-reconnaissance, airborne early warning, and search and rescue types, frequently use the term 'endurance' instead of 'range', since minimum speed is used to obtain long spells over a certain patrol area. Another term, for combat aircraft, is 'radius of action'—the 'there, attack, and back' range, in fact.

Rapide, D.H.89 Dragon: *see under* D.H.84 Dragon, de Havilland.

Rapier: a highly mobile surface-to-air missile built by the British Aircraft Corporation for the British Army and the Royal Air Force, the Rapier is primarily intended for use against very low-flying aircraft, and is a single-stage solid fuel rocket-powered missile, with radio-command guidance, carrying a high-explosive warhead.

RB-: U.S.A.F. designation for reconnaissance versions of bomber (*see* B-) aircraft types, including the Martin RB-57 Canberra (*see* B-57).

R.E.8, Royal Aircraft Factory: a reconnaissance-bomber biplane, the Royal Aircraft Factory R.E.8, usually known by its nickname 'Harry Tate', was powered by a single 150 hp Royal Aircraft Factory engine, allowing an endurance of up to four hours and a warload of up to 260 lb. The best aircraft of its type in Royal Flying Corps service during World War I, the R.E.8 was a tandem twin-seat aircraft with two rear-mounted Lewis guns and a single forward-firing gun.

Re.2001 Falco II, Reggiane: a development of the earlier Reggiane Rc.2000 Falco, the Re.2001 Falco II was extensively modified in the light of experience with the earlier aircraft, which had first flown during the late 1930s. The airframe was strengthened and a 1,175 hp Daimler Benz in-line engine replaced the Italian radial engine of the earlier aircraft. First operational use of the Re.2001 was in May 1942.

reaction propulsion: the term 'reaction propulsion' relates to any engine which depends for its propulsive effect on the rearward projection of hot gases, as in a rocket engine or a jet engine.

Read, Lieutenant Commander A. C., U.S.N.: the captain of the first aircraft to fly across the Atlantic, Lieutenant Commander A. C. Read flew the Curtiss NC-4 flying-boat which made the west–east crossing in stages between 8 and 31 May 1919. A crew of four accompanied Reed, who subsequently enjoyed a successful career in the United States Navy, retiring with the rank of Admiral.

recce: slang term for reconnaissance.

reciprocal agreements: the term 'reciprocal agreements' arises in the context of international civil aviation. Such agreements permit air services to be operated between two or more countries, usually with controls over the total number of flights, which are shared between the nations involved. *See also* air traffic agreement.

reconnaissance: *see* aerial reconnaissance *and* maritime reconnaissance.

Redeye: the General Dynamics F1M-43A Redeye surface-to-air missile is used for defence against low-flying aircraft, primarily in forward combat areas, and is used extensively by the U.S. Army and Marine Corps, and by the Australian and Swedish armies. A small missile of only four feet in length, capable of being fired by an infantryman using a shoulder launcher, Redeye uses a solid-fuel rocket motor and infra-red homing devices, while carrying a high-explosive warhead.

Red Top: a British air-to-air missile built by Hawker Siddeley Dynamics, Red Top has replaced the earlier Firestreak (q.v.) in the Royal Navy and Royal Air Force service, and is a second-generation missile with the ability to home on to the target without the necessity of being fired in clear visibility or by an aircraft directly in pursuit of the target. A solid-fuel rocket motor is used, while Red Top has an infra-red homing device and a high-explosive warhead.

reduced take-off and landing: a comparatively new concept, reduced take-off and landing, or R.T.O.L., has been evolved as an intermediate stage for large aircraft unable to achieve economic vertical or short take-off and landing within the limits of existing technology. Basically, reduced take-off can be achieved by modifications to existing designs, with additional flaps and slots and jetstream ducting, or by variations on existing design concepts, instead of the radical changes necessary for S.T.O.L. In addition, R.T.O.L. holds

the possibility of being allied with quiet take-off and landing, or Q.T.O.L., in a way which is not yet possible with V./S.T.O.L.

re-entry vehicle: the term 're-entry vehicle' applies to the last stage of a rocket or a guided missile when it is intended that this should return to the earth's surface in order to complete its mission. Features of a re-entry vehicle usually include provision for withstanding the considerable heat which results from the friction of the atmosphere on re-entry.

Regente, Neiva: a successor to the earlier Neiva L-6, the L-41 Regente all-metal, high-wing monoplane first flew in September 1961, and is currently in service with the Brazilian Air Force on training duties. A single 180 hp Lycoming 0-360-AIA piston engine provides the four-seat aircraft with a maximum speed of 165 mph and a range of 725 miles.

Reggiane: an Italian aircraft manufacturer which came into prominence during the late 1930s and early 1940s with the Re.2000 Falco and the Re.2001 (q.v.) Falco II fighters. Reggiane did not remain in existence after the end of World War II.

reheat: the term 'reheat', or 'afterburning', refers to the provision, in supersonic aircraft, of additional burners in the exhaust pipe of the jet engine to burn any unburnt gases, thus considerably boosting thrust. The provision of afterburners or reheat can be fairly described as being the same in practice as fitting a ramjet (q.v.) behind the turbojet.

Reims Aviation: a French aircraft manufacturer, Rheims Aviation produces a substantial part of the Cessna (q.v.) range under licence for European customers, as well as providing product support for the Max Holste 1521M Broussard.

Republic Aviation: a part of the Fairchild Corporation (*see* Fairchild-Hiller) since 1965, Republic Aviation dates from the early 1930s, when it was formed out of the Seversky (q.v.) concern. One of the first Republic designs to reach major production status, the F-47 (q.v.) Thunderbolt fighter of World War II, bore strong signs of its

Seversky ancestry. Post-war designs included an experimental four-engined photographic-reconnaissance aircraft, the XF-12 Rainbow, of which only two were built, while plans for a small amphibian aimed at the private-owner market were not fulfilled. However, the company returned to prominence with its first jet design, the F-84 (q.v.) Thunderjet, and its developments, the RF-84F Thunderflash and the F-84F Thunderstreak, all of which were supplied to many NATO countries in addition to the United States. The F-84 series was followed during the 1950s by the first supersonic fighter to enter regular operational service, the F-105 (q.v.) Thunderchief.

research and development: the term 'research and development', more usually referred to as 'R. & D.', covers both specific items of research and the more general and academic type of investigation. A large part of any research and development project associated with a specific aircraft is concentrated on the mainplane, if only because other items, such as engines and avionics, are sometimes standard for a number of projects. The mounting cost of developing new aircraft has led to massive state support in the more advanced countries, and even the United States has of late been forced to abandon its traditional policy of leaving the aircraft manufacturers to raise the development cost of a new civil project on the money market, notably in the case of such projects as the Boeing 2707 (q.v.) and the Lockheed Tri-Star (q.v.). On the other hand, military projects have always included research and development in the contracts for the aircraft, often with attendant fall-out into related civil projects. Government R. & D. assistance is usually claimed back in the form on a levy of sales.

reserve: the term 'reserve' in its military usage refers to a force of volunteers who have either spent a period of time serving with the service to which they still belong, or received training on a part-time basis; it is in any event usual for reservists to undertake periodic refresher courses and exercises. The object of any reserve is to provide a force of at least partially-trained men capable of rapid mobilisation in the event of the regular forces being placed under strain. Some countries have armed forces which are dependent on reserves to a very great degree, and maintain only a nucleus of regulars; these include Sweden, Switzerland, and South Africa. Other bodies, including the Royal Air Force, have only nominal reserves, although generally the United Kingdom has reserve forces of what may be described as intermediate size. Sometimes reserve forces of the kind described here are known as auxiliary forces, as is, for example, the Hong Kong Auxiliary Air Force.

An alternative meaning is that of regular forces held in reserve at some convenient point, ready for immediate dispatch to reinforce other forces as and when necessary.

reserve fuel: the range of an aircraft usually includes an allowance of fuel in reserve in case of a diversion from the planned landing point due to inclement weather or other problems, in which case the reserve equals the fuel necessary to fly to the diversionary airfield. The reserve can also be used for in-flight diversions or emergencies. An adequate reserve can sometimes be treated as an unnecessary luxury in military aviation, but civil authorities consider adequate reserves to be a vital necessity in air transport operations.

reverse thrust: reverse thrust is used by turbojet and turbofan aircraft to supplement the aircraft's brakes when landing, and is useful due to the very much higher landing speeds required for the safe operation of such aircraft types. Reverse thrust is obtained by deflecting the normal jet-stream forward.

RF-: U.S.A.F., U.S.N., and U.S.M.C. designation for reconnaissance developments of fighter (*see* F-) aircraft.

RF-4 Phantom II, McDonnell Douglas: *see* F-4 Phantom II, McDonnell Douglas.

RF-101 Voodoo, McDonnell: *see* F-101 Voodoo, McDonnell.

Richtofen, Rittmeister Baron Manfred von (1882–1918): the famous Red Baron of Germany, so-called because of his preference for a bright red Fokker Dr.1 Triplane, although this was only one of several types flown during his career. Baron Manfred von Richthofen was the most successful of the Central Powers' pilots during World War I, with eighty confirmed victories. Originally a cavalry captain (Rittmeister), Richthofen transferred to the Military Aviation Service in May 1915, and served in second-line squadrons until September 1916, when he was transferred to Oswald Boelcke's squadron. He received his own squadron in early 1917, and later that year was placed in command of several squadrons, which he organised into the first of his famous 'flying circus' fighter formations. In spite of his own considerable skill and the success of 'Richthofen's Flying Circus', he nevertheless had the reputation of being a fair and honourable opponent.

Richthofen was killed on 21 April 1918, after engaging two R.F.C. Sopwith Camels in combat. The exact circumstances of his death remain a mystery; officially he was shot down by one of the Camels, but it also seems possible that he was accounted for by ground fire from an A.N.Z.A.C. unit.

Rickenbacker, Captain Edward Vernon: a racing driver, Edward (or Eddie) Rickenbacker became interested in flying before joining the United States Army in 1917. He served initially as a staff car driver before being transferred to the Aviation Section and did not join an operational fighter unit until March 1918. Rickenbacker's squadron was Raoul Lufbery's (q.v.) Escadrille Lafayette, with which he flew the first American patrols over the Western Front. Eventually he had twenty-six confirmed victories before the Armistice, gaining the Congressional Medal of Honour, America's highest award in the process.

After the war Rickenbacker made an unsuccessful attempt at motor-car manufacture before returning to aviation, and subsequently being largely responsible for the creation of Eastern Air Lines (q.v.). During World War II he toured U.S.A.F. bases, and on one such tour spent three weeks on a life raft before being rescued after his aircraft ditched. Returning after the war to civil aviation, he became President of Eastern Air Lines in 1953.

Rio Treaty of Reciprocal Assistance: *see* Inter-American Treaty of Reciprocal Assistance.

Robin: *see* Avions Pierre Robin.

Roe, Sir Alliot Verdon (1877–1958): an Englishman, Alliot Verdon Roe designed a number of the first British aircraft, starting in 1907. He was also the second Englishman to fly (after J. T. C. Moore-Brabazon – see Brabazon of Tara). Roe's early designs included biplanes based upon the French Goupy, some early cabin biplanes, and the highly-successful Avro 504, which first flew in 1913. Sir Alliot Verdon Roe eventually left Avro (q.v.), and with S. E. Saunders formed the Saunders-Roe (q.v.) concern, which largely concentrated on flying-boats, in 1930.

roll: the term roll refers to, as it suggests, the movement of the aircraft about an axis which is both longitudinal and horizontal, usually indicating that the aircraft has undertaken the complete manoeuvre and returned to its original position.

Rollason: a British aircraft manufacturer, Rollason's main production effort is the Druine D.62 Condor, for which a production licence is held. The aircraft is marketed in the United Kingdom as the Rollason Condor – a two-seat, low-wing aircraft with a single engine and of basic construction and finish.

Rolls-Royce: one of the top three aero-engine manufacturers in the world, Rolls-Royce's origins are in motor car manufacture, and the company really came to prominence in aviation during World War I. The series of in-line engines produced by the company between the wars led ultimately to the Merlin, which powered, amongst many other aircraft, the Supermarine Spitfire (q.v.) fighter, the Avro Lancaster (q.v.) bomber, and the de Havilland D.H.98 (q.v.) Mosquito fighter-bomber during World War II. It was also produced under licence by Packhard in the United States for the North American F-51 (q.v.) Mustang, as well as being used after the war in the Canadair DC-4M Argonaut and the Spanish-built Messerschmitt Bf. 109.

After the war, Rolls-Royce went from strength to strength, establishing a strong position in turbojet and turboprop development and production with engines such as the Nene, the Avon, the Spey, the Conway, the Dart, and the Tyne. In 1961 Napier, another British aero-engine manufacturer, was acquired, and in 1968 Rolls-Royce merged with the sole remaining British aero-engine manufacturer, Bristol Siddeley Engines (q.v.), acquiring in the process the Pegasus vectored thrust engine for vertical take-off, the Olympus engine for the Concorde supersonic airliner, and the Gnome shaft turbine, used primarily as a helicopter powerplant. Both Rolls-Royce and Bristol were involved in producing marine versions of their aero-engines, while Rolls-Royce had developed lift jets, as opposed to vectored thrust, for vertical take-off and landing (q.v.).

Problems with the development of the advanced technology RB.211 engine for the Lockheed TriStar (q.v.) airbus led to the bankruptcy of Rolls-Royce in 1971, followed by a government acquisition of the company. Since that time, the non-aviation interests and the licence-production of Continental engines have passed to a new publicly-quoted company, Rolls-Royce Motors (q.v.).

Current Rolls-Royce production includes the Olympus engine for the Concorde, the RB.211 for the Lockheed TriStar, the Pegasus for the Harrier vertical take-off fighter, the Spey for a number of airliners and military aircraft (including the B.A.C. One-Eleven, the Trident, the Gulfstream II, the Phantom II, F-4K and M, and the Buccaneer), the Dart for the HS 748 and the Fokker Friendship, the Viper for executive and training aircraft, a number of Anglo-French engines (including the M45H for the V.F.W. 614), and helicopter engines and lift jets.

Rolls-Royce is a minority shareholder in Short Brothers and Harland (q.v.), and a member of the Turbo-Union consortium which is building the engines for the Panavia 200 Panther multi-role combat aircraft.

Rolls-Royce Motors: Rolls-Royce Motors is the former motor car, diesel engine, and light aircraft engine division of Rolls-Royce, and was formed as a separate publicly-quoted company in 1972 after the collapse and subsequent nationalisation of Rolls-Royce. The aviation interest is the licence-production of Continental light aircraft engines for British and European manufacturers.

Rosatelli, Celestino: one of the most notable Italian aircraft designers of the inter-war period, Celestino Rosatelli worked for Fiat, and was responsible for a number of successful and attractive designs. Prominent amongst the list of Rosatelli aircraft were the C.R.20 biplane fighter of the late 1920s and its successors, the C.R.30 and C.R.32 biplanes of the early and mid-1930s. These were followed by the B.R.20 Cicogna medium bomber monoplane of the latter half of the decade, as well as a number of fighter monoplanes which served in World War II.

rotary engine: the rotary engine is a type of internal combustion engine in which the cylinders rotate around a static crankshaft, and the propeller is actually attached to the cylinders. Invented by an Australian, Lawrence Hargrave (q.v.), in 1887, the rotary engine was developed for the aeroplane by Laurent Seguin in 1907, and first appeared as the production 50 hp Gnome engine. The idea was taken up by a number of manufacturers, and rotary engines powered almost all the main aircraft types in service during World War I and for a period afterwards, until radial and in-line engines came into prominence.

Although the rotary engine played a highly significant part in the development of the aeroplane, it suffered from problems of control and stability, particularly as hp ratings increased, which made the concept unacceptable as aircraft grew in size and performance.

The term 'rotary engine' is also used to describe the new Wankel engine, which replaces the reciprocating piston with a roughly triangular revolving piston, and holds some promise of application in the light aircraft field in the future.

rotary wing: the rotary wing is a wing which combines propulsive with lifting and control functions, and is found in helicopters, for which reason an alternative name for these aircraft is rotorcraft (q.v.).

Rotodyne, Fairey: the first true vertical take-off transport aircraft, as opposed to helicopter, the Fairey Rotodyne was flown extensively during the late 1950s and attracted considerable interest, including that of British European Airways, the Royal Air Force, and the United States Air Force, although the aircraft was later abandoned. Two 3,000 shp Napier Eland turboprops powered the aircraft in flight, while a rotor was placed amidships and in line with the fairly short span high-wing mainplane to provide lift during take-off; the rotor being powered by tip-mounted jets. Up to seventy passengers could be carried.

rotor blades: rotor blades differ from propeller blades in providing control and lift as well as propulsion, and therefore act as a wing replacement or alternative.

rotorcraft: a rotorcraft is an aircraft powered by rotor blades and receiving its main lift from rotor blades, making the term an alternative name for helicopter or gyroplane.

route structure: the term 'route structure' is used in any examination of a scheduled airline's activities, and refers to the type, frequency, and stage length of services operated.

Royal Aircraft Factory: a British government-owned concern, the Royal Aircraft Factory was established before World War I at Farnborough and produced a number of fighter, bomber, and reconnaissance aircraft designs before and during the war. Notable aircraft included the B.S.1., which was effectively the prototype for the World War I fighter aircraft; the B.E.3 'Bloater' reconnaissance aircraft, which was noted for its poor spin characteristics, but was the first British combat aircraft to land in France after the outset of World War I; the S.E.5 (q.v.) fighter of 1916; the R.E.8 (q.v.) reconnaissance-bomber of the same year; and the F.E.2a fighter and its development, the F.E.2b bomber. All R.A.F. designs were biplanes.

Royal Netherlands Aircraft Factory: *see* Fokker *and* V.F.W.-Fokker.

R.T.O.L.: *see* reduced take-off and landing.

RTV-2 Gargoyle, McDonnell: an experimental surface-to-air guided missile, the McDonnell RTV-2 Gargoyle project was initiated during the closing stages of World War II. In appearance the Gargoyle resembled a reduced-scale jet fighter aircraft, although rocket-powered, with a low-wing and butterfly 'V' tailplane. A 1,000 lb high-explosive warhead could be carried, and this warhead was fitted with a proximity fuse.

Rumpler: the Rumpler Flugzeug-Werke produced a number of designs for the German Military Aviation Service during World War I, including the Rumpler-Etrich Taube (q.v.) (Dove) fighter monoplane, which was built under licence by many other German firms, and the 'C' series of biplane fighters, of which the C.III and C.V. were probably the most notable.

runway visual range: the term runway visual range is self-explanatory; it means the distance along the runway which can be clearly seen by the pilots of aircraft landing or taking off. Visual range is usually determined by observations from the runway ends, using the landing lights as an indication of distance.

R.V.R.: *see* runway visual range.

S

S-2 Tracker, Grumman: a high-wing, twin-engined, carrier-borne anti-submarine aircraft, the Grumman S-2 Tracker first flew in December 1952, and the first deliveries to the United States Navy followed in February 1954. Early versions included the S-2A, also operated by Brazil, Italy, Japan, the Netherlands, Argentina, Taiwan, Thailand, and Uruguay, and the training TS-2A. The S-2C incorporated an enlarged weapons bay, the S-2D was fitted with improved equipment and an enlarged wingspan, and the S-2E had improved avionics. S-2As modified to the S-2E standard were designated S-2B. In addition to earlier models, the Royal Netherlands Navy also received some de Havilland Canada-built aircraft, as did the Royal Canadian Navy. Another S-2 operator is the Royal Australian Navy. Two 1,525 hp Wright R-1820-82 piston engines provide the S-2D with a maximum speed of 280 mph and a range of up to 1,350 miles, while a warload of twenty-five depth charges may be carried, with float lights and sonar buoys. An airborne-early-warning-development is the E-1B (q.v.) Tracer and a carrier onboard delivery development is the C-1 (q.v.) Trader.

S-3A Viking, Lockheed: a successor to the Grumman S-2 Tracker (q.v.), the Lockheed S-3A Viking first flew in Spring 1972. A twin-engined high-wing monoplane intended for carrier-borne operations, the S-3A entered service with the United States Navy in 1973. Two 9,000 lb thrust General Electric TF34 turbofans provide a maximum speed of 500 mph and a range of up to 2,000 miles, while a warload of depth charges, anti-submarine torpedoes, or missiles, and sonar buoys can be carried.

S.5, Supermarine: developed as a racing seaplane, the Supermarine S.5, a low-wing monoplane, was designed by Reginald Mitchell (q.v.) for Supermarine and won the 1927 Schneider Trophy for the United Kingdom with a maximum speed of 281 mph. The aircraft also established an unofficial world airspeed record of 284·4 mph. A 1,050 hp Napier engine powered the S.5, from which the later S.6 and S.6B (q.v.) were developed.

S.6/S.6B, Supermarine: the Supermarine S.6 in 1929 and the S.6B in 1931 were further developments on the monoplane seaplane theme by Reginald Mitchell (q.v.). After the success of the S.5 (q.v.) in 1927, the victory of the S.6 in 1929 and S.6B in 1931 won the Schneider Trophy outright for the United Kingdom. The S.6 established two world airspeed records, an unofficial 370 mph on 7 September 1929 and an official 357·75 mph on 12 September; while the S.6B established three speed records, an unofficial 388 mph on 13 September 1931, and 415 mph on 29 September, and an official 407 mph, also on 29 September in each case the pilot was Flight Lieutenant G. H. Stainforth, A.F.C., R.A.F. A 2,300 hp Rolls-Royce 'R' engine powered the S.6B.

S.7 Longhorn, Farman: *see* Longhorn, Maurice Farman.

S.11 Instructor, Fokker: a low-wing primary training monoplane, the Fokker S.11 Instructor first flew in 1947, with deliveries to the Royal Netherlands Air Force and the Israeli Defence Force/Air Force following. The aircraft has been built under licence in Italy, by Aermacchi, and in Brazil. A single 190 hp Lycoming O-435-A piston engine powers the twin-seat aircraft, giving a maximum speed of 130 mph and a range of up to 400 miles.

S-51, Sikorsky and Westland: one of the first practical helicopters to enter operational service with military air arms, the Sikorsky S-51 first flew in prototype form in February 1946, and subsequently entered service with the American armed forces as the U.S.A.F.'s H-5 and the U.S.N.'s HO2S-1. Westland built the S-51 under licence in the United Kingdom as the Dragonfly for the Royal Air Force and the Royal Navy, with a first flight in 1948 and deliveries starting in 1950. A number of Westland-built helicopters were also exported, including some of a developed version, the Widgeon, with a five-seat cabin and a modified rotor head. Few S-51s of any kind remain. The S-51, with four seats, used a single 450 hp Pratt and Whitney R-985 piston engine, while the Dragonfly and Widgeon used a 520 hp or 540 hp Alvis Leonides 50 piston engine, for a maximum speed of 110 mph and a range of up to 310 miles.

S-55, Sikorsky and Westland: the Sikorsky S-55 helicopter entered service during the early 1950s with the American armed forces, usually with an 800 hp Wright R-1300-3 piston engine, although the U.S. Army version, known as the UH-19 Chickasaw, used a 600 hp Pratt and Whitney R-1340. Performance of the various versions of the H-19 was generally around 112 mph maximum speed with a maximum range of up to 360 miles, while up to ten passengers could be carried in addition to the crew of two. The S-55 was also built under licence in Japan by Mitsubishi and in the United Kingdom by Westland as the Whirlwind, the latter versions being developed considerably, starting with the substitution of a 750 hp Alvis Leonides Major piston engine. Later versions, to which standard some of the earlier machines were also modified, used a single 1,050 shp Bristol Siddeley Gnome H.1000 shaft turbine, giving a maximum speed of 106 mph and a range of 300 miles. In addition to military users, the S-55 and the Whirlwind were used by civil operators.

S-58, Sikorsky and Westland: the Sikorsky S-58 helicopter was developed, starting in June 1952, in order to meet a U.S. Navy requirement for an anti-submarine helicopter, and the first flight of a prototype took place in March 1954, entry into service starting in August 1955. The U.S. Navy designation was SH-34 Seabat, while the U.S.M.C. used the designation UH-34D Seahorse, and the U.S. Army the designation CH-34 Choctaw. The S-58 was supplied to many of America's allies, as well as being built under licence in France by Sud Aviation and in the United Kingdom by Westland as the Wessex. Standard

U.S. helicopters used a single Wright R-1820-84 piston engine of 1,525 hp, giving the twelve-seat S-58 a maximum speed of 123 mph and a range of up to 260 miles. More recently, a number of shaft turbine conversions have been produced by Sikorsky.

The first flight of a Westland Wessex was in May 1957, and this differed from the standard S-58 in having a single Rolls-Royce Napier Gazelle N.Ga.11 shaft turbine of 1,100 shp, while the production Wessex 1 used a 1,450 shp Gazelle 161, the equipment including dipping sonar. Later up-rated versions with 1,600 shp engines were produced. The troop-carrying Wessex 5 appeared later still, with two coupled Bristol Siddeley Gnome turbines and strongpoints for rockets and air-to-surface missiles. Performance of the Wessex includes a maximum speed of 130 mph and a range of up to 600 miles. Civil versions have been produced by both Sikorsky and Westland.

S-61, Sikorsky and Westland: the prototype Sikorsky S-61 anti-submarine helicopter first flew in March 1959. Deliveries to the U.S. Navy started in September 1961 with the designation SH-3A Sea King, which used two coupled 1,250 shp General Electric T58-GE-88 shaft turbines. A number of SH-3As have since been modified to RH-3A standard, while a later development has been the SH-3D with up-rated 1,400 shp T58-GE-10 engines, which first appeared in 1965 and has a maximum speed of 159 mph and a range of up to 625 miles. U.S.M.C. and U.S. Army versions are used for assault duties, and the S-61 has been produced under licence in Italy by Agusta, and in Japan by Mitsubishi, while Westland has also built the Sea King in the United Kingdom for the Royal Navy, the Royal Norwegian Navy, and the Federal German Navy.

The Westland SH-3D Sea King uses twin coupled 1,500 shp Rolls-Royce Gnome H.1400 shaft turbines for a maximum speed of 143 mph and a range of up to 1,105 miles, while twenty-seven troops can be carried in addition to anti-submarine equipment which includes long-range sonar. Deliveries from Westland started in

1969. Westland also have on offer a troop-carrying version, the Commando.

Civil versions have been produced by Sikorsky only, with the S-61L and the amphibious S-61N; deliveries of these twenty-eight-passenger helicopters started during the early 1960s. The civil versions use two 1,350 shp General Electric CT58-110 shaft turbines for a maximum speed of 140 mph and a range of 275 miles.

A U.S.A.F. general-purpose version, with rear-loading ramp, is the S-16R, designated H-3C; this is also used now by the U.S. Coast Guard on search and rescue duties.

S-62, Sikorsky: a successor to the Sikorsky S-55 (q.v.), the S-62 uses a single 1,250 shp General Electric CT58 shaft turbine for a maximum speed of 100 mph and a range of up to 450 miles. Accommodation is for a crew of two and up to ten passengers. Production has now ended.

S-64 Skycrane, Sikorsky: the Sikorsky S-64 Skycrane helicopter first flew in prototype form in May 1962 and consists mainly of a control cabin and a backbone fuselage below which containers can be carried or other loads slung. Two 4,050 shp Pratt and Whitney JFTD12-A1 shaft turbines drive a single main rotor and provide a maximum speed of 117 mph and a range of 190 miles, while a payload of up to 20,000 lb may be carried; more recent versions have up-rated engines of 4,800 shp and a payload of 22,400 lb.

S-65A, Sikorsky: the Sikorsky S-65 is a cabin development of the S-64 (q.v.) Skycrane and uses the Skycrane rotor system. Developed for the U.S.M.C., first flight of the S-65 was in October 1964, and two 2,850 shp General Electric T64-GE-6 shaft turbines provide for a maximum speed of 195 mph and a range of up to 280 miles. Up to forty passengers can be carried, and the helicopter has the designation of CH-53 D Sea Stallion, although the U.S.N. uses the CH-53A designation and the U.S.A.F. the HH-53B/C designation. A three-engined civil version with accommodation for up to eighty-eight passengers is under offer to prospective operators.

S-67 Blackhawk, Sikorsky: a development of the S-61 (q.v.), the Sikorsky S-67 Blackhawk is a private-venture gunship helicopter which first flew in August 1970, and has since undergone successful trials with the U.S. Army. Tandem twin-seating is provided and a stub-wing is fitted for stability and for weapons strongpoints. Two coupled 1,500 shp General Electric T58-GE-5 shaft turbines provide for a maximum speed of more than 180 mph and a range of up to 1,100 miles, while a 6,000 lb warload can be carried. A fifteen-seat cabin version has been offered to the U.S. Army. On 14 and 19 December 1970 the S-67 established world speed records for a helicopter of 188·3 mph and 191·8 mph respectively.

SA-2: see 'Guideline'.

SA-3: see 'Goa'.

SA-4: see 'Ganef'.

SA-5: see 'Griffin'.

SA-6: see 'Gainful'.

S.A.321 Super Frelon, Aerospatiale: originally developed as the Sud Frelon (Hornet), the prototype Aerospatiale S.A.321 Super Frelon first flew in December 1962, using three coupled Turbomeca Turmo III C3 shaft turbines of 1,320 shp each, and incorporating a number of Sikorsky design features, including the boat-type hull and the rotor systems. Production models use three 1,515 shp Turmo III C5 shaft turbines. Among them are the S.A.321G for military users (who include the Aeronavale and the South African, Israeli, and Norwegian armed forces), the S.A.321F transport with accommodation for up to thirty-seven passengers, and the S.A.321J with tail ramp and sling. Maximum speed is 165 mph, and a range of up to 310 miles is possible while carrying a three-ton payload.

S.A.330 Puma, Aerospatiale: developed to meet an Armée de la Terre need for a utility helicopter of medium size, the Sud-designed S.A.330 Puma first flew in prototype form in April 1965, and has since become part of an Anglo-French helicopter production programme in which the two main partners are Aerospatiale and Westland. (The other helicopters involved are the S.A.341

S.A.341 Gazelle, Aerospatiale

(q.v.) and the W.G.13 (q.v.) Lynx.) The S.A.330 has entered service with the French Army and the Royal Air Force, and in Portugal, South Africa and Brazil. Two coupled Turbomeca Turmo III C4 shaft turbines of 1,300 shp each drive a single main rotor and provide a maximum speed of 174 mph and a range of up to 390 miles, while up to twelve fully-equipped troops can be carried.

S.A.341 Gazelle, Aerospatiale: a Sud Aviation design, the Aerospatiale S.A.341 Gazelle was developed as a successor to the Alouette II helicopter and the prototype, the S.A.340, first flew in August 1967. A further prototype first flew in 1969, and since then the S.A.341 has become a part of the Anglo-French (Westland-Aerospatiale) helicopter production programme. A single 600 hp Turbomeca Astazou shaft turbine provides the five-seat Gazelle with a maximum speed of 168 mph and a range of up to 425 miles.

SAAB: *see* Svenska Aeroplane A.B.

SAAB-17: a single-engined, mid-wing monoplane, the SAAB-17 was designed as a reconnaissance aircraft for the Royal Swedish Air Force during World War II, but eventually most aircraft of this type were modified for dive-bombing operations. A number of ex-R.Sw.A.F. SAAB-17s were supplied to Ethiopia after the war.

SAAB-18: the high-wing, twin-engined SAAB-18 was originally designed as a reconnaissance aircraft during World War II, but eventually was used on bombing, ground-attack, and torpedo-bombing duties. Two Swedish-built Pratt and Whitney Wasp radial engines powered the SAAB-18 at first, but these were later replaced by licence-built Daimler Benz DB.605s.

SAAB-21: the SAAB-21 fighter, or in R.Sw.A.F. service the J21, has the distinction of being the first and only fighter aircraft to be operational in both piston-engined and jet-powered versions. A single-seat twin-boom design, the piston-engined SAAB J21 was powered by a Daimler Benz engine built under licence in Sweden, driving a pusher propeller, but in later models a turbojet was substituted. Some

three hundred piston-engined and sixty turbojet-engined SAAB 21s were built during the closing stages of World War II and immediately after.

SAAB-29 Tunnan: holding the distinction of being Europe's first swept-wing operational jet fighter, the SAAB-29 Tunnan (aptly enough, in view of the fuselage shape, meaning 'Barrel') first flew in prototype form in September 1948. The aircraft entered R.Sw.A.F. service on interceptor duties in May 1951 with the designation J29A, and was followed by an improved version, the J29B, before the introduction of the A29 attack aircraft and the S29C photographic-reconnaissance aircraft. A number of further developments were built, ending with the J29E with modified wing leading-edge and reheat. The J29F used a single Svenska Flygmotor-built de Havilland Ghost turbojet of 6,170 lb thrust with reheat, providing a maximum speed of 658 mph and a range of up to 1,600 miles; two 550 lb bombs, or rockets, could be carried on underwing strongpoints, supplementing four fuselage-mounted 20 mm cannon.

SAAB-32 Lansen: the SAAB-32 Lansen first flew in November 1952 as a prototype and was intended for attack duties with the Royal Swedish Air Force, with which it entered service in 1955 as the A32A; this was followed by the J32B night and all-weather fighter and the S32 reconnaissance aircraft. In common with all modern Swedish aircraft, the Lansen was intended for operation from airstrips comprised of stretches of main road. A single Svenska Flygmotor RM.6A (licence-built Rolls-Royce Avon) turbojet of 14,690 lb thrust with reheat gave the A32A a maximum speed of 700 mph and a range of up to 2,000 miles, while rockets, bombs, and air-to-air missiles could be carried on underwing strongpoints, in addition to the four fuselage-mounted 30 mm cannon. All SAAB-32s were twin-seat aircraft.

SAAB-35 Draken: first flown as a prototype in October 1955, the SAAB-35 Draken (Dragon) single-seat fighter can be easily identified by its double-delta configuration wing. Deliveries to the R.Sw.A.F started in 1958 with the designation J35A interceptor. Successive versions have included the J35B, J35D, and J35F; a twin-seat conversion trainer, the Sk35C; and a reconnaissance version, the S35E. The Royal Danish Air Force obtained the SAAB-35XD version, and the Finnish Air Force has been another customer. A single Svenska Flygmotor RM.6C (Rolls-Royce Avon) of 17,200 lb thrust with re-heat provides a maximum speed of 1,320 mph (Mach 2·0) and a range of up to 1,600 miles, while two 30 mm cannon can be fitted and either four Sidewinder air-to-air missiles or 2,200 lb of bombs or rockets can be carried on underwing strongpoints.

SAAB-37 Viggen: the latest Swedish aircraft, the SAAB-37 Viggen was designed from the outset as a multi-role combat aircraft rather than having any particular leaning towards attack or interceptor duties. An unusual design, with the canard configuration of a small foreplane with flaps in addition to the delta mainplane, the SAAB-37 has a fairly short take-off run. The first flight of a prototype took place in February 1967, and production is now well advanced, early versions including the AJ37 strike aircraft, the JA37 interceptor, the twin-seat Sk37 trainer, and the S37 reconnaissance aircraft. A single Svenska Flygmotor RM.8 (licence-built Pratt and Whitney JT8D-22 turbofan) of 26,450 lb thrust with reheat provides a maximum speed in excess of Mach 2·0 and a range of up to 1,600 miles. A wide range of warloads can be carried on underwing and under-fuselage strongpoints.

SAAB-91 Safir: a two-to-three-seat low-wing monoplane for training duties, the SAAB-91 Safir first flew as a prototype in November 1945, and used a single 145 hp de Havilland Gipsy Major 10 piston engine. It entered service with the R.Sw.A.F. and the Ethiopian Air Force as the Tp91 or SAAB-91A. A 190 hp Lycoming O-435-A engine was fitted to the SAAB-91B and SAAB-91C, the latter having four seats. The most recent version is the SAAB-91D, with a 180 hp Lycoming O-360-AIA engine for a maximum speed of 165 mph and a range of 660 miles, and this has been put into service with the Austrian,

Finnish, and Tunisian air arms. In R.Sw.A.F. service the SAAB-91 is being replaced by the Scottish Aviation Bulldog (q.v.).

SAAB-105: a multi-purpose light twin-jet aircraft, the SAAB-105 has been primarily designed for basic jet training, although the aircraft can also be used for light attack and counter-insurgency duties, and for communications and liaison work, with an extra two seats behind the crew seats. First flown in June 1963, entry into service with the Royal Swedish Air Force as the Sk60A trainer followed in May 1966. Sk60B and Sk60C versions delivered later were equipped to light attack standard; several Sk60As were also converted to Sk60Bs. The Austrian Air Force now uses the SAAB-105 version as a SAAB-27 replacement, and the basically-similar SAAB-105XT is available on the export market. The standard aircraft uses two 1,600 lb thrust Turbomeca Aubisque turbofans for a maximum speed of 478 mph and a range of up to 1,100 miles, while a warload of up to 1,543 lb may be carried. The SAAB-105 XT uses two 2,850 lb thrust General Electric J85-17B turbofans.

SABENA–Société Anonyme Belge d'Exploitation de la Navigation Aérienne: the Belgian airline, SABENA, was formed in 1923 to succeed the Société Nationale pour l'Étude de Transports Aériens, which dated from 1919 and had been operating services to London, Paris, and Amsterdam. The new airline immediately started to develop the European network further, and started route-proving flights to the Belgian Congo in 1925, although it was not until 1935 that a Brussels–Leopoldville service could be operated on a regular basis with a Fokker F-VII. Douglas DC-3s were introduced during 1939, and the airline managed to get a number of its aircraft to the United Kingdom after the German invasion of 1940, enabling the aircraft to be used by the Allies.

Services were re-started in 1946, using Douglas DC-3s on the European routes, and DC-4s on a new service across the North Atlantic and on the African services. In 1947 the airline was the first in Europe to operate the Douglas DC-6, and in 1953 started the world's first regular international helicopter services – although the helicopter network has since been abandoned. During the 1950s Convair 440 Metropolitans and Douglas DC-7s were introduced, followed in 1960 by the airline's first jets, Boeing 707s, which were joined during 1961 by Sud Caravelles. Since that time, Fokker F.27 Friendships and Boeing 727s have been added to the fleet, the most recent additions being Boeing 737s and 747s and McDonnell Douglas DC-10-30CFs, to supplement the earlier 707s and Caravelles. Two Aerospatiale-B.A.C. Concorde options are held by the airline.

Sabre, North American F-86: *see* F-86 Sabre, North American.

Sabreliner, North American T-39: *see* T-39 Sabreliner, North American.

S.A.C.: S.A.C., or Strategic Air Command, is the U.S.A.F. command responsible for the maintenance and operation of America's land-based nuclear deterrent forces. Originally it used Boeing B-29 (q.v.) Superfortress and Convair B-36 (q.v.) bombers, and in more recent years Boeing B-52 (q.v.) Stratofortress and Convair B-58 (q.v.) Hustler jet bombers, although the Hustler is now in reserve and the B-52 fleet has been largely involved in conventional bombing duties in Vietnam. The main force of the S.A.C. today is its intercontinental ballistic missile strength of Minuteman and Titan rockets, although some B-52s with stand-off bombs remain, and a new manned bomber, the North American Rockwell B-1 (q.v.), is being developed for the mid-1970s.

'Saddler': the Soviet SS-7 'Saddler' intermediate range ballistic missile is capable of carrying a nuclear warhead in the 5–10 megaton bracket, and large numbers of these missiles are known to be fixed on important military and industrial targets in Western Europe. The range is in the region of 2,000 miles, and an inertial guidance system is used.

Safeguard: the United States anti-ballistic missile system, Safeguard, is largely aimed at the defence of the deterrent forces and the air defence system, with a secondary function of providing cover for the major centres of population and industry. Basically, Safeguard consists of perimeter acquisition radar (q.v.) to detect any intrusion by incoming missile warheads, either from over the North Pole by intercontinental ballistic missiles or from the oceans by submarine-launched ballistic missiles. Once attack is detected, Spartan (q.v.) long-range interceptor missiles are used for destruction of the warheads while still outside the atmosphere, or Sprint (q.v.) terminal interceptor missiles are used for last-minute destruction, although this involves nuclear fall-out over the area being defended. *See also* anti-ballistic missile defences *and* intercontinental ballistic missile.

Safir, SAAB-91: *see* SAAB-91 Safir.

SAGE: the strangely-named semi-automated ground environment system, or SAGE, consists of the defence radar and interceptor aircraft and missiles of NATO, including the NORAD (q.v.) air defence network. SAGE detects any intrusion by enemy aircraft and directs interceptors or surface-to-air missiles to the target.

Salamander, Heinkel He.162: *see* He.162 Heinkel.

SALT: the Strategic Arms Limitation Talks, or SALT, have been taking place between the United States and the Soviet Union over a number of years without any progress until mid-1972, when agreement was reached on the number of intercontinental ballistic missiles and submarine-launched ballistic missiles to be deployed by either side. The advantage in terms of the number of missiles has been given to the Soviet Union, reflecting the greater number of American warheads due to the greater use of multiple warheads (*see* multiple independently-targeted re-entry vehicle) on American Poseidon (q.v.) and Polaris (q.v.) missiles; but the balance of power may well be tipped against the United States in the long run since the agreement does not prevent modernisation of warheads or missiles, thus allowing the Soviet Union the opportunity to catch up. Strangely, neither the United Kingdom nor France was a party to the talks, but the Soviet Union considers British and French submarine-launched missiles as being an integral part of the American force, even though the British Polaris missiles have British-built warheads and the French missiles have no United States involvement at all in their construction and operation.

S.A.M.: *see* surface-to-air missile.

Samson, Air Commodore Charles Rumney, C.M.G., D.S.O., A.F.C., R.A.F., (1883–1931): originally a regular Royal Navy officer, having joined the service in 1898, the then Lieutenant Charles Rumney Samson, R.N., underwent a course of flying instruction in 1911, and by December of that year had become the first British pilot to take off from a warship, when he flew a Short S.27 from a platform built over the bows of the battleship H.M.S. *Africa*. The following May, Samson was probably the first pilot to take off from a ship under way when he flew a Short pusher biplane from another battleship, H.M.S. *Hibernia*. (The first flight from a ship under way is believed by some to have been on 9 May, by Lieutenant R. Gregory, R.N. (q.v.).) Promoted to Commander, Samson became Officer Commanding the Naval Wing of the Royal Flying Corps in October 1912.

Samson's record in World War I was distinguished as well as pioneering, and included leading the first bombing raid on Germany, and later commanding a force of three seaplane carriers in the Middle East. He followed this by experimenting with flights from barges towed by destroyers. While a Wing Captain, Royal Naval Air Service, he was transferred in 1918 to the then newly-formed Royal Air Force, being given the rank of Colonel before being made a Group Captain in 1919, the year in which he became a Companion of the Order of St Michael and St George (C.M.G.) and was awarded the Air Force Cross; he had been awarded the Distinguished Service Order in 1914.

'Sandal': a simple medium-range ballistic missile, the Soviet 'Sandal' uses a liquid-fuel rocket motor for a range of about 1,200 miles, and can carry a one megaton nuclear warhead. Based on German V-2 (q.v.) technology, the 'Sandal' has small tail fins and control vanes in the rocket efflux, while there is an inertial guidance system for control in the descent stage. A single-stage rocket, 'Sandal' was first displayed in 1960 and was deployed in Cuba for a brief period in 1962.

Santos-Dumont, Alberto (1873–1932): a rich expatriate Brazilian resident in Paris at the time of his involvement in aeronautics, Alberto Santos-Dumont enjoyed the much-sought-after distinction of making the first heavier-than-air powered flights in Europe in his '14-bis', a tail-first biplane of roughly boxkite structure. The '14-bis' used a 24 hp Antionette engine on its first flight at Bagatelle on 13 September 1906; this covered a distance of only seven metres, but on 12 November, using a 50 hp Antionette engine, it made four flights, of which the longest was 220 metres. Some credit a Dane, J. C. H. Ellehammer, with making the first powered flights in Europe, but those of Ellehammer's flights which preceded the Santos-Dumont flights were of short duration in an aircraft which was tethered.

Alberto Santos-Dumont had in fact been involved in aeronautics since 1898, when he flew his first airship, and made a considerable number of airship journeys over Paris with the minimum of formality – literally landing for coffee! Although the Santos-Dumont airships were less advanced than those of the Lebaudy brothers and Count Ferdinand von Zeppelin, he was nevertheless able, in his VI, to make a thirty-minute flight from St Cloud on 19 October 1901, flying round the Eiffel Tower and returning to his starting point. Santos-Dumont's contributions to airship and aviation development consisted largely in publicising what might be possible with these new concepts, rather than making any significant technological advances.

S.A.R.: *see* search and rescue.

'Sark': an obsolescent Soviet submarine-launched missile, 'Sark' can only be launched by the submarine while on the surface, and has a range of only 400 miles, thus indicating considerable restrictions on use.

SARO: *see* Saunders-Roe.

Sarti, Vittorio: an Italian, Vittorio Sarti was one of the early designers of helicopters, in this case a co-axial rotor design in 1828, but in common with others, no attempt was made to build, let alone fly, the machine.

S.A.S.–Scandinavian Airlines System: the Scandinavian Airlines System dates from 1946 and the merger of three airlines, the Danish D.D.L., the Swedish A.B.A., and the Norwegian D.N.L. The ownership of S.A.S. was split between Sweden, Denmark, and Norway in the proportions 3:2:2, with each national share being half state-owned and half private enterprise. D.D.L. dated from 1919, A.B.A. from 1924, and D.N.L. from 1927. Initially S.A.S. were only concerned with transatlantic operations and was known as O.S.A.S. (Overseas Scandinavian Airlines System), but the merger of the European networks in 1948 led to the formation of E.S.A.S. (European Scandinavian Airlines System), which merged with S.A.S. in 1951 to form the present airline. This exists under an agreement due for renewal in 1975.

It did not take long for the route network to expand into Africa and Asia, and S.A.S. made commercial air transport history in 1954 by inaugurating the first air service over the North Pole, initially with Douglas DC-6s on a service to the West Coast of the United States. This was followed with a DC-7 service in 1957 which cut the Europe–Japan journey-time in half. Ten years later, S.A.S. became the first European airline to operate via Tashkent in the U.S.S.R., cutting 1,350 miles off the Copenhagen to Bangkok and Singapore journey. The first jet airliner, Sud Caravelles, had been ordered in 1957.

Today, S.A.S. operates a worldwide network of international services. The airline does not operate domestic services since these are left to airlines operating within each Scandinavian country, although S.A.S. has a 50 per cent interest in the Swedish domestic airline, Linjeflyg, and a 25 per cent interest in Greenlandair. A 30 per cent interest in Thai International Airways is also held. The current fleet includes Boeing 747s, McDonnell Douglas DC-10-30s, DC-8s (including the 'Super-Sixty' series), and DC-9s, Sud Caravelles, and Convair 440 Metropolitans.

'Sasin': a Soviet ballistic missile, the SS-8 'Sasin' is believed to be deployed as an intermediate-range weapon for use against Western Europe. A warhead of between five and ten megatons can be carried, and the missile is thought to use a liquid fuel rocket. The two-stage 'Sasin' has been operational since 1963.

satellite: the term satellite refers to any object in orbit around a larger object in space, and strictly speaking this includes the relationship between moons and their planets, as well as the more usual sense of a man-made object in orbit round a planet.

The first man-made satellite to enter earth orbit was the Soviet Sputnik I (q.v.), launched on 4 October 1957, which remained in orbit until January 1958. A Russian dog entered orbit in a satellite launched on 3 November 1957; this was the second Russian satellite. The first human to be carried into earth orbit was the Russian Flight Major Yuriy Gagarin (q.v.), who was launched in his Vostok 1 spacecraft on 12 April 1961. Numerous other satellites have been launched over the intervening years, to the extent that the space around the earth is cluttered with man-made objects, since only the earliest satellites crashed back to the surface. The uses to which such satellites are put include communications, of which the first example was the American Telstar, meteorology, and military observation (including the 'spy-in-the-sky' concept). An advanced concept is that of the orbiting solar observatory (q.v.). There are Soviet and American plans for space stations in orbit around the earth which would fulfil a number of functions, including acting as staging posts for space exploration of the planets.

satellite airfield: as the term implies, a satellite airfield is one which cannot be considered an airfield in its own right but which must be taken as subservient to some way to a larger near-by airfield. The way in which this is done can vary, including the allocation of certain traffics or other functions to the satellite to relieve pressure on the major airfield. Often the satellite will be within the same terminal air traffic control system as the major airfield, and its capacity will be limited by the precedence given to the air traffic control requirements of the major airfield. Satellite airfields frequently perform an important poor-weather diversionary function.

Saudi Arabian Airlines: the Saudi Arabian Airlines Corporation is a state-owned organisation dating from 1945, when it was formed with assistance from Trans World Airlines (q.v.). Charter operations were undertaken prior to the first scheduled flights in 1947, and during the late 1940s a network of domestic and international services was established, using Douglas DC-3s. Bristol Wayfarers were introduced in 1951, followed by other aircraft in the Douglas Commercial series in the decade which followed, Boeing 720 jet airliners being introduced in 1962.

The T.W.A. management and technical assistance contract continues to operate. Today Saudi Arabian Airlines operates a fleet of Boeing 707s, 720s, and 737s, McDonnell Douglas DC-9s, Convair 340s, and Douglas DC-3s on an extensive domestic and international network covering the Middle East and major points in Europe, Africa, and Asia.

Saunders-Roe Aircraft: a British company, Saunders-Roe was formed when Sir Alliot Verdon Roe (q.v.) left Avro during the late 1920s and joined forces with the S. E. Saunders boatbuilding concern, from which partnership a number of flying-boats and amphibians were produced, starting with the twin-engined Cutty Sark of 1929. Although never a large company, Saunders-Roe (or SARO as it was frequently called) produced Britain's largest aircraft after World War II in the Princess (q.v.) flying-boat, and at around the same time produced the world's only jet-powered flying-boat fighter, the SR.A1 (q.v.); but both aircraft were cancelled after a number of test flights.

A light helicopter which entered R.A.F. and British Army service during the 1950s was the Skeeter, a twin-seat design for liaison and A.O.P. duties. This was after the acquisition of Cierva (q.v.), the gyro-plane manufacturer, in 1950. A further helicopter development ultimately appeared as the Westland Scout (q.v.).

During the late 1950s, Saunders-Roe was chosen by the government-sponsored National Research and Development Corporation to build the first full-sized hovercraft, under the direction of its inventor, Sir Christopher Cockerell (q.v.). The series of hovercraft which followed, starting with the SR-N1 (q.v.) of 1958, all carried the SR-N designation, meaning 'Saunders-Roe, Nautical'. However, in 1960 Saunders-Roe became a part of Westland Aircraft (q.v.), and the hovercraft work passed to the joint Westland and Vickers-owned British Hovercraft Corporation (q.v.).

'Savage': one of the main Soviet intercontinental ballistic missiles, the SS-13 'Savage' is a three-stage solid-fuel propellant rocket with a range believed to be up to 6,000 miles, although it is possible that this might be less. A one-megaton warhead is carried

Savoia-Marchetti: an Italian aircraft manufacturer currently specialising in light aircraft, Savoia-Marchetti (sometimes known as SIAI-Marchetti) dates from the interwar years, when it was primarily a manufacturer of seaplanes, one of the earliest designs winning the first post-war Schneider Trophy (q.v.) race. Amongst the more notable aircraft from the company can be included the S.65, a twin-boom design with twin engines mounted fore and aft of the cockpit driving tractor and pusher propellers. This was entered for the 1929 Schneider Trophy race. Another notable design was the S.M.55X flying-boat, a twin-boom, twin-hulled aircraft with two engines mounted in tandem above the middle of the wing, which made a number of record-breaking mass flights during the early 1930s, including that by a formation of twenty-three aircraft from Rome to New York and Chicago in 1933.

The first landplanes from the firm included the S.M.79 Sparviero (Hawk) tri-motor bomber of World War II, credited with being the best Italian bomber of the war, and the S.M.95 transport, also of World War II, although a number were also built after the war for service with Alitalia.

Production today largely centres on the SF.260, a single-engined low-wing monoplane for training duties and for the private owner.

'Sawfly': the Soviet equivalent of the American Polaris (q.v.) submarine-launched ballistic missile, the 'Sawfly' first appeared in 1967, and is considered to be a two-stage solid-fuel propellant missile with a range of between 1,500 miles and 2,000 miles. 'Sawfly' is about forty-two feet in length with a diameter of about six feet.

SC-1, Short: a British experimental V.T.O.L. jet fighter, the Short SC-1 made a number of test flights during the early 1960s as part of a major test programme, although eventually the rival Hawker Siddeley P.1127 (*see* Harrier) was selected instead for development for R.A.F. service. Unlike the vectored-thrust Harrier, the SC-1 used separate lift and propulsive jets – Rolls-Royce RB.108s.

'Scamp': the Soviet 'Scamp' intermediate-range ballistic missile consists of the top two stages of the SS-13 'Savage' (q.v.) and has an estimated range of 2,500 miles, using solid-fuel propulsion and carrying a one megaton nuclear warhead. In service since 1965, the 'Scamp' is considered to be based mainly on the Sino-Soviet border. A high degree of mobility has been attained by mounting the missile on a tracked launch vehicle.

'Scarp': the largest Soviet I.C.B.M., the SS-9 'Scarp' is in service in a number of versions, including the earlier types with a large single warhead of around twenty megatons; a developed version for launching orbiting bombs; and the latest version, with a multiple re-entry vehicle dispenser containing three re-entry vehicles with five megaton warheads. It is believed that the most recent re-entry vehicles are independently targeted. 'Scarp' is a large two-stage rocket of about 120 feet in length, and is believed to use liquid-fuelled motors. Little is known about range, etc.

Schneider, Jacques: a wealthy Frenchman, Jacques Schneider was convinced that the ideal concept for speed in the air was the seaplane, and so in 1912 he inaugurated the Schneider Trophy seaplane race, with a $5,000 trophy and a $5,000 first prize, the trophy to be retained by any nation winning it three times in succession.

Schneider Trophy: the Schneider Trophy or, to be exact, the Jacques Schneider Air Racing Trophy for Hydro-Aeroplanes, was first competed for in April 1913 at Monaco and won for the first time by a Deperdussin float-monoplane powered by a 160 hp Gnome rotary engine flying at an average speed of 45·75 mph – this would have been higher had not there been some dispute requiring a further circuit of the course. The second race, and the last before the competition was suspended during World War I, was won by a floatplane conversion of the Sopwith Tabloid, which flew at an average speed of 86·78 mph.

The first post-war race was won by a Savoia biplane, and other notable winners of the race after the war included the Curtiss R3C-2 biplane of 1925, with a speed of 232 mph, the Macchi monoplane of 1926, and the Supermarine S.5, S.6, and S.6B monoplanes of 1927, 1929, and 1931 respectively, winning the Trophy outright for the United Kingdom and, on the last occasion, producing an average speed of 340 mph

Schweitzer Aircraft: a small United States manufacturer, specialising in glider production, but also undertaking production and marketing of the Thurston Teal amphibian.

Scottish Aviation: a British aircraft manufacturer, Scottish Aviation entered aircraft production on its own account during the 1950s with the Pioneer and Twin Pioneer S.T.O.L. transports for the R.A.F., while also undertaking overhaul and refurbishing work for the R.A.F. and other operators. Further work during the late 1960s consisted of subcontract work on the Handley Page Jetstream and the Lockheed C-130K Hercules, but towards the end of the decade the company acquired the production jigs and rights for the Beagle Bulldog (q.v.) after Beagle's bankruptcy. Production rights and jigs for the Jetstream (q.v.) were acquired in 1972 from Jetstream Aircraft, which had previously acquired Jetstream production rights from Handley Page on that company's collapse. Bulldog and Jetstream production now comprise the main part of Scottish Aviation's workload.

Scout, Bristol: one of the first operational reconnaissance aircraft in service with the Royal Flying Corps during the early stages of World War I, the Bristol Scout single-seat biplane was unarmed, except for a revolver carried by the pilot, and was later relegated to training duties rather than upgraded to fighter operations. A single 80hp Le Rhône rotary engine powered the aircraft.

Scout, Morane: the Morane Scout biplane was used in small numbers by the British and French air arms during the early stages of World War I, mainly on reconnaissance duties, but both the pilot and the observer were provided with Lewis guns later. A 110 hp Le Rhône rotary engine powered the aircraft.

Scout, Nieuport: the Nieuport Scout single-seat biplane was in service with the Allies at the start of World War I, and at first was completely unarmed, although a Lewis gun was fitted later. A single 80 hp Le Rhône or Gnome rotary engine was fitted.

Scout, Westland: originating as a Saunders-Roe (q.v.) design, first flying in prototype form in July 1958, the Westland Scout light liaison helicopter entered production for the British Army in 1960, and a navalised version, the Wasp, with a wheeled undercarriage instead of skids, entered production for the Royal Navy in 1962. Scout and Wasp helicopters are now out of production having been superseded by the Westland W.G.13 (q.v.). Wasps are also in service with the navies of South Africa, the Netherlands, New Zealand and Brazil, operating from anti-submarine and general-purpose frigates. A single 710 shp Rolls-Royce Nimbus 103 or 104 shaft turbine provides a maximum speed of 120 mph and a range of up to 270 miles, while three passengers can be carried in addition to the crew of two, or two anti-submarine torpedoes (Wasp) or air-to-surface missiles (Scout).

'Scrooge': a Soviet intermediate-range ballistic missile, the 'Scrooge' has a range of some 3,500 miles and is highly mobile, being mounted on a tracked launch vehicle. Large numbers of 'Scrooges' are believed to be deployed on the Sino-Soviet border. Little is known about this missile, although a nuclear warhead would certainly be carried.

'Scud': a Soviet battlefield support missile, the 'Scud' has a range of fifty miles and uses an inertial guidance system. Nuclear or conventional warheads can be carried, and the missile is single-stage with a liquid propellant rocket. 'Scud' is approximately thirty-three feet in length and thirty inches in diameter, and is mounted on a converted tank chassis.

SE-: international civil registration index mark for Sweden.

S.E.5, Royal Aircraft Factory: a twin-seat fighter biplane of the latter half of World War I, the Royal Aircraft Factory S.E.5 was undoubtedly one of the best fighters of the war, and entered Royal Flying Corps service in large numbers. Propeller-synchronised Lewis guns were fitted, and Hispano-Suiza 'V' engines of 150 or 200 hp were used, depending on the version. An improved variant, the S.E.5a, was in service before the Armistice.

Seaboard World Airlines: originally formed in 1946 as Seaboard and Western Airlines, Seaboard World operated charter freight services until 1956, when scheduled transatlantic services were started. Canadair CL-44 'swing-tail' freighters were introduced in 1962, and replaced in 1966 by the first of a fleet of McDonnell Douglas DC-8-55CFs, which were later joined by DC-8-63CFs. Three Boeing 747QCs have recently been added to the fleet, which consists of these aircraft and Douglas DC-8-63Fs. Scheduled freight services are operated from New York to most major European destinations.

Seacat: one of the most widely-used shipboard surface-to-air missiles today, the Short Seacat is a short-range weapon with a secondary surface-to-surface capability and a wide range of guidance systems available, although radio command guidance is probably the most common. A two-stage solid propellant rocket is used, with a high-explosive warhead, and the missile is normally mounted on a quadruple launcher. Missile length is about five feet. A version for use by land forces is the Tigercat (q.v.).

Sea Dart: a medium-range surface-to-air missile for use aboard warships, the Hawker Siddeley Dynamics Sea Dart entered service in 1972 aboard the Royal Navy's Type 82 destroyers, and will also equip Royal Navy and Argentinian Navy Type 42 destroyers. A secondary surface-to-surface capability is provided, in addition to interception of aircraft and surface-to-surface missiles at a wide range of altitudes. A semi-active radar homing system of guidance is used, and the missile is ramjet-powered, with a solid-fuel rocket booster first stage. Maximum range is in excess of twenty miles.

Seafire, Supermarine: the carrier-borne version of the highly-successful Supermarine Spitfire (q.v.) fighter of World War II, the Seafire was far less successful and less common, even allowing for the lower numbers of naval aircraft. A number of shortcomings included an insufficiently strong undercarriage for carrier deck landings.

Sea Hawk, Hawker: developed during the immediate post-World War II period, the Hawker Sea Hawk carrier-borne jet fighter first flew in 1947 as the Hawker P.1040, from which the Hunter (q.v.) was also later developed. Initial production was by Hawker, but after the first thirty-five aircraft were delivered to the Royal Navy, production and further design work was transferred to Armstrong-Whitworth. The aircraft was also later supplied to the Federal German Navy and to the Indian Navy. A straight leading-edge wing, unlike the related Hunter, was characteristic of the aircraft, and plans for a swept wing development were dropped. A single-seat aircraft, the Sea Hawk used a single 5,400 lb thrust Rolls-Royce Nene 103 turbojet in the F.(G.A.) Mk.6 version for a maximum speed of 590 mph and a range of around 600 miles, while four 20 mm can-non were fitted in the fuselage, and bombs, rockets, or drop tanks could be carried from four underwing strongpoints.

Sea Hornet, de Havilland: a carrier-borne version of the de Havilland Hornet (q.v.) for the Royal Navy, this was the last piston-engined combat aircraft to enter service with the Fleet Air Arm.

Sea King, Sikorsky S-61/SH-3: *see under* S-61, Sikorsky and Westland.

Sea Knight, Boeing-Vertol Bv. 107: *see* Bv. 107 Sea Knight, Boeing-Vertol.

seaplane: the term 'seaplane' refers to an aircraft which lands on water and uses floats in place of an undercarriage, differing from the flying-boat in not having a hull immersed in the water, and from the amphibian in not having the capability of landing on land. The predecessor of the seaplane was the float-plane (q.v.), and hence the pioneers in this field of aviation were Henri Fabre (q.v.) and Glenn Curtiss (q.v.).

Subsequent development of the seaplane was given considerable incentive before and after World War I by the Schneider Trophy (q.v.) races which started in 1913 and ended in 1931. During World War I a number of seaplane types were put into service, including the Fairey Campania and the Short 184, and supported by seaplane carriers or tenders. Significant producers of seaplanes included Britain's Supermarine, Short Brothers, Fairey, and Hawker, America's Grumman and Curtiss, Italy's Savoia-Marchetti and Aermacchi, and Heinkel in Germany. Not all seaplanes were single-engined types – most were, but exceptions included the twin-engined Heinkel He.59, the trimotor Z.506B Airone, and the four-engined Short Mercury (*see under* Mayo, Short). While European manufacturers tended to favour twin floats, American manufacturers produced designs with a large single float.

The seaplane, in common with the flying-boat and the amphibian, was of decreasing importance after the start of World War II (during which the only significant seaplane was the Heinkel He.115 twin-engined torpedo-bomber), largely due to the ever-increasing range of landplanes and the improved performance of carrier-borne aircraft, followed by the advent of the helicopter. There are no modern aircraft designed as seaplanes, although a number of landplane designs are available with floats.

search and rescue: one of the many tasks falling to aviation over the years has been that of search and rescue or air-sea rescue, for which the aeroplane can be invaluable since even at low height an extended horizon is provided which, with the increased speed over surface vessels, means that a larger area can be searched more quickly and thoroughly than might otherwise be possible. Limitations on the aircraft include adverse weather, but it has not been unknown for aircraft to be able to take off while ships have been kept in port due to the perils of navigating a difficult harbour entrance.

Search, however, is only a part of the operation, and the aim has always been to design, develop, and produce an aircraft capable of effecting the rescue as well as the search function – although at worst search aircraft can always drop inflatable life rafts and supplies. Few aircraft have in fact been built specifically for search and rescue duties, since this role has often fallen to maritime-reconnaissance (q.v.) aircraft and units, while helicopters (q.v.), flying-boats (q.v.) and amphibians (q.v.) have in general had obvious advantages for these tasks, albeit with limitations of speed in the case of the helicopter. Suitable flying-boats have been available since the early 1920s.

Specific search and rescue aircraft have, however, included the Supermarine Walrus (q.v.) and the Consolidated Catalina (q.v.) of the late 1930s, both of which remained in service throughout World War II and, in the case of the Catalina, for some years afterwards as well. Helicopters, from the Sikorsky S-51 (q.v.) or Westland Dragonfly onwards have always been used for rescue duties, specially-designed machines including the Kaman Husky and Seasprite, while there have also been special versions of the S-61 (q.v.).

A modern search and rescue operation requires a degree of co-ordination and control not unlike that of an air defence system, the need for prompt response furthering the comparison. Systems vary, but in most countries the armed forces are involved to a greater or lesser degree, usually in the provision of aircraft. In the United Kingdom, H.M. Coastguard Service co-ordinates rescue, which can be effected by the voluntary Royal National Lifeboat Institution, the Royal Air Force or the Royal Navy or, in certain areas without service helicopter cover, by civil helicopters leased by the Coastguard Service. In the United States the Coastguard Service has its own ships and aircraft, but can call on U.S.A.F. and U.S.N. units if necessary.

Seaslug: a medium-range surface-to-air missile for shipboard applications, the Hawker Siddeley Seaslug is now obsolescent. Both Mk.1 and Mk.2 versions of the Seaslug use radar beam riding guidance and have solid fuel propellant rocket motors with solid fuel boosters, but the Mk.2 has a longer range and improved capability when used against low-flying targets, as well as having a secondary surface-to-surface capability.

Seasprite, Kaman UH-2: *see* UH-2 Seasprite, Kaman.

seat pitch: *see* pitch, seat.

SEATO: the South East Asia Treaty Organisation, or SEATO, was formed as a result of the Manila Treaty of 1954, and itself dates from 1955. It is a multilateral alliance and the founder-members included Australia, France, New Zealand, Pakistan, the Philippines, the United Kingdom, and the United States, of which Pakistan left in late 1972. There is no formal command organisation, and members are only bound to consult in the event of an attack upon the forces of a member state. Control is through meetings of a Council, and although there are no forces committed to the Organisation, British and American units in the area, and elsewhere, are nevertheless allocated a SEATO role. Apart from SEATO, many of the member states are bound by other treaties, notably ANZUS (q.v.), ANZAM (binding together the forces of Australia, New Zealand, and the United Kingdom in Malaysia), and a separate Anglo-Malaysia Treaty.

Sea Vampire, de Havilland: *see under* Vampire, de Havilland.

Sea Venom, de Havilland: *see under* Venom, de Havilland.

Sea Vixen, Hawker Siddeley: developed during the early 1950s as the de Havilland D.H.110, the Hawker Siddeley Sea Vixen carrier-borne interceptor first flew in production form in March 1957, and entered Royal Navy service in late 1958 as the F.(A.W.)Mk.1. The F.(A.W.)Mk.2 entered service in 1962, with Red Top missiles instead of the Firestreaks of the earlier aircraft, and increased fuel capacity. A twin-seat, twin-boom design, the Sea Vixen used two 11,250 lb thrust Rolls-Royce Avon 208 turbojets for a maximum speed of about Mach 1·0, and an armament of either Red Top or Firestreak air-to-air missiles in the interceptor role, or rocket pods, Bullpup air-to-surface missiles, or 500 lb bombs on six underwing strongpoints in the ground-attack and fighter-bomber roles. The Sea Vixen was retired from Fleet Air Arm service in 1972 during the rundown of the F.A.A.'s fixed-wing element.

Seawolf: little information has been released so far about the British Aircraft Corporation's Seawolf shipboard surface-to-air missile, other than that it will be a fast-reaction device for use against supersonic aircraft and surface-to-surface missiles, while also possessing a secondary surface-to-surface capability of its own. A single-stage rocket with solid-fuel propellant, guidance will be by radio command.

second line: the expression 'second line' generally relates to aircraft used for an air arm's own internal support functions, such as communications duties, target towing, navigational aids duties, and so on. Combat and transport aircraft are not generally included in the designation, although towards the end of their lives they can be relegated to second line operations.

second strike capability: in terms of modern nuclear warfare, second strike capability is the capacity to mount a second strike which, in this case, is regarded as being made after suffering a nuclear attack since, the theory is that a first strike will be made as soon as a nuclear attack is detected as being on its way. It is, in theory at any rate, possible that no first strike would be mounted for fear that this action would be precipitate; however, not only would a nuclear attack by missiles or aircraft be obvious, but failure to make a first strike would also result in the risk of discovering that the second strike capability is non-existent. The means of mounting a second strike include specially hardened silos for I.C.B.M.s, such as the American Minuteman, and possibly some highly-mobile means of counter-attack able to escape detection and attack, such as a nuclear submarine fleet equipped with missiles.

semi-active radar homing: a means of surface-to-air missile guidance, semi-active radar homing requires only a simple tracking radar in the missile (hence the term semi-active radar) to enable it to home on to emissions bounced off the target by ground radar, which is frequently a part of the launch equipment.

semi-automatic ground environment system: *see* SAGE.

Seneca, Piper: the Piper PA-34 Seneca is essentially a twin-engined development of the PA-28E Cherokee (q.v.) Arrow, and was introduced in 1971. Two 200 hp Lycoming 10-360-AIA piston engines provide a maximum cruising speed of 190 mph and a range of up to 1,160 miles, while seven seats are fitted.

Sentinel: a U.S. anti-ballistic missile system first mooted during 1967, the Sentinel project was replaced by the Safeguard (q.v.) system, although using the same component parts – missiles and radar. The difference between Sentinel and Safeguard lies in the fact that Sentinel was intended mainly as a defence against I.C.B.M.s coming over the North Pole, while Safeguard provides cover against submarine-launched ballistic missiles and orbital bombs as well.

'Serb': a Soviet submarine-launched missile, the 'Serb' is an intermediate weapon between the 'Sark' (q.v.) and the later

'Sawfly' (q.v.), having a range of between 400 and 700 miles, according to estimates, and being capable of launch while the submarine is submerged. A two-stage liquid-fuel missile of some thirty-three feet in length and with a diameter of five feet, 'Serb' can carry a one-megaton warhead, and was first publicly displayed in 1964.

Sergeant: a single-stage battlefield support missile, the Sergeant, built by the Univac Salt Lake City Company, has replaced the earlier Corporal, and is in service with the United States and West German Armies. Approximately thirty-four feet in length and with a diameter of thirty-one inches, the Sergeant has a range of seventy-five miles and uses a solid propellant rocket motor and an inertial guidance system. Either nuclear or high-explosive warheads may be fitted.

sesquiplane: a sesquiplane is a biplane with one wing, usually the lower, markedly smaller than the other. Sometimes the smaller wing is only half the area of the larger.

7P-: international civil registration index mark for Lesotho.

7Q-: international civil registration index mark for Malawi.

7T-: international civil registration index mark for Algeria.

Seversky Aircraft: formed by Alexander de Seversky, a Russian who, like Igor Sikorsky (q.v.), had fled the Bolshevik Revolution, Seversky Aircraft built a number of monoplane designs during the inter-war period, including the three-seat SEV-3MWW low-wing float amphibian and the U.S.A.A.F.'s P-35 fighter, which entered service during the late 1930s and used a single 900 hp Pratt and Whitney Wasp radial engine. Later the firm was absorbed into Republic Aviation (q.v.).

Shackleton, Avro: developed from the Avro Lincoln heavy bomber, itself a development of the earlier Lancaster, the Avro Shackleton was built primarily for long-range maritime-reconnaissance operations, and first flew in March 1949. Entry into R.A.F. service started towards the

end of 1951, with deliveries of the Mk.1. An improved version, the Mk.2, followed in 1952, joined later by the Mk.3, with further improvements including the substitution of a nosewheel for the tailwheel of earlier versions. The Mk.3 was also supplied to the South African Air Force. Replacement of the Shackletons in the maritime-reconnaissance role has been by Hawker Siddeley Nimrods, but a number of aircraft remain in service after conversion to the airborne-early-warning radar role. A mid-wing aircraft with a twin-fin tailplane, the Shackleton uses four 2,455 hp Rolls-Royce Griffon 57A piston engines and two 2,500 lb thrust Bristol-Siddeley Viper 203 turbojets for a maximum speed of over 300 mph and a range of up to 3,660 miles. Before conversion to A.E.W. depth charges and torpedoes could be carried.

'Shaddock': now obsolete and possibly out of service, the Soviet 'Shaddock' tactical cruise missile uses a turbojet with rocket boosters, and has a range of 250 miles with a one-kiloton warhead.

SHAPE: SHAPE (Supreme Headquarters, Allied Powers in Europe) is, with ACLANT one of the two major NATO (q.v.) Commands. Based in Belgium and coming under the Supreme Allied Commander, Europe (SACEUR), SHAPE operates through the Allied Command, Europe (*see* ACE) to its subordinate commands, AFNORTH (q.v.), AFCENT (q.v.), and AFSOUTH (q.v.). So far, SACEUR has been an American general. SHAPE responsibilities include the defence of the territory of all European NATO members, with the exception of the United Kingdom and of Portuguese territorial waters.

Shin Meiwa: the aircraft production interest of the Japanese Shin Meiwa concern centres around the PS-1 (q.v.) maritime-reconnaissance flying-boat for the Japanese Maritime Self-Defence Force, which is the only flying-boat in production today.

Short Brothers and Harland: a British Company formed before World War I by Horace and Eustace Short, Short Brothers and Harland was known for many years simply as Short Brothers. The company made an early mark on aviation history in

1912 with experiments in arming aircraft and in deck take-offs, with the Short S.27, from platforms fitted over the bows of British battleships. The Short Type 38 was the first aircraft to be both radio-equipped and armed with a machine gun in pre-World War I experiments by the Royal Flying Corps. During World War I the Short 184 and 225 float-biplanes played an important part in naval reconnaissance work with the Royal Naval Air Service, while after the war the Silver Streak (q.v.) biplane was the first aircraft to employ a stretched skin all-metal monocoque construction, and the Short-built version of the Seaplane Experimental Station's F.5 flying-boat was the first military flying-boat to have a metal hull.

A number of ultra-light aircraft designs were produced after the war and during the early 1920s, including both landplanes and flying-boats, but the main preoccupation of the company was with the design, development, and production of a series of large flying-boats for military and civil users, with some landplane development. Amongst the first of these were the Short Calcutta (q.v.) for Imperial Airways, the Rangoon for the Royal Air Force, the Scipio four-engined flying-boat and its landplane development, the Scylla, and the famous Empire (q.v.) flying-boat of the late 1930s for Imperial Airways and Qantas. Unusual aircraft included the Short Mayo (q.v.) composite aircraft, while a pressurised landplane project, the Stratoliner, was abandoned on the outbreak of World War II.

A large Short flying-boat, the Sunderland (q.v.), developed from the Empire series, figured prominently in maritime-reconnaissance operations during World War II, and Shorts also produced Britain's first real heavy long-range bomber, the Stirling (q.v.). A stretched version of the Sunderland, the Shetland, was developed at the end of World War II, but cancelled, although a new civil flying-boat, the Solent, went into service with B.O.A.C. Early post-war aircraft from he company also included the Sturgeon, a twin-engined target tug for the Royal Navy, and the Sealand twin-engined amphibian.

A production facility had previously been established with the shipbuilders Harland and Wolf, and this was known as Short-Harland until 1947, when it was merged with the parent company to form Short Brothers and Harland.

During the post-war period Short Brothers and Harland have moved completely away from their flying-boat image, with licence-production of Britannia (q.v.) transports for the Royal Air Force, and the development from the Britannia of the giant Belfast (q.v.) strategic heavy transport. They have also built and conducted an experimental programme with a vertical take-off jet fighter, the SC-1 (q.v.). The company has also made its mark in the guided weapons field with the highly successful Blowpipe (q.v.), Seacat (q.v.), and Tigercat (q.v.) missiles, and is currently producing these, and the new Skyvan (q.v.) and Skyliner transport aircraft. Subcontract work in recent years has included building pods for the Rolls-Royce RB.211 engines used by the Lockheed TriStar airbus. Non-aviation activities include armoured car production.

Short Brothers is owned mainly by the British Government, with Rolls-Royce (q.v.), now also state-owned, holding a minority interest.

short haul: civil transport aircraft with a range of up to 1,000 miles, or scheduled air services of less than 1,000 miles in distance, are generally covered by the 'short-haul' description.

Shorthorn, Farman: a biplane with a single 70 hp Renault engine driving a pusher propeller, the Maurice Farman Shorthorn was so called because of the very short and stubby landing skids, which were in marked contrast to those of the Longhorn (q.v.). A tandem twin-seat design, the Shorthorn was used by the French Aviation Militaire for artillery spotting and light observation duties at the start of World War I.

short take-off and landing: increasing aircraft speeds and sizes have required ever-increasing runway lengths over the years, and in an attempt to find aircraft which can operate from reduced lengths of runway for their size or speed, the short take-off and landing (S.T.O.L.) aircraft has been evolved. The helicopter's poor performance has also been behind the evolving of S.T.O.L. fixed-wing aircraft. S.T.O.L. can be achieved in one or more of three ways: by increasing wing area, which tends to reduce aircraft speed as well; by increasing engine size, which means having costly excess power in flight; and by the complication of providing extra flaps and slots, perhaps with thrust from the engines bled over them. A fourth system, only found so far in high performance military aircraft, is variable geometry (q.v.) or swing wings.

A number of aircraft manufacturers have either produced S.T.O.L. aircraft of their own or modified existing designs to achieve near-S.T.O.L. performance. Some manufacturers, such as de Havilland Canada, Pilatus, and Britten-Norman, have specialised in this field. Other important S.T.O.L. aircraft from manufacturers with other aircraft in production include the Short Skyvan (q.v.), the Dornier Do.28D (q.v.) Skyservant, the Breguet 941S (q.v.), and a version of the Lockheed C-130 (q.v. Hercules.

Shrike: a U.S. air-to-surface missile manufactured by Texas Instruments, the AGM-45A Shrike is the first to be designed exclusively for use against enemy radar installations, which it attacks by homing on to their emissions. The range of the missile is about two miles. It is some ten feet in length, uses a solid propellent rocket, and carries a high-explosive warhead. In U.S. Navy service since 1964, the Shrike is now also in U.S.A.F. and Israeli Defence Force/Air Force service.

Sidewinder: the AIM-9 Sidewinder air-to-air guided missile has been in production since 1956, and is in service with the American armed forces and with the armed forces of most of America's allies. Production has been by Philco-Ford or Raytheon, with some licence-production in West Germany. Sidewinder 1A entered operational service in 1956, and uses infra-red homing, which was coupled with a range of just over half a mile, and poor performance except in good visibility and when fired in direct pursuit of an opponent. Sidewinder 1C uses semi-active radar homing and possesses a range of just about two miles,

while Sidewinder 1D uses a much-improved wide-angle infra-red homing device, and has a range of more than two miles. All Sidewinders use solid-fuel propellants and are just under ten feet in length.

Sikorsky, Igor (1889–1972): a Russian, Igor Sikorsky was undoubtedly the greatest aircraft designer from that country. His more important designs included the world's first four-engined aircraft, the Bolschoi (q.v.) of 1913; this was followed during World War I by the giant four-engined Ilya Mourometz (q.v.) heavy bomber, which operated successfully with the Imperial Russian Air Service. Sikorsky left Russia after the Bolshevik Revolution and settled in the United States, establishing Sikorsky Aircraft (q.v.), which produced a number of good flying-boat and amphibian designs between the wars, before developing the first successful helicopter to be flown outside of Germany, and following this with the first of a series of highly-successful and practical helicopter designs.

Sikorsky Aircraft: formed in 1923 as the Sikorsky Aero Engineering Corporation by the Russian *émigré* Igor Sikorsky, Sikorsky Aircraft specialised for many years in the manufacture of flying-boats and amphibians, including the S-39 of 1931, a five-seat, single 400 hp Pratt and Whitney radial-engined amphibian; the S-42 four-engined flying-boat of 1934 and the S-43 twin-engined amphibian, both for Pan American World Airways; and the VS-44, another four-engined flying-boat. The connection with United Aircraft, of which Sikorsky is today an important part, started in 1929.

Sikorsky's record in flying-boat manufacture was creditable enough, but the company's real claim to fame was in the development and production of the first operational helicopters, following on from Igor Sikorsky's work with the VS-300 experimental helicopter in 1939, the first successful helicopter design produced outside of Germany. The VS-300 led to the R-4, which undertook trials with the British and American armed forces, paving the way for the first production helicopter, the R-5 or Sikorsky S-51 (q.v.), which was also built under licence in the United Kingdom by Westland as the Dragonfly.

The S-51 was followed by the series of helicopters which has ensured a strong position for Sikorsky Aircraft in this field until the present time, even after the entry of many other manufacturers into helicopter development and production. Sikorsky helicopters have included the S-55 (q.v.) and the S-58 (q.v.), both with piston engines as well as the turbine S-61 (q.v.), S-62, S-64 (q.v.), and S-65 (q.v.). Many of these have been built under licence outside of the United States, mainly by Westland, but also by Dornier, Sud Aviation, Agusta, and Kawasaki. Apart from further helicopter development, including an armoured battlefield development, Sikorsky have also undertaken studies into various kinds of stowed rotor and contra-rotating rotor V.T.O.L. aircraft.

Silver Streak, Short: the first aircraft in the world to have a stressed skin all-metal monocoque fuselage structure, the Short Silver Streak (or Swallow as it was initially called) was a single-engined, single-seat biplane built of duralumin, and first flew in 1920. The aircraft was far ahead of its time, and it was to be some years before this type of construction was employed on standard production aircraft, particularly those coming from the British industry.

simulator: modern airline and (to a considerable extent) military pilots receive a considerable part of their training on flight simulators, which are replicas of flight decks or cockpits, usually being based on definite aircraft types, and with handling characteristics built in. A high degree of realism is produced with the assistance of film and sound tapes, enabling training to take place in greater safety and at lower cost than would be the case if all training was conducted on an aircraft – although actual flying cannot be dispensed with altogether. Modern simulators can also be used for conversion training of qualified aircrew, or for safety checks on aircrew.

The flight simulator evolved between the wars, largely due to the invention of the Link trainer, named after its inventor. During World War II Link trainers, which were a very basic type of simulator, were used to give novice pilots the feel of aircraft controls.

Sioux, Bell 47: see Bell 47 Sioux.

Sirius, Lockheed: the first Lockheed aircraft to use a low-wing layout, the Sirius was built in 1929 as a single-engined monoplane with tandem twin-seating and a fixed undercarriage. Flown by Colonel Charles Lindbergh, it established a record for the West–East U.S.A. flight of $14\frac{3}{4}$ hours. A development, the Altair, with a retractable undercarriage, made the first non-stop East–West flight across the Pacific Ocean in November 1934.

60S-: international civil registration index mark for Somalia.

6V-: international civil registration index mark for Senegal (see also 6W-).

6W-: international civil registration index mark for Senegal.

6Y-: international civil registration index mark for Jamaica.

'Skean': a development of the 'Sandal' (q.v.) surface-to-surface missile, the Soviet 'Skean' has a range of some 2,000 miles and can be regarded as being in the I.R.B.M. category, using liquid propellant and carrying a nuclear warhead.

Skycrane, Sikorsky S-64: see S-64 Skycrane, Sikorsky.

Skyhawk, McDonnell Douglas A-4: see A-4 Skyhawk, McDonnell Douglas.

Skyliner, Short: see under Skyvan, Short.

Skymaster, Douglas DC-4/C-54: see DC-4 and C-54 Skymaster, Douglas.

Skyraider, Douglas A-1: see A-1 Skyraider, Douglas.

Skyservant, Dornier Do.28D: see Do.28D Skyservant, Dornier.

Skyvan, Short: developed as a simple utility aircraft with a box-like fuselage, a high wing, and a non-retractable undercarriage, the Short SC.7 Skyvan first flew in prototype form in January 1963, using two Continental GTS10-520 piston engines of 390 hp each, although it was soon decided to substitute turboprops, and the prototype was re-engined with Turbomeca Astazou II turboprops. Early production models used 690 shp Tubromeca Astazou XII turboprops, but the standard specification is now for two 715 shp Garrett-AiResearch TPE331-201 turboprops, giving a maximum cruising speed of 207 mph, a range of up to 690 miles, and a maximum payload of 4,600 lb. The original aircraft could accommodate eighteen passengers, but executive and Skyliner versions are now available, with improved soundproofing and interior comfort, which can carry up to twenty-two passengers. Further development work is taking place and will result in a thirty-passenger SH-3D version with a generally 'scaled-up' appearance, modified tailplane, and retractable undercarriage.

slot: a slot is a form of high lift device resulting from the fitting of a slat to the leading edge of an aircraft's wing; in flight the slat is normally retracted into or against the leading edge, closing the slot. The slot augments the lift by providing a smooth airflow over the upper surface of the wing. The first slotted wings were developed by Handley Page (q.v.) during the 1920s.

SNECMA: the main French aero-engine manufacturer, SNECMA (Société National d'Étude et de Construction de Moteurs d'Aviation), dates from 1945 when it was formed from the amalgamation of a number of small aero-engine manufacturers, of which the most famous were Gnome et Rhône, which had done much to develop the rotary engine during World War I, and the aero-engine division of Renault. Since then SNECMA has acquired Bugatti and, more recently, Hispano-Suiza (q.v.). SNECMA is owned by the French Government.

Currently, SNECMA is collaborating with Rolls-Royce (q.v.) on a number of projects, including the Olympus engine for the Concorde supersonic airliner and the M45H engine for the V.F.W. 614, as well as building Rolls-Royce Tyne turboprops under licence for the Franco-German C-160 (q.v.) Transall and Br. 1150 (q.v.) Atlantique projects. Pratt and Whitney (q.v.) and General Electric (q.v.) engines

are built under licence for the Caravelle and the A.300B airbus respectively. SNECMA also builds its own Atar series of engines, based on an original German design and first run in 1948, for the Mirage series of aircraft. Collaboration with the other French engine manufacturer, Turbomeca (q.v.), is proceeding on the engine for the Alphajet trainer, and a 'ten ton' (i.e. 22,000 lb thrust) advanced technology project with General Electric is under consideration.

Soko: the main Yugoslav aircraft manufacturer, Soko is state-owned, and is currently engaged in producing the G2-A (q.v.) Galeb basic jet trainer and its derivative, the J-1 (q.v.) Jastreb light attack aircraft. The smallest aircraft in the range is the Kraguj light attack aircraft, which uses a piston engine.

Sol-Sol Balistique Stratégique: an Aerospatiale product, the French Armée de l'Air has at least eighteen Sol-Sol Balistique Strategique, or S.S.B.S., missiles deployed, currently in the S-2 version, but an improved S-3 is under consideration for the late 1970s. An atomic warhead of only 150 kilotons is fitted, although a one megaton hydrogen warhead is under development. A two-stage rocket of forty-eight feet in length and five feet diameter, the S.S.B.S. is deployed in hardened concrete silos and has a range of 1,800 miles. Control is by gimballed nozzles and inertial guidance is used.

sonar: an acronym for sound navigation ranging, sonar's main application is in the detection of submarines, and the workings of the device can best be described as being an acoustic version of radar, since signals are bounced off submerged objects. Sonar is less exact than radar and much more limited, since performance depends on the temperature and salinity of the water. Sonar can also be used as a means of navigation and as a homing device for guided torpedoes, and also as a means of underwater communication. Use of sonar by aircraft for submarine detection consists in either dunking a sonar buoy in the water from a helicopter hovering overhead, or dropping sonar buoys from an aircraft unable to use the dunking facility, and receiving radio signals from the sonar buoys.

sonic boom: the sonic boom results from the considerable pressure wave generated by an aircraft flying at supersonic (q.v.) speed, and the size of the pressure wave depends on the speed, size, and height of the aircraft, and on the temperature of the atmosphere. If an aircraft is flying at a considerable height and only just above Mach 1 (the speed of sound), it is unlikely that the boom will reach the ground. On the other hand, if the aircraft is large enough and flying low enough and at considerable speed, the pressure wave felt on the ground can be such as to cause considerable damage.

Sopwith: founded before World War I by T. O. M. Sopwith, Sopwith Aircraft built a number of excellent designs before and during World War I, giving the Royal Flying Corps and the Royal Naval Air Service some of their best fighter and scout aircraft. Although the firm was mainly a landplane manufacturer, the Sopwith 'Bat Boat' was the first real amphibian, and a float conversion of the Sopwith Tabloid (q.v.) won the second Schneider Trophy contest in 1914. The Tabloid in its landplane form was later one of the first fighter biplanes. The next few years saw a succession of Sopwith designs, including the Pup (q.v.), the 1½-Strutter (q.v.), the Dolphin, and the Camel (q.v.), the latter being the first aircraft to fly from barges towed behind destroyers. Almost all Sopwith designs were biplanes, the exception being the triplane, or 'Tripe', which flew with the Royal Naval Air Service. Towards the end of the war the company built the Snipe fighter and the Cuckoo torpedo-bomber.

The company ceased business after World War I ended, only to have its factory taken over by the newly-formed Hawker Aircraft (q.v.), with which T. O. M. Sopwith remained associated, although there were no financial links between Sopwith Aircraft and Hawkers.

sound barrier: the so-called sound barrier is in effect the speed of sound, probably gaining this popular title because of the pressure wave resulting from an aircraft flying at supersonic speed, which is usually most apparent on the ground as a double bang or sonic boom (q.v.). The speed of sound is 760 mph at sea level, decreasing to 660 mph in the stratosphere because of the lower temperature prevailing at high altitudes. Regardless of the actual height or speed, the speed of sound is expressed as Mach 1·0.

South African Airways: formed in 1934 when the Union Government acquired Union Airways, a privately-owned airline unable to meet the capital demands of a rapidly-growing airline, South African Airways expanded further in 1935 with the acquisition of South West African Airways, another privately-owned company dating from 1932. The mainstay of the fleet at this time consisted of Junkers F-13s, but these were later supplemented by Junkers Ju.52/3m trimotors and Ju.86s, with a sizeable fleet of Lockheed Lodestars also pressed into service before the start of World War II. All commercial operations had ceased by 1940, and the aircraft were pressed into the South African Air Force, although limited air services, necessary in a large country, were re-started in 1944.

A service to London was started in 1945, and during the late 1940s the airline operated a fleet of Douglas DC-3s and DC-4s, Avro Yorks, and Lockheed Lodestars, to which Lockheed Constellations were added in 1950, until replaced by Douglas DC-7Bs in 1956. De Havilland Comet Is were leased from B.O.A.C. in 1953, and flown with South African crews, but the airline entered the jet age permanently in 1960 with Boeing 707s, although Vickers Viscount turboprops had been operated for some four years prior to this. Some re-routing of services had to take place in 1963 when a number of newly-independent African states banned overflying of their territory by South African aircraft, but the new routes to Europe, via the Cape Verde Islands, have in fact proved to be popular with passengers.

The present fleet consists of Boeing 747s, 707s, 727s, and 737s, with some Hawker Siddeley 748s, operating an extensive domestic network and international services to Europe, Brazil, the United States, and Australia.

South East Asia Treaty Organisation: *see* SEATO.

South Pole: although the first flight across the South Pole by an aeroplane was made on 28–29 November 1929 by Rear Admiral R. Evelyn Byrd, U.S.N. (q.v.), in a Ford Trimotor, the South Pole has figured less in aviation than the North Pole – largely-because of its greater distance from civilisation and the more hostile territory. No airline services are operated over the South Pole, although tentative plans exist for South American–South African services via this route – but without any date for inauguration.

SP-: international civil registration index mark for Poland.

space shuttle: in order to eliminate the high cost and waste of using conventional rockets with their single-launching life, and to keep an orbiting space station or laboratory supplied, the National Aeronautics and Space Administration evolved the space shuttle concept. After a long period of evaluation of different proposals, North American Rockwell was chosen in mid-1972 to be the main contractor. Basically, the shuttle will comprise two parts, both of which will be re-usable: a booster and the actual vehicle to transport men and supplies into earth orbit and make the return journey. A further advantage of the concept arises from the substantial cost savings possible from starting voyages of exploration from a space station, rather than from the earth's surface.

Spad Aircraft: a French manufacturer prominent during World War I, Spad (Société pour Aviation et ses Dérives) produced a number of fighter designs which were used by the British, French, and American air arms, with some supplementary licence-production in the United Kingdom of the Spad S.7 biplane.

span: the span, or more correctly wingspan, of an aircraft is the measurement along the wing from tip to tip, regardless of whether or not the wing is 'broken' by the fuselage.

The largest wingspan of any aircraft to date is that of the Hughes Hercules (q.v.) flying-boat.

Sparrow III: the Raythcon AIM-7E Sparrow air-to-air missile is the standard armment for the McDonnell Douglas F-4 Phantoms used by the United States and her principal allies. Semi-active radar homing is for all-weather performance, and a version with improved performance, the AIM-7F, is used for the F-14 and F-15 air superiority fighters. Reputed to be a reliable and effective weapon, the Sparrow III is more than twelve feet in length and a foot in diameter, and uses a solid-fuel propellant rocket motor for a maximum speed of well over three times the speed of sound and a range of some seven miles. The warhead, of high-explosive, is exceptionally large for a missile of this variety.

Spartan: the first line of defence in the American Safeguard (q.v.) anti-ballistic missile system, the Spartan, for which the main contractors are Bell Telephone and McDonnell Douglas, is a three-stage missile intended to intercept an incoming enemy ballistic missile while outside the earth's atmosphere, destruction of the ballistic missile being by blast and neutron kill from the five-megaton nuclear warhead. Unlike Sprint (q.v.), the other Safeguard missile, Spartan would not inflict any damage on the earth's surface, and is undoubtedly the lesser evil compared with the effects of the detonation of a ballistic missile warhead over its target. In one series of tests, Spartan showed an interception success rate of more than 80 per cent. The maximum range of the Spartan is 400 miles, and the missile is fifty-four feet in length and has a diameter of ten feet. Guidance is by radar command linked to the missile site radar.

Spider, Fokker: Anthony Fokker's (q.v.) first aircraft, built when he was nineteen, the Spider first flew in 1911. The aircraft's main feature was the use of a high degree of dihedral for lateral control in the absence of ailerons or wing warping.

spin: the term 'spin' covers the condition in which an aircraft falls vertically while autorotating; it is a manoeuvre which can be induced on purpose, but in the early days of aviation was more often accidental and final. The first pilot to recover from a spin was an Englishman, Frederick Langham, while flying an Avro biplane in 1911. The following year another Englishman, Lieutenant Wilfred Parke, R.N., flying an Avro cabin biplane, was able to put the aircraft into a spin and demonstrate recovery.

Spitfire, Supermarine: developed by R. J. Mitchell for Supermarine from his Schneider Trophy seaplanes, the S-5, S-6, and S-6B, the prototype Supermarine Spitfire first flew in March 1936, and was followed by some 23,000 production models throughout World War II. The earlier versions used Rolls-Royce Merlin engines, (early models using the 1,030 hp Merlin II/III), but later models used Packhard-built Merlins and the Rolls-Royce Griffin; all in all there were some thirty main production variants of the Spitfire and the related Seafire, the latter being for carrier-borne operations, and developed from the Spitfire Mk.V.B, with a strengthened fuselage.

In spite of its ancestry, the Spitfire was a landplane, and only a handful of floatplane versions were produced. Duties included fighter, fighter-bomber, and photographic reconnaissance work, while a tandem twin-seat conversion trainer was produced during the latter years of the war. The earlier versions used eight machine guns, but later these were superseded by four 20 mm cannon, and racks for bombs and rockets were sometimes fitted. Maximum speed depended on version, with the prototype having a maximum speed of 346 mph, but was generally in excess of 400 mph. Early models had the then fashionable 'hump back' fuselage form, although this was left off later models and a teardrop canopy substituted for the earlier blister.

Although the Spitfire was used in all theatres of war, the most famous action was undoubtedly the Battle of Britain in the summer of 1940, when the Spitfire proved superior to a larger number of German aircraft, of which the Messerschmitt Bf.109 was the most important. It lacked the speed of the Bf.109, but the Spitfire's greater manoeuvrability and strength, and the pro-

vision of light armour-plating around the cockpit, proved to be more important.

Sprint: the last-ditch stage of the American Safeguard (q.v.) anti-ballistic missile system, the Sprint, on which the prime contractors are Bell Telephone and Martin Marietta, intercepts the incoming enemy ballistic missile warhead within the atmosphere. A two-stage rocket, the Sprint has an astonishing rate of acceleration well in excess of 100g, and a range of twenty-five miles, while guidance is by radar control linked with the missile site radar. Control of the conical-shaped missile is by fluid injection in the first stage, and aerodynamic fins in the second stage. The warhead is small (of only one or two kilotons) in an attempt to minimise the inevitable fall-out over friendly territory. Early tests indicate a success rate in excess of 50 per cent.

Sputnik I: the first man-made satellite to enter earth orbit, the Soviet Sputnik I was launched on 4 October 1957, and weighed 184 lb. Consisting of a metal sphere some two feet in diameter, the Sputnik orbited the earth once every ninety-six minutes and completed some 1,400 orbits before dropping back into the earth's atmosphere and burning up on 4 January 1958. Tasks of the satellite were to measure the density and temperature of the upper atmosphere, to measure the concentration of electrons in the ionosphere, and to transmit the information back to earth.

spy satellite: the so-called 'spy-in-the-sky' surveillance or spy satellite is a man-made earth satellite equipped with a variety of aids for gathering military information, which is coded and transmitted back to earth. The two main differences between a spy satellite and a weather satellite are the more exacting information required and its coding before transmission to the satellite's control – meteorological satellites 'broadcast' their information in a form available to any listening station. The more important aids in the spy satellite include moving and still cameras with the capability of penetrating cloud, vegetation, and camouflage, and equipment to monitor communications and broadcasts, and to detect radar networks by their emissions. Major troop movements, fleet actions, and the construction of airfields and missile sites can all be detected. It is the use of spy satellites which has enabled the United States vastly to improve its intelligence system – hitherto a one-way flow of information with regard to the Soviet Union – and to drop its insistence on inspection of Soviet missile sites as part of any arms limitation agreement (see SALT).

squadron: the standard form of unit in any air arm is the squadron, although the strength of a squadron varies considerably from as few as four aircraft in a French Mirage IV bomber squadron to as many as twenty-five fighters in some air forces. It is not essential that a squadron should use only one aircraft type, and with transport and communications units in particular two or more aircraft types may be operated. Squadrons are normally broken down into a number of flights. Usually three squadrons make a wing, and six squadrons a group, although here too there is considerable scope for variation.

The rank of squadron leader in many air forces, primarily those linked to the United Kingdom, no longer has any great direct significance, and is today really just another rank, as indeed is the naval rank of captain. The larger and more sophisticated aircraft of today often require a squadron to be commanded by a wing commander. The equivalents of the ranks of squadron leader and wing commander in army terms and the terms of some air forces are major and lieutenant colonel respectively; or lieutenant commander and commander in naval terms. It is normal to speak of a squadron commander or squadron C.O.

The first recorded 'squadron leader' in history was the Frenchman Jean-Marie-Joseph Coutelle, who commanded a balloon unit at the Battle of Fleurus in 1794.

SR-71, Lockheed: the fastest military aircraft in service, although in many ways rendered obsolescent by spy satellites, the Lockheed SR-71 was originally developed from the A-11, and first flew in 1961 as a U-2 (q.v.) replacement. The reconnaissance version, of which seventeen were built, carries the SR-17 designation, while there were three interceptors, designated YF-12A. Two 32,500 lb thrust Pratt and Whitney J58 turbojets with afterburners give the aircraft a maximum speed of Mach 3·3 and an endurance of ninety minutes at Mach 3·0.

SR.A1, Saunders-Roe: the only flying-boat jet fighter ever built, the Saunders-Roe SR.A1 first flew during the immediate post-World War II period and used two Metrovick turbojets. Extensive trials were undertaken and the aircraft performed well, but the project was abandoned since carrier-borne jet fighters held considerably more potential.

Sraam 75: the Hawker Siddeley Dynamics Sraam 75 air-to-air guided missile is intended to equip a wide variety of aircraft from the mid-1970s onwards, although the missile is primarily to be used on the Panavia 200 Panther multi-role combat aircraft. Infra-red homing will be used and a high degree of reliability is anticipated over very short ranges. Little else is known about the missile.

Sram: the Boeing AGM-69A Sram air-to-surface missile is designed to operate over short ranges of less than 100 miles, but with the capability of delivering a nuclear warhead in spite of encountering advanced defensive measures. The missile can be carried in large numbers by the Boeing B-52H and the General Dynamics FB-111, and will also equip the North American Rockwell B-1A. Fourteen feet in length and eighteen inches in diameter, Sram uses a solid propellant and inertial guidance.

SR-N1, Saunders-Roe: the world's first full-sized hovercraft, the SR-N1 was built by Saunders-Roe in 1958 for the National Research and Development Corporation, to the design of the inventor of the hovercraft, Sir Christopher Cockerell. In common with the test models, the SR-N1 used a single motor for lift and propulsion and, starting in 1959, was used extensively on a number of tests, including a cross-Channel run. Maximum speed was 25 knots.

SR-N2, Saunders-Roe: the first passenger-carrying hovercraft to enter service, only one SR-N2 was built by Saunders-Roe, and this used two air propellers and two lift fans driven by two aircraft engines. Up to sixty passengers could be carried at a speed of up to 75 knots. Initial operations were on the Bristol Channel.

SR-N3, Saunders-Roe: a large development of the SR-N2 (q.v.) and another one-off design, the Saunders-Roe SR-N3 is still in service with the British Interservice Hovercraft Trials Unit, undertaking military transport and assault exercises.

SR-N4, British Hovercraft Corporation: originating as a Saunders-Roe design, the SR-N4 is in limited production for civil users, and has operated across the English Channel since 1968. Although up to six hundred passengers could be carried in an all-passenger version, the general tendency is to use the SR-N4 as a mixed passenger and car craft, for up to twenty-five cars and three hundred passengers. Four Rolls-Royce Proteus turboprops of 3,400 shp each drive four air propellers and four lift fans, giving a maximum speed of 75 knots.

SRN-5, British Hovercraft Corporation: although a later design than the large SRN-4 (q.v.), the Saunders-Roe-designed SR-N5 was put into full production by the British Hovercraft Corporation well in advance of the SR-N4, and in fact has become the first hovercraft in the world to achieve production-line status, although being delivered primarily to military users. The SR-N5 has been built under licence in the United States by Bell Aerosystems as the Sk-5 for the United States armed forces, and is also in service with the British, Iranian, and Saudi Arabian armed forces, as well as with other government bodies and with some civil users. A number of pioneering journeys in Africa and South America have been undertaken by the craft, which has an eighteen-passenger cabin and uses a single 900 shp Rolls-Royce Gnome turboprop engine for a maximum speed of 50 mph. A stretched version is the SR-N6 (q.v.).

SR-N6, British Hovercraft Corporation: a stretch of the Saunders-Roe-designed SR-N5 (q.v.), the British Hovercraft Corporation SR-N6 uses the same 900 shp Rolls-Royce Gnome engine, but can accommodate thirty-eight passengers, and a number of craft have been returned for further stretching, to carry fifty-six passengers. Primarily a civil craft, the SR-N6 was first used on services across the Solent in 1965.

SS-11: the French surface-to-surface missile, the Aerospatiale SS-11 is wire-guided and has a range of under two miles, using a two-stage rocket motor. Maximum speed is 360 mph, and the SS-11 is four feet in length and twenty inches in diameter. A helicopter-mounted version is the AS-11.

SS-12: the Aerospatiale SS-12 is a development of the SS-11 (q.v.) surface-to-surface missile, but with spin stabilizers to augment the wire guidance system, a larger warhead, and an improved propulsion system giving a range of more than three miles, and a maximum speed of 425 mph. The SS-12 is six feet in length and two feet in diameter. A helicopter-borne version is the AS-12.

S.S.B.S.: *see* Sol-Sol Balistique Stratégique.

ST-: international civil registration index mark for the Sudan.

stability: aircraft stability may be longitudinal or lateral, with the absence of the former leading to pitching, and of the latter leading to rolling. A stabilizer is fitted to some aircraft; this is an automatic device to help improve stability, although the term can also refer to the tailplane. Lateral stability is assisted by flaps and ailerons, or by a high degree of dihedral.

stagger: in a biplane, the placing of one wing slightly ahead of the other is known as stagger; generally the more forward of the wings is the upper.

stall: the term stall refers to the condition when an aircraft's wings lose lift and the aircraft starts to fall – usually because of too little speed, although too high an angle of incidence has the same effect.

Stampe et Vertongen: a Belgian company in business between the wars, J. Stampe et M. Vertongen produced a number of aircraft designs, including some fighters, but the company's fame rests upon its training and aerobatic aircraft, some of which remain in use today. The most famous Stampe was the SV-4, which was also built under licence by Nord, and after the war was built in Belgium by Stampe et Renard.

Standard: the American General Dynamics RIM-66A and RIM-67A Standard missiles are respectively single-stage Tartar (q.v.) replacement and two-stage Terrier (q.v.) replacement shipboard surface-to-air guided missiles, now entering service with the United States Navy. The RIM-66A is fifteen feet in length and the RIM-67A twenty-one feet, both with a diameter of one foot. Ranges are respectively ten and thirty miles, and maximum speeds in excess of Mach 2·5. Components common to both are used, including solid fuel propellants, semi-active radar homing, control by fins, and high-explosive warheads.

stand-off missile: *see* air-to-surface missile.

Starfighter, Lockheed F-104: *see* F-104 Starfighter, Lockheed.

Starlifter, Lockheed C-141: *see* C-141 Starlifter, Lockheed.

Stinson Aircraft: an American aircraft manufacturer largely specialising in light aircraft for private and military owners, Stinson Aircraft was formed in 1925 and continued as a separate entity until after World War II, when it became a part of General Dynamics (q.v.). Early Stinson aircraft included the Detroiter four-seat mailplane, followed by the SM-1 and SM-2 light aircraft. Exceptions to the light aircraft production line of Stinson included the Tri-motor of 1930, with three 240 hp Lycoming radial engines and accommodation for up to ten passengers, and its successor, the 'A' trimotor of the mid-1930s, which was also one of the few low-wing Stinsons – most aircraft from this company were high-wing monoplanes. Also before World War II, the Reliant range of light aircraft was first introduced.

During World War II Stinson produced a number of observation and liaison types, including the L-1 and L-5 for the United States Army, and the AT-10 navigational trainer.

Stirling, Short: Britain's first four-engined heavy bomber, the Short Stirling first flew in 1939 and entered R.A.F. service in 1940, using four 1,600 hp Bristol Hercules radial engines for a maximum speed of 280 mph and a bomb load of up to eight tons. A mid-wing monoplane with a single-fin tailplane, the Stirling was fitted with nose, dorsal, and tail machine gun turrets, but was not a particularly successful design.

S.T.O.L.: *see* short take-off and landing.

strategic air war: a traditional form of aerial warfare of the kind developed during World War I and World War II, a strategic air war generally involves heavy conventional bombing on targets behind the front line, such as docks, factories, railways, etc., but without the use of nuclear weapons. American bombing raids over North Vietnam fall into this category.

Strategic Arms Limitation Talks: *see* SALT.

strategy: the course of producing an overall policy for the attainment of an important national or international objective may be regarded as strategy, implying long-term policies of a more extensive nature than would be in the case if, in military terms, one was considering the role of just one arm of the armed forces; a strategy in fact needs to go beyond just one service and, indeed, beyond the armed forces, which are merely the tools of national and foreign policy.

Stratocruiser, Boeing 377: *see* Boeing 377 Stratocruiser.

Stratofortress, Boeing B-52: *see* B-52 Stratofortress, Boeing.

Stratolifter, Boeing C-135: *see* C-135 Stratolifter, Boeing.

Stratoliner, Boeing 307: *see* Boeing 307 Stratoliner.

Stratotanker, Boeing KC-135: *see under* C-135 Stratotanker, Boeing.

strike: the term strike is more specific than 'attack', since it implies a form of attack which is both sharp and sudden, and also selective.

strike-fighter: basically a modern development of the fighter-bomber (q.v.) concept, the term strike-fighter differs little from fighter-bomber in meaning and really can only be said to reflect the greater and heavier variety of weapons now likely to be carried by such an aircraft. The term strike aircraft, which is also sometimes used, means a light bomber or an inter-diction bomber, i.e. a bomber aircraft able to operate against selected military targets rather than undertake heavy blanket bomb-ing or long-range nuclear attaack.

Stringfellow, John (1799–1883): an Englishman, John Stringfellow was an engineer who first became involved in aviation development when he collaborated with his friend W. S. Henson (q.v.), in building a model steam-powered aeroplane in 1843. Later, Stringfellow built a model of his own based on the Henson design, but with a much improved engine. In 1868 Stringfellow won an Aeronautical Society prize for the steam engine with the greatest power-to-weight ratio. The Stringfellow and Henson designs had a considerable effect on the public imagination, but lacked the promise of practicality, although the layout of the models could be described as being 'modern'.

Stuka, Junkers Ju.87: *see* Ju.87 Stuka, Junkers.

'Styx': a very basic surface-to-surface missile for shipboard use, the threat of the Soviet 'Styx' has, nevertheless, prompted much development in this field by the Western powers. 'Styx' has a range of about twenty miles and can carry a high-explosive warhead. The missile has to be aimed at the target before firing, and its radar homing system is only effective for last-minute course alterations, although the launcher aiming system can receive trans-ponder signals to allow it to differentiate between friend and foe. Maximum speed is Mach 0·9. Unlike most other missiles, 'Styx' has been fired in anger: three missiles sank the Israeli destroyer *Eilat* in October 1967, after being fired from an Egyptian gun-boat.

SU-: international civil registration index mark for the United Arab Republic.

Su-7B 'Fitter', Sukhoi: the standard Soviet

Bloc ground-attack fighter, the single-seat Sukhoi Su-7B, NATO code-named 'Fitter', has now been in service for a number of years. A single 22,050 lb thrust TRD-31 turbojet with reheat provides the swept-wing aircraft with a maximum speed of 1,060 mph. Two 30 mm cannon are fitted in the wing roots, and bombs, rockets, napalm tanks, or missiles can be carried on underwing strongpoints.

Su-7UT1 'Moujik', Sukhoi: a two-seat conversion training development of the Su-7B (q.v.), the Sukhoi Su-7UT1, NATO code-named 'Moujik', has tandem seating but is otherwise identical to the combat version.

Su-9 'Fishpot', Sukhoi: an interceptor development of the Su-7B (q.v.), the Sukhoi Su-9, NATO code-named 'Fishpot', has a longer range than the MiG-21. The Su-9 has an Su-7B fuselage, but with a delta wing not unlike that of the MiG-21. A single 22,050 lb thrust TRD-31 turbojet with reheat gives a maximum speed of Mach 1·8. A single-seat aircraft, the Su-9's standard armament is the 'Alkali' air-to-air missile.

Su-11 'Flagon A', Sukhoi: the latest Sukhoi design to enter service, the Sukhoi Su-11, NATO code-named 'Flagon A', is an interceptor using two 22,000 lb thrust turbojets with reheat for a maximum speed of Mach 2·5. There is also an experimental V./S.T.O.L. development with lift jets.

subsonic: the term subsonic refers to air-craft flying at below the speed of sound (760 mph at sea level, falling to 660 mph in the upper atmosphere).

Sudan Airways: the state-owned airline of the Sudan, Sudan Airways was formed in 1946, and started operations during 1947 with assistance from Airwork, a British company, using a fleet of de Havilland Dove light transports on domestic services. Steady expansion followed, to the extent that, by the late 1950s, the airline was operating to major points in the Middle East, Africa, and Europe with a fleet of Doves and Douglas DC-3s, and a Vickers Viscount. Fokker F-27 Friendship airliners were introduced in 1962, and in 1963 the airline received its first jet airliners, de Havilland Comet 4Cs.

The current route network, which covers domestic, African, Asian, and European destinations, is served by a fleet of leased Boeing 707s, Fokker F-27 Friendships, and de Havilland Canada Twin Otters, and one Douglas DC-3.

Sud Aviation: the main predecessor of Aerospatiale (q.v.).

Sukhoi: the Sukhoi design bureau is the Soviet Union's newest, the first design of importance to be noted in the West being the Su-7B (q.v.) ground-attack aircraft, and its interceptor development, the Su-9 (q.v.).

Sunderland, Short: the only large long-range flying-boat operated by the Royal Air Force during World War II, the Short Sunderland was used for maritime-reconnaissance and search and rescue duties. Basically a development of the Empire (q.v.) flying-boats, the Sunderland used four Bristol Pegasus radial engines for a maximum speed of 210 mph and a range of 3,000 miles, and had nose, dorsal, and tail gun-turrets. Sunderlands remained in service with the R.A.F. and some Commonwealth air forces for some years after the end of World War II with 741 aircraft built.

Super Constellation, Lockheed: *see under* Constellation, Lockheed.

Super Courier, Helio U-10: *see* U-10 Super Courier, Helio.

supercritical aerodynamics: a concept aimed at improving aircraft performance at speeds of around Mach 1·0, supercritical aerodynamics are based on thick aerofoils with a flat upper surface, and hold considerable promise for civil and military applications, particularly if used in conjunction with area ruling (q.v.). Benefits are expected to include a general improvement in performance and reduced structural weight, with less stress on the airframe. Extensive tests have been carried out by the National Aeronautics and Space Administration in the United States, using a modified Ling-Temco-Vought F-8 Crusader.

Super Frelon, Aerospatiale S.A.321: *see* S.A.321 Super Frelon, Aerospatiale.

Super Magister, Potez: *see under* CM.170 Magister, Potez.

Supermarine Aircraft: the first Supermarine designs appeared during World War I, including the Baby biplane flying-boat, and were followed during the 1920s and 1930s by a series of flying-boats, including the Sea Eagle, Southampton, and Walrus (q.v.), all of which were biplanes. Some of these aircraft were the work of the company's famous chief designer, R. J. Mitchell (q.v.). However, in spite of its undoubted capacity to produce good flying-boats (primarily for military users), the company's real fame during the period came from the design and development of cantilever monoplanes for the Schneider Trophy races, from the ill-fated S-4 of 1924 to the successful S-5, S-6, and S-6B, which won the Trophy outright for the United Kingdom. These seaplanes were also important milestones in Mitchell's work, and led directly to the famous Supermarine Spitfire (q.v.) fighter of World War II.

After the war the company built its first jet fighter for the Royal Navy, which was the first carrier-borne jet fighter designed as such for that service. A supersonic jet fighter for the Royal Air Force, the Swift (q.v.), entered service during the early 1950s and established a world air speed record before being withdrawn from service due to aerodynamic defects. This was the first British swept-wing jet fighter, however. Later in the same decade the Scimitar carrier-borne jet fighter was put into service with the Royal Navy. Supermarine and its parent company since the inter-war period, Vickers, became a part of the British Aircraft Corporation in 1960.

Super Mystère, Dassault: the Dassault Super Mystère single-seat interceptor and strike-fighter was a development of the earlier Mystère (q.v.) jet fighter, with a number of modifications for supersonic performance. First flown in prototype form in March 1955, the Super Mystère initially used a Rolls-Royce Avon turbojet, but in the production version, first flown in February 1957, the SNECMA Atar 101G turbojet of 9,700 lb thrust with reheat was substituted, giving a maximum speed of Mach 1·13 and a range of 600 miles. Two 30 mm cannon were fitted, and two 1,100 lb bombs or rockets, napalm tanks, or air-to-air missiles could be carried on underwing strongpoints. Few if any of these aircraft, which were used by the French and Israeli air forces, remain in service.

Super Porter, Pilatus: *see* PC-6 Porter/ Turbo-Porter, Pilatus.

Super Sabre, North American F-100: *see* F-100 Super Sabre, North American.

Super Skymaster, Cessna 337: *see* Cessna 337 Super Skymaster.

supersonic: supersonic flight is flight above the speed of sound, or Mach 1·0. The first aircraft to attain the speed of sound was the Bell XS-1, launched from a Boeing B-29 Superfortress in October 1947. The first aircraft to reach the speed of sound after taking off normally was the Douglas Skyrocket, while the first production aircraft capable of such speed was the North American F-86 Sabre, although it was not until the F-100 Super Sabre appeared that an aircraft taking off under its own power was able to break the sound barrier in level flight.

The first British supersonic aircraft was the de Havilland 108 tailless experimental aircraft, while the first operational supersonic aircraft were the Supermarine Swift and the Hawker Hunter. The first Russian supersonic aircraft was the Mikoyan-Gurevich MiG-17, although the first to be able to break the sound barrier in level flight was the MiG-19.

Commercial supersonic aircraft include only the Tupolev Tu-144 and the B.A.C.–Aerospatiale Concorde at present, the Russian aircraft being the first to fly – although it is believed that technical problems may delay its entry into service.

Super VC.10, Vickers: *see under* VC.10, Vickers.

surface-to-air missile: development of the surface to air missile was made necessary because of the greater speed and altitude of attacking military aircraft made possible by the jet engine; against these conventional anti-aircraft artillery was at a marked disadvantage. Although manned,

the forerunner of the modern surface-to-air missile was in many ways the Messerschmitt Me.163 Komet of World War II. No missiles as such were developed and put into service during the war, although Fairey (q.v.) developed such a weapon starting in 1944, primarily to defend warships against Kamikaze attack; but the missile entered service too late to see action, and was used instead, as the Stooge, on test duties.

The first American surface-to-air missile was the Western Electric Nike-Ajax, the development of which had started in 1945, while the Boeing Bomarc was the first long-range missile. The first operational British missiles were the Bristol Bloodhound (q.v.), which entered service with the R.A.F. in 1958, and the English Electric Thunderbird (q.v.), which has been used by the British Army since 1959.

surface-to-surface missile: the term 'surface-to-surface missile' embraces a wide range of weapons, including battlefield missiles of very short range, such as the French SS-11, shipboard missiles, and the very long range intercontinental ballistic missiles (I.C.B.M.s). The first surface-to-surface missile to enter service was the German V-2 (q.v.) designed by Dr Wernher von Braun during World War II, although it should be remembered that the Chinese had rockets for battlefield use for many years, and these were brought to Britain and re-introduced into European warfare by Sir William Congreve (q.v.).

The first American surface-to-surface missile was the Firestone Corporal, introduced into the United States and the British armies during the early 1950s, while the U.S.A.F. received the longer-range Martin Matador – essentially a flying bomb, although these, including the German V-1 (q.v.), can also be fairly described as surface-to-surface missiles. A British ballistic missile, the Blue Streak (q.v.), was developed by Hawker Siddeley, but cancelled in 1960; it has since been used as a satellite launch vehicle, establishing an enviable reputation for reliability. The American Polaris (q.v.) was the first submarine-launched surface-to-surface missile.

Many surface-to-air missiles, particularly those for shipboard use, retain a secondary surface-to-surface capability.

surveillance satellite: *see* spy satellite.

SV-4, Stampe et Vertongen: a single-engined, two-seat biplane for training and aerobatics, the SV-4 was put into production by Stampe et Vertongen in Belgium during the early 1930s, and was also built under licence by Nord for the French Aeronavale. A 130 hp de Havilland Gipsy Queen engine powered the original Belgian-built versions, but post-war Belgian aircraft by Stampe et Renard and the Nord aircraft used 140 hp Renault engines. Maximum speed of the aircraft, of which many examples survive, is 120 mph.

Svenska Aeroplane A.B.: the Swedish SAAB, or Svenska Aeroplane A.B., was formed in 1937 and has established a reputation both in aviation and as a motor car manufacturer. Today it is a part of the SAAB-Scania Group, in which heavy commercial vehicles are also an important product. During World War II SAAB built a number of aircraft for the Royal Swedish Air Force, founding a tradition of allowing the R.Sw.A.F. to buy most of its frontline combat aircraft from domestic sources, instead of incurring the foreign exchange cost and foreign policy limitations of being dependent on an outside arms supplier. SAAB wartime aircraft included the SAAB-17 (q.v.) light bomber, the SAAB-18 (q.v.) bomber, and the SAAB-21 (q.v.) fighter, which was the only fighter in the world to be produced in both piston and jet-engined versions when, after the war, a jet-engined model was put into production and service.

After the war the SAAB-29 (q.v.) Tunnan and SAAB-91 (q.v.) Safir trainer were put into production, followed by the SAAB-32 (q.v.) Lansen during the mid-1950s, and the SAAB-35 (q.v.) Draken and SAAB-105 (q.v.) during the 1960s. The most recent aircraft, supplementing Draken production, has been the SAAB-37 (q.v.) Viggen, which will probably be the last all-Swedish combat aircraft. SAAB is a member, with B.A.C. and M.B.B., of the European consortium which is examining the possibility of building a reduced take-off airliner.

Swearingen: the American Swearingen Aircraft Corporation's main product to date has been the Merlin, a twin-engined executive and feeder-liner, production of which has passed to Fairchild-Hiller.

Swift, Supermarine: the first British swept-wing aircraft, the Supermarine Swift entered R.A.F. service during the early 1950s, although aerodynamic failings led to its prompt withdrawal. However, the aircraft did establish a world air speed record on 25 September 1953, of 735 mph.

Swissair–Swiss Air Transport Company/Schweizerische Luftverkehr A.G.: the Swiss Air Transport Company, or Swissair, was formed in 1931, by the merger of Balair of Basle and Ad Astra Aero of Zürich, dating from 1925 and 1919 respectively. Most of Swissairs aircraft at the time of its formation were of Fokker manufacture, but these were soon supplemented and eventually replaced by Lockheed Orions in 1932, Curtiss Condor biplanes in 1934, and Douglas DC-2s in 1935. Operations had to be suspended on the outbreak of World War II in 1939, although Switzerland was neutral. By this time Douglas DC-3s had joined the fleet.

Post-war expansion included additional DC-3s and the introduction of the airline's first four-engined airliners, Douglas DC-4s, permitting the start of transatlantic services in 1949. The airline had by this time been recognised as the national (but not the nationalised) airline of Switzerland, under a 1947 agreement which allocated a 30 per cent interest to the state, leaving the remainder for private investors. During the 1950s a fleet of Douglas DC-6s and DC-7s was operated, with Convair 440 Metropolitans for the shorter routes. Swissair received its first jets, Douglas DC-8s, in 1960, and later supplemented these with Convair 990 Coronado, Sud Caravelle, and McDonnell Douglas DC-9 jets.

The current Swissair fleet consists of Boeing 747s, Douglas DC-8s (including some of the 'Super Sixty' series), DC-10-30s, and DC-9s, and Convair 990 Coronado jet airliners, with some Piaggio

P.149Es and other light aircraft. A 56 per cent shareholding is held in the charter airline Balair, and a maintenance agreement, KUSS or K.S.S.U. (q.v.), has been entered into with S.A.S. (q.v.) and other airlines. The route network includes domestic, European, African, Asian, and North and Latin American services.

Swordfish, Fairey: first flown in 1934, the Fairey Swordfish tandem twin-seat biplane was used operationally by the Royal Navy as a torpedo-bomber during the first part of World War II, notable actions including the attack on the Italian Fleet at Taranto in November 1940, and the action leading to the sinking of the *Bismark*. Mainly a carrier-borne aircraft, some float-plane versions were also put into service.

SX-: international civil registration index for Greece.

Syrian Arab Airlines: formed in 1961 by the Government of Syria, after Syria had left the United Arab Republic of Syria and Egypt, Syrian Arab Airlines soon acquired jet aircraft in the form of Sud Caravelles, supplementing these with older piston-engined types such as Douglas DC-3s, DC-4s, and DC-6s. Services were operated domestically, throughout the Middle East, and to Europe. The current fleet has changed little, consisting of Caravelles and DC-6Bs, and there has been little further expansion of the route network.

T

T-: U.S.A.F. and U.S.N. designation for training aircraft, including basic, advanced, and navigational trainers, but excluding conversion training developments of fighter and bomber aircraft.

T.1, Fuji: the first Japanese-designed jet aircraft, apart from certain wartime Kamikaze types, the T.1 was designed and built by Fuji as an intermediate jet trainer, and made its first flight in January 1958. The first forty production models were fitted with a single 4,000 lb thrust Bristol Siddeley Orpheus 805 turbojet, and have been designated T.1A by the Japanese Air Self-Defence Force. These were followed by twenty T.1Bs, with a single 2,645 lb thrust Ishikawajima-Harima J3-IHI-3 engine. A tandem twin-seat aircraft, the T.1A version has a maximum speed of 575 mph and a range of up to 1,210 miles, while a 0·50 in. machine-gun pack and up to 1,500 lb of bombs, rockets, or napalm tanks may be carried on underwing strongpoints.

T.2, Mitsubishi: first flown during the autumn of 1971, the Mitsubishi T.2 has been designed as an advanced trainer and ground-attack aircraft for the Japanese Air Self-Defence Force, and some two hundred are now in the course of production. A maximum speed of Mach 1·6 and a range of up to 1,500 miles is available from two 6,950 lb thrust Rolls-Royce/Turbomeca Adour turbofans with reheat.

T-2 Buckeye, North American: developed using the wing of the North American FJ-1 Fury carrier-borne fighter and the control system of the North American T-28C Trojan trainer, the T-2 Buckeye carrier-borne jet trainer first flew in January 1958, in standard production form since the prototype stage was avoided. Initial production versions were designated the T-2A, and used a 3,400 lb thrust Westinghouse J34-WE-36 turbojet. The T-2B, first flown in May 1965, uses two 3,000 lb thrust Pratt and Whitney J60-P-6 turbojets, giving a maximum speed of 530 mph to this tandem twin-seat aircraft. An up-rated version of the T-2B, designated the T-2C, entered service in 1970.

T-5B Freedom Fighter, Northrop: see under F-5A/B Freedom Fighter, Northrop.

T-6 Harvard/Texan, North American: first introduced in 1938, the North American T-6 basic trainer aircraft remained in production throughout World War II and the Korean War until 1954, and the aircraft has been operated not only by the American armed forces, but also by those of many allies and by many neutral and Latin American nations as well. A low-wing tandem twin-seat monoplane, the T-6 used a single 550 hp Pratt and Whitney R-1340-AN-1 engine for a maximum speed of 212 mph and a range of up to 870 miles. A considerable number of Harvards remain in service at the present time.

T-11 Kansan, Beech: see Beech 18.

T-28 Trojan, North American: winner of a 1948 U.S.A.F. design competition for an advanced piston-engined trainer, the North American T-28 Trojan first flew in prototype form in September 1949, as the T-28A with an 800 hp Wright R-1300 piston engine. Developments of this tandem twin-seat low-wing monoplane have included the T-28B for the U.S.N. with an up-rated engine, the T-28C with arrester hook, and the T-28D for counter-insurgency duties. Maximum speed of the T-28B/C/D is 380 mph and the range, with a weapon-load of bombs, rockets, napalm tanks, or machine gun packs, is 500 miles.

T-33, Lockheed: a tandem twin-seat version of the Lockheed F-80 (q.v.) Shooting star jet fighter, retaining the same engine as the fighter aircraft, although intended as an advanced jet trainer rather than just a conversion trainer.

T-34 Mentor, Beech: a development of the Beech Bonanza (q.v.) using the same aerodynamic surfaces, the T-34 Mentor military basic trainer first flew in December 1948. The first U.S.A.F. order was placed in 1950, followed by orders from the U.S.N. and many of America's allies, with licence-production in Japan by Fuji. A single 225 hp Continental 0-470-13 piston engine gives the tandem twin-seated T-34 a maximum speed of 189 mph and a range of 735 miles.

T-37, Cessna: the first and (until the Citation (q.v.)) only Cessna jet, the T-37 basic trainer first flew in October 1954, after winning a U.S.A.F. design competition. The initial production version, the T-37A, used two Continental J69-T-9 turbojets; this was followed by the up-rated T-37B with 1,025 lb thrust Continental J69-T-23 turbojets, to which standard the T-37As have now been converted. The T-37C is an armed version for counter-insurgency duties. Maximum speed is 510 mph. A twin-seat aircraft, the T-37's cockpit has a side-by-side seating layout.

T-38 Talon, Northrop: developed from the F-5B Freedom Fighter (*see* F5A/B), the Northrop T-38 Talon advanced jet trainer retains the same fuselage structure, albeit with some simplification of the control surfaces. First flight of the T-38 was in April 1959, with the first deliveries of production aircraft to the U.S.A.F. in March 1961. The aircraft is also used by the Luftwaffe. Two 3,850 lb thrust General Electric J85-GE-5A turbojets with reheat give a maximum speed of 860 mph (Mach 1·3) and a range of 1,140 miles.

T-39 Sabreliner, North American: originally built as a private venture navigational trainer and utility transport, the North American T-39 Sabreliner first flew in September 1958, the prototype using two tail-mounted 2,500 lb thrust General Electric J85 turbojets. Production aircraft for the U.S.A.F. and U.S.N. use two 3,000 lb thrust Pratt and Whitney JT12 engines for a maximum speed of 550 mph and a range of up to 2,000 miles, while up to ten passengers may be carried. The Sabreliner is now also marketed as a civilian executive jet aircraft.

T-41A, Cessna: the military development of the Cessna 172 (q.v.) light aircraft, the T-41A is used for basic training for U.S.A.F. pilots by civilian contractors who undertake this phase of the U.S.A.F.'s flying training. T-41As have also been supplied to a number of other countries.

T-42, Beech: *see* Baron, Beechcraft.

TA-: U.S.A.F. and U.S.N. designation for conversion training developments of attack (A-) aircraft.

TA-4 Skyhawk, McDonnell Douglas: *see under* A-4 Skyhawk, McDonnell Douglas.

T.A.A.–Trans Australia Airlines: the Australian domestic state airline, Trans Australia Airlines has operated since 1946, and today has an extensive domestic service network operated in competition with Ansett Airlines of Australia (q.v.) and the other Ansett subsidiaries, although competition in Australia is strictly and extensively regulated, to the extent of affecting departure times, frequency, aircraft types, and seating capacity. Flying doctor services in parts of Australia are operated by T.A.A. on behalf of the Royal Australian Flying Doctor Service and the Northern Territory Aerial Medical Service. The current fleet includes Boeing 727 and McDonnell Douglas DC-9 aircraft, Fokker F-27 Friendships, de Havilland Canada DHC-6 Twin Otters, and Douglas DC-3s.

Tabloid, Sopwith: produced in both floatplane and landplane versions, the Sopwith Tabloid in its former guise had the distinction of winning the 1914 Schneider Trophy race for the United Kingdom, with a speed of 86 mph, and later of raising the world speed record for floatplanes to 92 mph. A single-seat biplane powered by an 80 hp Gnome rotary engine, the Tabloid was developed as a landplane, and most of the aircraft built were landplanes. First flown in 1913, it was based upon the Royal Aircraft Factory's B.S.1 of the previous year, which had been designed by Geoffrey de Havilland and the B.S.1. was in fact the forerunner of the World War I fighter.

tactics: a much more limited term than strategy (q.v.), tactics can generally be described as the science of conducting a battle. Tactics are used within a strategy, and while the general nature and the object of an attack may be described as a strategy, the problems arising from its execution will be tactical rather than strategic.

tail-first: a concept of aircraft design in which the tailplane is placed before the mainplane, although the fin and rudder may still be in the conventional position. Many early designs, including those of the Wright brothers, used the tail-first layout, but the concept fell from favour before World War I. It has only been revived again in recent years with such aircraft as the SAAB-37 Viggen; more correctly such aircraft today are described as canards.

tailless: an attempt to minimise the risk of stalling and to improve the controllability of the early aeroplane, the first tailless (i.e. without a tailplane, but not necessarily without a rudder and fin) designs were produced in the United Kingdom before World War I by John William Dunne. These inspired Professor G. T. R. Hill to design his series of tailless Pterodactyls for Westland, the first of which flew in 1926. Another exponent of the tailless aircraft during the inter-war years was Dr Alexander Lippisch in Germany, whose research aircraft which first flew in 1931. Lippisch was concerned with discovering the ideal shape for high-speed flight rather than with the behaviour of the aircraft at low speed, which was the main interest of Dunne and Hill.

Lippisch's work in fact led ultimately to the delta wing (q.v.) aeroplane, although this configuration did not always lack a tail. The first operational tailless aeroplane was in fact the Messerschmitt Me.163 Komet (q.v.) rocket-powered interceptor of World War II, while the first operational delta wing aircraft (in this case also tailless) was the Gloster Javelin (q.v.). The largest was the purely experimental Northrop Flying-Wing. Other notable tailless aircraft have included the Convair F-102 and F-106, the Fairey F.D.2, the Dassault Mirage III, IV, and V, the Hawker

Siddeley Vulcan, the Convair Hustler, and the B.A.C.-Aerospatiale Concorde (q.v.).

Talon, Northrop T-38: see T-38 Talon, Northrop.

Talos: the Bendix RIM-8G Talos is an American shipboard surface-to-surface missile designed for use over long ranges, the medium range being in excess of sixty miles. Some difficulties have been encountered with the missile, in spite of the success scored by the U.S.S. *Long Beach* against two MiGs over North Vietnam in 1968. In service only with the U.S. Navy, the Talos is twenty-one feet in length and has a diameter of two feet. It uses a solid propellant booster and a liquid-fuelled sustainer, while guidance is by a combination of beam-riding and semi-active radar homing. Control is by moving wings. A nuclear or conventional warhead may be carried, and the maximum speed is about Mach 2·5.

tandem seating: the seating of the occupants of an aircraft behind one another, tandem seating is found most often in the smaller military aircraft, such as fighters and fighter-bombers, if these have a two-man crew; the layout is favoured in such cases because it permits a narrow, low-drag fuselage. In combat aircraft the second seat is normally occupied by an observer or navigator, and in training aircraft the pupil sits in front of the instructor.

Tank, Professor Kurt: the chief designer for the German Focke-Wulf (q.v.) concern before and during World War II, Kurt Tank's most famous designs were the Fw.190 fighter and its development, the Ta.152H, which was fitted with a pressurised cockpit and other equipment for high altitude operations. After the war Tank worked in India for a period, designing the Hindustan HF-24 (q.v.) Marut interceptor, which was the first jet combat aircraft to be designed in Asia.

Tarom–Transporturile Aeriene Romane: in common with many East European airlines, Tarom was formed in 1946 with Russian assistance, the Soviet Union holding a 50 per cent interest in the airline. The initial fleet consisted of Lisunov Li-2s. A pre-war Rumanian airline was LARES, formed in 1932 by the Rumanian Government, which had ceased operations in 1939.

Ownership of Tarom passed completely to Rumania in 1954, and after this the airline began a steady expansion, assisted by the introduction of Illyushin Il-14 airliners. The airline's first turboprop airliners were introduced in 1962 – these were Ilyushin Il-18s – and in 1968 the airline introduced its first jets, B.A.C. One-Elevens. The present fleet includes the One-Elevens, Ilyushin Il-14s and Il-18s, and Antonov An-24s, operating a network of domestic and European services.

Tartar: in use with the United States Navy and with the navies of a number of other NATO countries as well as with the Royal Australian Navy, the General Dynamics RIM-21 Tartar shipboard surface-to-air missile is currently being replaced in American service by the Standard (q.v. missile. Approximately fifteen feet in length and about two feet in diameter, the Tartar has a range of ten miles, and is a two-stage solid propellant missile with semi-active homing radar guidance and control by folding rear fins. A high-explosive warhead is carried, and maximum speed is in the region of Mach 2·5.

Taube, Rumpler-Etrich: developed from a glider designed by the Austrian Etrich concern, the Taube (Dove) was built in Germany by the Rumpler factory. At the outbreak of World War I half of the Military Air Service's strength of 250 reconnaissance aircraft were Taubes, built by either Rumpler or its licensees, including Albatros, Aviatik, A.E.G., D.F.W., Euler, Gotha, L.V.G., Otto, and the Jeannin concern, which built an all-steel version known as the Stahltaube. A monoplane, the Taube used various engines of about 100 hp.

TC-: international civil registration index mark for Turkey.

Teal, Thurston-Schweizer: designed by Colin Thurston and now manufactured by the American Schweizer Aircraft, the Teal is the lowest-priced amphibian on the market, and is a two-to-three-seat, single-engined aircraft aimed at the private-owner market. A high-wing monoplane, apart from having the engine drive a pusher propeller, the Teal has the other notable feature of 'retracting' the undercarriage only for water landings – at all other times, even in flight, the undercarriage is in the down position, since there is no internal accommodation in the wing or fuselage for it.

telecommunications satellite: perhaps one of the best-known uses for space satellites in earth orbit, telecommunications satellites are today in fixed orbits allowing military and civil users the opportunity of good-quality and reasonably reliable rapid telecommunication around the world. The satellites can receive and relay radio and television broadcasts and radio-transmitted telephone calls.

Tempest, Hawker: one of the fastest piston-engined fighter aircraft, the R.A.F.'s Hawker Tempest was one of Sir Sydney Camm's designs, and was developed as a fighter from the slightly earlier Typhoon (q.v.), which was primarily a ground-attack aircraft. Early versions of the Tempest used a Bristol Centaurus engine, but later models used Napier Sabres of 2,180 hp and even 2,500 and 2,825 hp. A single-seat, low-wing monoplane, the Tempest had a maximum speed well in excess of 450 mph and was the only fighter to be able to operate successfully against the German V-1 flying bombs. A carrier-borne development for the Royal Navy was the Hawker Sea Fury.

Terrier: the General Dynamics RIM-2 Terrier shipboard surface-to-air missile's development started in 1951 and the original missile, which had encountered a number of difficulties during its service, was taken out of service in 1968. The Advanced Terrier, first introduced in 1963, continues to be used by the United States Navy, and by the navies of the Netherlands and Italy. A medium-range weapon, Terrier has a range of twenty miles and a maximum speed of Mach 2·5, using a two-stage solid propellant rocket and a guidance system which combines beam-riding and semi-active radar homing. A high-explosive warhead is carried.

Test Ban Treaty: originally framed and signed by the United Kingdom, the United State, and the Soviet Union in 1963, the Test Ban Treaty is an agreement not to conduct nuclear weapons tests in the atmosphere, in order to avoid any build-up of damaging radio-active fallout; it has since been signed by about a hundred nations. The two main non-signatories are France and Communist China, an omission which is made the more serious by the primitive and 'dirty' nature of their weapons, and by the fact that their testing grounds tend to be closer to populated areas than those used by other nations in the past.

TF-: U.S.A.F. and U.S.N. designation for conversion trainer versions of fighter (F-) aircraft.

TF-: international civil registration index mark for Iceland.

TF-104 Starfighter, Lockheed: *see* F-104 Starfighter, Lockheed.

TG-: international civil registration index mark for Guatemala.

Thai Airways International: formed in 1959 with a 30 per cent Scandinavian Airlines System interest, Thai Airways International started operations in 1960, acquiring the fleet and the services of the former Thai Airways Company. Since that date a network of services throughout the Far East, and to Australia and Europe, has been created, using a fleet which now consists entirely of McDonnell Douglas DC-8s.

third-level: not to be confused with third-rate, the term 'third-level' applies to that increasingly important category of air transport operation which consists of scheduled air taxi services (i.e. scheduled services with aircraft of below 12,500 lb weight – although this is likely to be increased to 14,000 lb or more in the future). In many countries, third-level services are actively encouraged by the regulatory authorities, and these include Australia and the United States; such

services are also being developed in Canada, the United Kingdom, and many parts of Europe.

3A-: international civil registration index mark for Monaco.

Thunderbird: the English Electric-designed Thunderbird surface-to-air missile has been built in Mk.1 and Mk.2 forms by the British Aircraft Corporation for the British Army. The Mk.1 first became operational in 1960. Thunderbird is in fact more of a missile system than simply a missile, normally being available with associated surveillance and height radar, and also with the short-range Rapier (q.v.) missile as a back-up weapon. Thunderbird itself, in Mk.2 form, is twenty-one feet long and twenty-one inches in dia- and is a single-stage solid propellant rocket with boosters. Guidance is by beam-riding, radar control, or infra-red homing, while wings are used for control. Maximum speed is Mach 2·5, and while details of the maximum range are unknown, it is believed to be in the region of forty to fifty miles. A highly mobile and reliable system, Thunderbird carries a high-explosive fragmentation warhead, and the missile and its associated weapons and radar can be air-transported.

Thunderbolt, Republic F-47: *see* Thunderbolt, Republic.

Thunderchief, Republic F-105: *see* F-105 Thunderchief, Republc.

Thunderjet, Republic F-84: *see* F-84 Thunderjet, Republic.

THY–Turk Hava Yollari: originally formed in 1933 by the Ministry of Defence as Devlet Hava Yollari, Turk Hava Yollari came into being in 1956 and initially operated a small network of services in the Middle East with Douglas DC-3s. The new airline was largely owned by the Turkish Government, but B.O.A.C. had a small interest. Vickers Viscounts were introduced in 1957, followed by Fokker F-27 Fellowships in 1960, and the first jets, McDonnell Douglas DC-9s, in 1968.

The present fleet consists of Boeing 707-320s, McDonnell Douglas DC-9s, and Fokker F-27s and F-28s, operating on a network of domestic services and international services to major centres throughout Europe. The British Overseas Airways Corporation still holds a 2 per cent interest in the airline.

TI-: international civil registration index mark for Costa Rica.

Tigercat: a short-range surface-to-air missile, the Short Tigercat is a land-based version of the Seacat (q.v.), and is a fully mobile system, using a director trailer and a launcher trailer, both of which are towed by a Land-Rover.

Tiger Moth, de Havilland D.H.82: *see under* D.H.60 Moth, de Havilland.

Tips, Ernest: Managing Director of Avions Fairey (q.v.) during the 1930s, Ernest Tips designed a number of light aircraft for the amateur pilot, including the Tipsy Trainer and Belfair, and the Tipsy Nipper, the last-named also being intended to be a 'homebuilt' design.

Tissandier, Albert and Gaston: the two Tissandier brothers constructed the first electric-powered dirigible airship in 1883. Possessing an exceptionally poor power-to-weight ratio, the airship's Siemens motor had an output of only 1½ hp, but at the first flight at Auteuil, in France, in October 1883, it still proved possible to effect some control in spite of there being a stiff breeze.

Titan: originally the main component of the American intercontinental ballistic missile force, the Martin LGM-25C Titan 2 first became operational in 1963, and some fifty-four remain in service, replacing the earlier Atlas and Titan 1, from which the Titan 2 was evolved. The Titan 2 is 103 feet in length and 10 feet in diameter, and is a two-stage liquid-fuelled rocket using inertial guidance and control by gimballed engines. A nuclear warhead of up to eighteen megatons can be carried over a range of up to 9,000 miles.

TJ-: international civil registration index mark for the Cameroons.

TL-: international civil registration index mark for the Central African Republic.

T.M.A.: *see* **Trans Mediterranean Airways.**

TN-: international civil registration index mark for the Republic of the Congo.

Tomcat, Grumman F-14: *see* F-14 Tomcat, Grumman.

TR-: international civil registration index mark for Gabon.

Tracer, Grumman E-1B: *see* E-1B Tracer, Grumman.

Tracker, Grumman S-2: *see* S-2 Tracker, Grumman.

tracking radar: alternatively known as narrow-beam radar, tracking radar is designed to lock on to and follow targets, while a narrow beam oscillates around the target to make evasion difficult. An essential part of any major air defence system, tracking radar is also important in directing anti-aircraft artillery and certain types of surface-to-air guided missile.

Trader, Grumman C-1A: *see* C-1A Trader, Grumman.

trailing edge: the trailing edge is, as the name implies, the rear edge of any aerodynamic surface, with particular reference to the mainplane.

Transall, C-160: *see* C-160 Transall.

transatlantic flights: *see* Atlantic flights.

Trans Mediterranean Airways S.A.L.–T.M.A.: formed in 1953 as an air freight charter airline, Trans Mediterranean Airways initially used Avro Yorks. Scheduled air freight services were started in 1960, and the airline is today the largest all-freight scheduled airline outside of the United States. The first round-the-world all-cargo service was introduced by T.M.A. in 1971. A fleet of Boeing 707-320Cs and Douglas DC-6s is used on a network which includes Europe, the Middle East, and the Far East.

transonic: transonic speed is achieved when an aircraft is travelling near the speed of sound (Mach 1·0), and the airflow over the upper surface of the wing is moving at supersonic speed, and that over the lower surface at subsonic speed.

transponder: an essential item of aircraft electronics (avionics), the transponder is activated by radar in an air traffic control system or an air defence system, and issues a signal which can be coded to provide identification of the aircraft. Certain Soviet warships are also equipped with transponders to avoid accidental attack by 'Styx' missile-carrying gunboats.

Transportes Aereos Portugueses–Portuguese Airlines: the Portuguese airline Transportes Aereos Portugueses (T.A.P.) was formed in 1944 by the Department of Civil Aviation, and services started in 1946 with a Douglas DC-3 service to Madrid, followed by services to Angola and Mozambique. During the years which followed, T.A.P. expanded its European network and introduced Douglas DC-4 and Lockheed Constellation and Super Constellation aircraft. A reorganisation in 1953 made T.A.P. a company structure, and today the airline is in fact largely privately-owned.

T.A.P.'s first jets, Aerospatiale Caravelles, were introduced in 1962, although before this pool agreements with Air France and B.E.A. had enabled the airline to use the Caravelles and Comets of these two airlines. The current fleet consists of Boeing 747-292Cs, 707-382Bs, and 727-82s, and Aerospatiale Caravelles, operating on a European and African network, with services to North and South America.

Trans World Airlines–T.W.A.: formed in 1930 as Transcontinental and Western Air Express from the merger of two other airlines, T.A.T. and Maddux and Western Air Express, which dated from 1925, T.W.A. initially concentrated on expanding its domestic network, with Douglas DC-2 and DC-3 airliners, and later became the first airline to operate a pressurised airliner, the Boeing 307 Stratoliner. During World War II many of the airline's aircraft (including the Stratoliners) and their crews were pressed into military service, and it was during this period that T.W.A. gained its first experience of operations on transoceanic routes.

The airline commenced international services in 1946, using Lockheed Constellations on a transatlantic service from New York to Paris via Shannon, and later introducing a service to London. The present title was adopted in 1950.

Trans World Airlines is today one of the largest airlines in the world, operating a fleet of Boeing 747s, 707s, and 727s, Lockheed TriStars, Convair 880 Coronados, and McDonnell Douglas DC-9s on an extensive domestic network and on services to Europe, the Middle East, and the Far East.

Trenchard, Marshal of the Royal Air Force Hugh Montague, 1st Viscount Trenchard, G.C.B., G.C.V.O., D.S.O., (1873–1956): known as the 'father of the Royal Air Force', Trenchard was originally an army officer who learned to fly in 1912, and became an instructor in the Royal Flying Corps. After the outbreak of World War I he was given the task of increasing the R.F.C.'s strength in the United Kingdom when the bulk of the force accompanied the British Expeditionary Force to France. In 1915 Trenchard was placed in command of the R.F.C.'s combat units, and under his leadership the R.F.C. progressively gained aerial supremacy.

The first Chief-of-Staff of the Royal Air Force on its formation in 1918, Trenchard quickly set about establishing the new service with ranks, uniforms, and colleges of its own, including the Royal Air Force College at Cranwell, in order to emphasise its independence of the Army and the Royal Navy. During the years before his retirement in 1929, he created an air force, placing great importance on the strategic bomber and on fighter defence, both of which policies were to be proved correct during World War II.

After his retirement from the R.A.F., Viscount Trenchard was Chief Commissioner of the Metropolitan Police until 1935, and established the Police College at Hendon.

Trident, Hawker Siddeley: originating as the de Havilland D.H.121 medium-haul jet airliner, the Trident was originally to have been built by the Aircraft Manufacturing Company, but was caught in the 1960 mergers of the British aircraft industry and has since been marketed as a

Hawker Siddeley product. First flight of a Trident 1C, with three rear-mounted 9,850 lb thrust Rolls-Royce Spey 505 turbojets, took place in January 1962, and this type entered service with British European Airways, to whose specification the aircraft was designed. The Trident 1E export version followed, with a first flight in November 1964; and the Trident 2E, with aerodynamic modifications and increased fuel-tankage, as well as 11,930 lb thrust Rolls-Royce Spey 512 turbofans, giving increased speed (maximum cruising speed of 610 mph) and range (maximum range with 115 passengers, 2,450 miles), first flew in July 1967. A further development, the stretched-fuselage Trident 3B, with a maximum seating capacity of about 150 passengers, first flew in 1970; a feature of this aircraft, which has a range of 1,600 miles, is the addition of a 5,250 lb thrust Rolls-Royce RB.162 lightweight turbojet below the centre Spey 512 to provide additional take-off thrust. An order for Tridents by the Civil Aviation Administration of China in late 1972 specified the Trident Super 3B, with a range of 2,300 miles, and some further stretching of the fuselage.

The Trident was the first aircraft to be designed specifically for automatic landings, and after extensive tests was certified for autoflare (q.v.) operations in June 1965, and for autoland (q.v.) in May 1967. The worst conditions in which a Trident can land are category 3a or 3b (see category, weather) depending on the version of the aircraft and on the airport.

trijet: the term 'trijet' refers to a three-engined airliner, usually with the engines tail-mounted, as with the Boeing 727, Hawker Siddeley Trident, and the Yakovlev Yak-40.

trimotor: although the term 'trimotor' obviously refers to a three-engined aircraft using radial or piston engines, strictly speaking the term is strangely limited to aircraft with one nose-mounted engine (and the other two wing-mounted or on the wing bracing struts), such as the Junkers Ju.52/3M, Fokker F.VIIB, and the Ford Trimotor of the inter-war years, rather than to aircraft, such as the Short Cal-

cutta or the Dornier Do.24, with three wing-mounted engines.

Trimotor, Ford: originally designed by William Stout, the Ford Trimotor was one of a number of single- and triple-engined Stout designs acquired by the American Ford Motor Corporation; numerous versions were built after its first appearance in 1925. Nicknamed the 'Tin Lizzie', largely because of the all-metal construction with corrugated fuselage, but also because of the spartan and metallic interior, the Ford Trimotor contributed much to the development of air transport, particularly in the United States. Early versions used 200 hp Wright Whirlwind radial engines, but on the final versions, built during the early 1930s, 450 hp Pratt and Whitney Wasp radials were used, while seating accommodation ranged from eight passengers to fourteen.

triplane: an aircraft with three wings, or sets of wings, mounted above one another, with or without stagger, is known as a triplane, appropriately enough. Interest in triplanes reached its zenith during World War I, with aircraft such as the Sopwith Tripe and the Fokker Dr.1, but declined rapidly afterwards, probably because of the considerable drag experienced.

Trislander, Britten-Norman: a development of the Britten-Norman BN-2 (q.v.) Islander, the Trislander retains much of the smaller aircraft's structure, and the main differences are the stretched fuselage, with accommodation for up to eighteen passengers, and the addition of a third engine on the fin. Three 260 hp Avco Lycoming 0-540-E465 piston engines give a maximum speed of 200 mph and a range of up to 700 miles. A stretched nose with luggage accommodation was introduced in 1972 to overcome the shortage of luggage space on initial versions of the Trislander.

TriStar, Lockheed L-1011: development of the Lockheed L-1011 TriStar medium-haul airbus started in 1968 with the then unprecedented order-book of 144 aircraft. The first flight took place in 1971, and the aircraft entered airline service during the Spring of 1972. Three Rolls-Royce

RB.211-22 triple-spool turbofans of 42,000 lb thrust provide a maximum cruising speed of 610 mph and a range of 2,800 miles in the Tristar 1, and development of a 4,000 mile range TriStar -2 with 45,000 lb thrust RB.211-24s is well under way. Maximum passenger capacity is 400, although this depends on class mix and pitch, and the figure is usually nearer to the 280 mark. Only one of the three engines is tail-mounted.

Troopship, Fokker F-27M: *see under* F-27 Friendship, Fokker.

TS-: international civil registration index mark for Tunisia.

TT-: international civil registration index mark for Chad.

TT, Martin: one of the earliest Martin designs, the TT first appeared before the United States entered World War I, and was used by the Army. A 90 hp Curtiss OX-2 engine was used.

TU-: international civil registration index mark for the Ivory Coast.

Tu-16 'Badger', Tupolev: the Tupolev Tu-16 jet bomber, NATO code-named 'Badger', first appeared in 1954. Two 20,950 lb thrust Mikulin AM-3M turbojets give a maximum speed of 590 mph and a range of up to 4,000 miles. The standard aircraft, the 'Badger A', can carry up to nine tons of bombs; the 'Badger B' can carry two 'Kennel' air-to-surface missiles; and 'Badger C' is employed on reconnaissance duties and carries a single 'Kipper' air-to-surface missile. The mid-wing Tu-16 is no longer an important part of the Soviet Union's offensive air power.

Tu-20 'Bear', Tupolev: the only turboprop heavy bomber ever put into operational service, the Tupolev Tu-20, NATO code-named 'Bear', first appeared in 1955. The original aircraft was designated 'Bear A', and a modernised version, the 'Bear B', remains in limited service. Four 14,795 shp Kuznetsov NK-12M turboprops give the Tu-20 a maximum speed of 500 mph and a range of 7,800 miles. Eleven tons of bombs can be carried internally, or

'Kangaroo' missiles can be fitted on underwing strongpoints. The high-wing 'Bear' is today normally used on electronic countermeasures and reconnaissance duties.

Tu-22 'Blinder', Tupolev: the only bomber in the world with tail-mounted engines and the only supersonic aircraft to have engines in that position, the Tupolev Tu-22, NATO code-named 'Blinder', first appeared in 1961, and has since been taking over many of the duties performed by the Tu-16 and Tu-20. A low-wing monoplane, the Tu-22 uses two 26,500 lb thrust turbojets with reheat for a maximum speed of Mach 1·5 (1,000 mph) and a range of up to 3,000 miles, while a 'Kitchen' air-to-surface missile can be carried.

Tu-104 'Camel', Tupolev: the first Soviet turbojet transport aircraft, the Tupolev Tu-104, NATO code-named 'Camel', first flew in 1955. The initial version of this low-wing monoplane carried the Tu-104 designation and could accommodate fifty passengers. Later versions included the seventy-seat Tu-104A and, in 1959, the stretched Tu-104B with one hundred seats. Two wing-mounted 21,385 lb thrust Mikulin AM-3M-500 turbojets provide a maximum speed of 500 mph and a range of 1,400 miles.

Tu-114 'Moss'/'Cleat', Tupolev: for many years the world's largest airliner, the Tupolev Tu-114, NATO code-named 'Cleat', was based on the Tu-20 (q.v.) bomber and first flew in late 1957, although it did not enter Aeroflot service until April 1961. Up to 220 passengers can be carried over a range of up to 5,560 miles at a maximum cruising speed of 478 mph. Four 14,795 shp Kuznetsov NK-12MU turboprops are used. A maritime-reconnaissance and airborne-early-warning development is NATO code-named 'Moss'.

Tu-124, Tupolev: a smaller development of the Tu-104 (q.v.) for short-haul operations, the Tupolev Tu-124 retains the appearance of the larger aircraft. First flown in June 1960, the Tu-124 entered service in October 1962. Two 11,905 lb thrust Soloviev D-20P turbofans provide a maximum cruising speed of 500 mph and a range of 760 miles. There is accommodation for about sixty passengers.

Tu-134, Tupolev: essentially a modified Tupolev Tu-124, the Tu-134 has had the engines moved from the wings to the tail, and a 'T' tailplane fitted, along with other minor modifications. The Tu-134 first flew in 1964 and entered service with Aeroflot in 1967. Two 15,000 lb thrust Soloviev D-30 turbofans provide a maximum cruising speed of 560 mph and a range of up to 1,500 miles, while between seventy and eighty passengers can be carried.

Tu-144, Tupolev: the first supersonic airliner to fly, the Tupolev Tu-144 first flew at the end of December 1968, but has since been delayed by numerous and unspecified difficulties, to overcome which a number of modifications have been made. It is not yet known whether the aircraft is in production, but entry into airline service is unlikely before mid-1974 at the earliest. Maximum speed is Mach 2·37, and the prototype can accommodate 126 passengers, although it is believed that a production version will carry nearer 150 passengers. Range so far appears to be in the region of 2,400 to 3,200 miles. Four 38,500 lb thrust Kuznetsov Nk.144 turbofans with reheat are used.

Tu-154, Tupolev: first flown in October 1968, the Tupolev Tu-154 is the Soviet Union's rival to aircraft in the Hawker Siddeley Trident 3B and Boeing 727 range, having three tail-mounted 21,000 lb thrust Kuznetsov Nk-8-2 turbofans giving a maximum cruising speed of 526 mph and a range of up to 2,500 miles, while up to 164 passengers can be carried. This appears to be the most promising Soviet airliner so far.

Tunis Air–Société Tunisienne de L'Air: Tunis Air was formed in 1948 with the support of Air France, which also held the majority interest in the airline until Tunisian independence in 1957. Today, a network of domestic services and international services throughout North Africa

and the Mediterranean area is operated by a fleet of Aerospatiale Caravelles and Boeing 727s and 737s, with some light aircraft.

Tunnan, SAAB-29- *see* SAAB-29 Tunnan.

Tupolev, Andrei Nikolaevich (1888–1972): one of the Soviet Union's leading airframe designers, Andrei Tupolev was instrumental in persuading Lenin to establish the Central Aerodynamics and Hydrodynamics Research Institute in Moscow after the Bolshevik Revolution. Tupolev's first opportunity to design an aircraft in his own right came while he was working at the factory established near Moscow in 1922 by the German Junkers concern – this was the single-seat ANT-1 monoplane. After this, Tupolev headed the design bureau which today bears his name, and started to produce the bomber and transport designs for which he has become famous; some of these were produced in prison, to which he was committed for political reasons in 1934. While he was in prison, his TB-7 strategic bomber design was taken over by his assistant, Petlyakov, who saw the design into production as the Petlyakov Pe-8, the only four-engined strategic bomber produced by Russia during World War II. However, Tupolev also designed the Tu-2 twin-engined bomber during this period.

After the war captured Boeing B-29 Superfortress bombers were copied, and entered service in 1946, being followed by a succession of turbojet and turboprop designs, including the Tu-16 (q.v.) and Tu-20 (q.v.) bombers, and Tu-104 (q.v.) and Tu-114 (q.v.) airliners. More recently, Tupolev's design bureau has produced the Tu-22 (q.v.) supersonic bomber, the Tu-144 (q.v.) supersonic airliner, designed by his son, Alexei, and the Tu-134 (q.v.) and Tu-154 (q.v.) airliners. A new Mach 3·0 bomber, NATO code-named 'Backfire', is under development for service during the mid-1970s.

In spite of the growing influence of other designers, including his son, Alexei, in the Tupolev Bureau, Andrei Tupolev retained an advisory function until his death in December 1972.

turbine: more commonly known as the gas turbine (q.v.) or jet (q.v.).

Turbo-Beaver, de Havilland Canada DHC-2: *see under* DHC-2 Beaver, de Havilland Canada.

turbofan: a development of the turbojet engine, the turbofan engine includes a ducted fan, larger than the compressor blades, which drives a part of the air intake past the compressor blades, thus giving the turbofan its alternative title of a by-pass engine. Advantages of the turbofan include good performance at low altitude, thus enabling many turboprop replacement aircraft to employ turbofans, and the shielding of the exhaust gases from the engine inside a circle of cooler air, thus reducing the noise which is created by hot exhaust gases hitting a cold atmosphere abruptly.

turbojet: the standard jet (q.v.) or gas turbine (q.v.) engine.

Turbomeca: Société Turbomeca was formed in 1938 and has been producing turbojets since 1947, including such famous designs as the Artouste and the Marbore. The company's production is largely concentrated at the smaller end of the field, and current production includes the Astazou, Bastan, Astafan, and the Turmo, while the Adour is a joint venture with Rolls-Royce for the B.A.C.-Breguet Jaguar and the Mitsubishi T.2. A new twin-spool military engine is the Larzac, and this is also the first engine to have been developed jointly by the private-enterprise Turbomeca and the state-owned SNECMA (q.v.).

turboprop: the turboprop engine is basically a gas turbine aero engine fitted with a propeller to provide better performance at lower speeds and altitudes than would be possible with a jet engine.

The first turboprop, the Rolls-Royce Trent of 1945, was in fact a modified Derwent turbojet, and after trials on a modified Gloster Meteor and a prototype Vickers Viscount, Rolls-Royce developed from it the highly successful Dart turboprop, which has been used on numerous civil and military aircraft. A more power-ful counterpart of the Dart has been the Rolls-Royce Tyne. Other important manufacturers of turboprop engines have included Bristol Siddeley, Allison, Turbomeca, Avco Lycoming, and Napier. For many applications the turboprop is now falling from favour, due to the superior performance of the turbofan, although increasingly shaft turbines are being used for helicopters and executive and third-level aircraft, in spite of the high costs compared with piston engines, which is balanced by higher performance.

Tutor, Canadair CT-114: *see* CT-114 Tutor, Canadair.

Twin Comanche, Piper: based on the Comanche (q.v.), the Piper PA-30 Twin Comanche first flew in prototype form in November 1962, and entered production the following year. Using the same fuselage and cabin arrangement as the single-engined aircraft, the Twin Comanche has two 160 hp Lycoming IO-320-B piston engines, giving this four-to-six-seat, low-wing monoplane a maximum speed of 200 mph and a range of 1,100 miles.

Twin Otter, de Havilland Canada DHC-6: *see* DHC-6 Twin Otter, de Havilland Canada.

Twin Pioneer, Scottish Aviation: bearing no resemblance to the Scottish Aviation Pioneer (q.v.), the Twin Pioneer first flew in August 1957, as a S.T.O.L. light transport for the Royal Air Force. Deliveries to the R.A.F. started in 1958, and the aircraft has also been sold to the Royal Malaysian Air Force; a few are used by civil air transport undertakings. Two 640 hp Alvis Leonides 531/8 radial engines give the twelve-to-sixteen-seat Twin Pioneer a maximum speed of 165 mph and a range of up to 700 miles. In appearance the Pioneer is a twin-engined, high-wing monoplane with a triple fin tailplane.

TY-: international civil registration index mark for Dahomey.

Type 38, Short: the Short Type 38 or S.38 biplane became the first aircraft to take off from a ship under way when Lieutenant Charles Rumney Samson flew a Type 38 from H.M.S. *Africa* in December 1911;

267

the ship had had a special platform constructed over the bows for the purpose. The Type 38 was a single-engined, pusher-propeller aircraft with tandem seating for two.

Typhoon, Hawker: primarily a ground-attack aircraft, the Hawker Typhoon, a single-seat, low-wing monoplane, carried rockets in addition to its four 20 mm cannon, and was used during the closing stages of World War II on anti-tank and train-busting duties. Designed by Sir Sydney Camm, the Typhoon had a maximum speed of 416 mph from its 2,240 hp Napier Sabre piston engine. A development was the Tempest (q.v.) Main user of the type was the Royal Air Force.

TZ-: international civil registration index mark for Mali.

U

U-: U.S.A.F. designation for utility aircraft, although it has been used for other types in order to maintain secrecy (*see* U-2). Helicopters in this category are designated UH-.

U-2, Lockheed: designated as a utility aircraft in order to maintain secrecy for as long as possible during development, the Lockheed U-2 high-altitude reconnaissance aircraft first entered U.S.A.F. service in 1954, and aircraft of this type made high-altitude flights over the Soviet Union until 1960, when one was shot down. Many of the aircraft have since been modified for weather flights, and designated WU-2, while the remainder have been supplied to Nationalist China for reconnaissance duties over the Chinese mainland. The initial production aircraft was the U-2A, with an 11,200 lb thrust engine, but most aircraft were built to U-2B standard, with a 17,000 lb thrust Pratt and Whitney J75-P-13 turbojet giving a speed of 520 mph and a range of more than 3,000 miles. Few U-2s remain in service.

U-8 Seminole, Beechcraft: *see* Queen Air, Beechcraft.

U-10 Super Courier, Helio: a light utility monoplane with a high wing, the S.T.O.L. Helio Courier was first ordered for trials by the U.S. Army in 1952. Orders followed in 1958 for the Super Courier, which was designated the U-10. A single 295 hp Lycoming G0-480 piston engine powers the five-seat aircraft, giving a maximum speed of 167 mph and a range of up to 615 miles.

U-21, Beechcraft: A turboprop development of the Queen Air (q.v.), the Beech U-21 utility transport is in U.S.A.F. service. It uses two 550 shp United Aircraft PT6A-20 turboprops for a maximum cruising speed of 250 mph and a range of up to 1,600 miles.

Udet, Oberleutnant Ernst: Germany's second ranking fighter pilot during World War I, Ernst Udet ended the war with sixty-two confirmed victories, gained while flying Fokker D.VII and Siemens D.III biplanes. During the inter-war period Udet was a stunt pilot and explorer, also managing to produce a design for a twin-engined, high-wing airliner, the 'Kondor', during the mid-1920s. He was granted a number of senior appointments after Adolf Hitler came to power in Germany, although he had little liking for the régime and eventually committed suicide in 1941.

UH-2 Seasprite, Kaman: the Kaman UH-2 Seasprite utility helicopter first flew in July 1959, and was the first Kaman design to use a conventional single main rotor instead of the intermeshing rotors used by the company until then. The UH-2A entered U.S. Navy service in December 1962, and this was followed by the UH-2B with simplified avionics. Up to eleven passengers can be carried, and the single 1,250 shp General Electric T58-GE-8B turbine provides for a maximum speed of 162 mph and a range of 670 miles.

undercarriage: although strictly the term 'undercarriage' relates to any form of aircraft landing gear, including skids, floats, skis, and wheels, the general use of the world relates solely to wheeled undercarriages.

Early aircraft used skids and often took off after running along a set of rails, although Alberto Santos-Dumont's 14-bis of 1906 used a wheeled undercarriage without skids, while many aircraft built before World War I used a wheeled undercarriage with skids. Pneumatic tyres appeared at a relatively early stage, on the Vuia monoplane of 1906. An American, Matthew Sellers, is reputed to have built an aircraft with a retractable undercarriage, while the second retractable undercarriage design appeared in 1911 on a German Wiencziers monoplane. The first practical retractable undercarriage was that of the Dayton-Wright high-wing monoplane for the Gordon-Bennett Trophy race of 1920.

Retractable undercarriages are the rule today, even for many light aircraft, since they reduce drag and assist economical cruising flight, although the extra cost of retractable undercarriages has meant that on some low cost and utility designs, fixed undercarriages are common. During World War II and shortly afterwards, the tailwheel undercarriage was replaced by the so-called tricycle undercarriage, with a nosewheel, offering better stability during the taxi and take-off runs, and giving a better forward view during ground movement – a more recent advantage is the greater ease with which such aircraft designs may be stretched.

undershoot: the term undershoot refers to an aircraft landing or falling short of the runway during the landing approach.

United Air Lines: second only to Russia's Aeroflot in size, United Air Lines is the largest genuine commercial airline in the world. United Air Lines itself dates from 1933, but the company's origins lie in the formation in 1926 of an airmail operator, Varney Air Lines, and in 1927 of three other airlines, Boeing Air Transport, National Air Transport, and Pacific Air Transport, all of which carried airmail for the U.S. Post Office. Boeing acquired Pacific in 1928, and Varney and National in 1930; the resulting airline was called United Aircraft and Transport when it was formed in 1931. The present airline came into existence after Congress forbade the grouping of manufacturing and operating interests, causing Boeing to dispose of its large interest in United.

In 1933 United was the first airline to introduce the Boeing 247, which with its twin-engined, all-metal, low-wing construction, was the forerunner of the modern airliner. The firm remained a mainland United States airline until after World War II, when a route to Hawaii was awarded to it in 1946. In 1947 the airline was the first

to introduce the Douglas DC-6 pressurised airliner, and other such firsts include the first Douglas DC-8 jet airliners in 1959, Super DC-8s (series 61) in 1967, Boeing 720s in 1960, and Boeing 727s in 1962; while in 1961 United became the only American operator of the Sud Caravelle jet airliner. A merger in 1961 with Capital Airlines, which had encountered financial difficulties, boosted United's revenue by a quarter, as well as giving the airline a large fleet of Vickers Viscount turboprop airliners.

Although a domestic airline, United operates throughout the United States, including Hawaii, and has a fleet which currently includes Boeing 747s, 720s, 727s and 737s, and McDonnell Douglas DC-10s, and DC-8s, including the 'Super Sixty' series.

U.T.A.–Union de Transportes Aériens: the French independent airline, U.T.A. was formed in late 1963, following a merger of U.A.T. (Union Aéromaritime de Transport), and T.A.I. (Transports Aériens Intercontinentaux), both of which dated from the immediate post-World War period and had a strong shipping connections. While U.A.T. had concentrated on scheduled services to Africa, T.A.I. had operated mainly as a charter airline until starting services to the Pacific in 1956. U.A.T. had been one of the airlines to operate the de Havilland Comet I airliner.

Today, U.T.A. continues the services of its predecessor airlines, with a strong African network and services to the Pacific, so that the airline operates in association, rather than in competition, with Air France. The fleet includes McDonnell Douglas DC-10s and DC-8s, including the 'Super Sixty' series, and Aerospatiale Caravelles. Subsidiary airlines include Air Niger, Air Hebrides, Air Ivoire, Air Polynesie and Air Volta, and Transgabon. There is also an association with Air Afrique.

V

V-1: the German V-1 'Flying Bomb', so-called because it was the first of the 'Vergeltungswaffen' (Revenge Weapons), was more properly designated the F2G-76. First used against targets in England during 1944, the V-1 was in effect a small aircraft powered by a pulse-jet motor and capable of carrying an 1,870 lb warhead for up to 150 miles at a speed of 400 mph. No guidance system was fitted, the course being pre-determined before the V-1 took off along its launching ramp, and maintained by a gyroscopic stabiliser. Some 20,000 V-1s were launched against England during World War II, mainly from the ground, though a few were air-launched after the Allies overran the launching sites.

V-2: the German V-2 rocket was the second of the 'Vergeltungswaffen' (Revenge Weapons) used by Germany against the British Isles during the closing stages of World War II. Officially known as the A-4, it was made possible by Goddard's work, starting in 1919, on liquid-fuel rockets, although the development of the V-2 was started by Dr Wernher von Braun in 1933. Trials with the V-2, which was the forerunner of the ballistic missile of today, started in 1942, but operational use against targets in France, Belgium, and England did not start until September 1944, some 1,100 rockets being used against England alone during the following six months. The V-2 could carry a 2,000 lb warhead for up to 200 miles.

VAK.191B, V.F.W.-Fokker: an experimental vertical take-off jet fighter, the V.F.W.-Fokker VAK.191B first flew in 1970 and three prototypes were built for an extensive test programme which lasted until late 1972, largely in connection with the Panavia 200 Panther development programme. No production VAK.191Bs were ever built, although the aircraft performed well on test. A single-seat, high-wing monoplane, the VAK.191B used a single 9,920 lb thrust Rolls-Royce/M.A.N. RB.193-12 vectored thrust turbofan and two 5,580 lb thrust Rolls-Royce/M.A.N. RB.162-81 lift jets. Maximum speed was about 600 mph.

Valiant, Vickers: the first British strategic nuclear bomber, or V bomber, the Vickers Valiant first flew as a prototype in May 1951. More than a hundred production aircraft followed, entering R.A.F. service in 1955 on strategic bombing, photographic reconnaissance and, towards the end of their careers, in-flight refuelling duties. Four 9,500 lb thrust Rolls-Royce Avon R.A.14 turbojet engines gave the high-wing Valiant a maximum speed of 560 mph and a range of up to 3,400 miles. It was necessary to withdraw the Valiant from service prematurely due to fatigue problems.

A Valiant dropped the first British atom bomb on 11 October 1956, and the first British hydrogen bomb on 15 May 1957.

Vampire, de Havilland: the second British jet fighter to enter production, the de Havilland Vampire single-seat, twin-boom fighter-bomber first flew in September 1943, and a few aircraft still remain in service around the world, mainly of the twin-seat trainer version. A large number of versions were built, including the Sea Vampire for operation from the Royal Navy's aircraft carriers, and twin-seat (side-by-side) night-fighters and trainers. The Vampire was also built under licence in France by Sud Aviation and in Italy by Fiat and Aermacchi. A single 3,350 lb thrust de Havilland Goblin 3 turbojet gave the mid-wing Vampire a maximum speed of 548 mph and a range of up to 1,200 miles, while four 20 mm cannon were fitted in the nose and 2,000 lb of bombs or rockets could be carried on underwing strongpoints.

Vanguard, Vickers: intended as a larger sister to the highly-successful Viscount (q.v.), the Vickers Vanguard was a commercial failure, only forty-three production aircraft being built for B.E.A., which took twenty, and Air Canada. First flight of a prototype took place in January 1959. This was followed by the Type 951 production model for B.E.A., while Air Canada received the Type 952, with uprated engines and increased weights; the British airline's order was in fact completed with the Type 953, with Type 952 standard weights and Type 951 engines. The Type 953 used four Rolls-Royce Tyne 512 turboprops of 5,545 shp each for a maximum cruising speed of 425 mph and a range of up to 1,830 miles with a maximum payload of 139 passengers. The British Vanguards have largely been converted to all freight operations, and renamed the Merchantman, while the Canadian aircraft are with new owners.

variable geometry: the term 'variable geometry', sometimes referred to as 'variable sweep' or 'swing wings', relates to the ability to vary the angle of sweep of an aircraft's wings during flight, allowing minimum angles of sweep for take-off, landing, and low speed flight, and a higher angle of sweep for very fast flight.

The first aircraft in the world to employ variable geometry (and variable incidence (q.v.)) was the Swedish Palson Type 1 of 1918, although it is generally doubted whether this design can ever have flown. The Grumman XF10F Jaguar jet fighter, which first flew in May 1953, employed variable geometry, but only two prototypes were built because of the complications associated with the design. The first operational variable-geometry aircraft was the General Dynamics F-111 (q.v.), while the Dassault Mirage G.8 (q.v.) was the first fighter design – these designs owing much to the work in Great Britain of Sir Barnes Wallis (q.v.) after World War II. A variable-geometry transport aircraft, the Boeing 2707 (q.v.) was abandoned due to design difficulties and heavy cost escalation.

A disadvantage of variable-geometry aircraft is the unproductive weight of the 'hinges', which on the F-111 add two tons to the airframe weight, while the design

complications, extra maintenance, and liability to drag near the wing roots are further problems.

variable incidence: the term 'variable incidence' is generally used to describe the ability to alter the angle of incidence between the wings and the fuselage of an aircraft.

The first known variable-incidence aircraft was the Swedish Palson Type 1 of 1918, which also employed variable geometry (q.v.), but is not thought to have flown. The Ling-Temco-Vought F-8 (q.v.) Crusader carrier-borne jet fighter, which first flew in March 1955, was the first operational variable-incidence aircraft, while the McDonnell Douglas F-15 (q.v.) has a variable-incidence tailplane.

Varig–Empresa de Viacao Aerea Rio Grandense: the Brazilian airline, Varig, was formed in 1927 by the German Condor Syndicate to operate their Brazilian services, initially with a Dornier Wal flying-boat. Varig remained a domestic airline until 1953, although in the meantime operations were expanded and another domestic airline, Aero Geral, was acquired in 1951. The first international service was to the United States, but the international expansion of the airline really got under way after the acquisition in 1961 of the R.E.A.L. consortium of airlines (which included Aerovias Brasilia, Nacionale, and Aeronorte), doubling the domestic network and quadrupling the international services. Services to Europe followed the acquisition of Panair do Brasil in 1965, after that airline encountered financial difficulties.

Today Varig, which is owned 85 per cent by employees and 15 per cent by management, operates a fleet of Boeing 707s and 727s, Hawker Siddeley HS 748s, and Lockheed L-188 Electras on an extensive domestic network, with international services to the United States, Europe, Africa, and the Far East, as well as throughout Latin America.

V bomber: the term given to the first generation of British strategic nuclear bombers, so-called partly because of the names of the aircraft, the Vickers Valiant (q.v.), the Avro Vulcan (q.v.), and the Handley Page Victor (q.v.), and partly because all the aircraft concerned had swept leading edges to the wings.

VC.10, Vickers: developed to meet a B.O.A.C. requirement for an aircraft capable of operating from hot and high airfields on the airline's African routes, but eventually being developed with full transatlantic capability, the Vickers VC.10 first flew in June 1962, followed by the Super VC.10, with up-rated engines, a fuselage stretch, and extended wingspan, in May 1964. Aircraft for British United Airways incorporated a forward freight door in the standard fuselage; R.A.F. aircraft also included this feature, but with the addition of Super VC.10 engines and wings. The standard aircraft can accommodate up to 150 passengers, and the Super VC.10 up to 178 passengers; the latter uses four 22,500 lb thrust Rolls-Royce Conway 43 Mk.550 turbofans for a maximum cruising speed of 570 mph and a range of up to 4,600 miles with maximum payload. A low-wing aircraft, the VC.10s has tail-mounted engines and a 'T' tailplane.

Venom, de Havilland: a development of the earlier Vampire (q.v.), the de Havilland D.H.112 Venom retains the single-engined, single-seat, twin-boom layout of the Vampire, but has a new engine, thinner and slightly swept wings, and the addition of wingtip tanks. Some Venoms remain in service today, and the design can certainly be counted as a success. First flown in September 1949, the Venom fighter-bomber was first delivered to the R.A.F. starting in 1951, and was followed by the Sea Venom for the Royal Navy and by licence-production in France and Switzerland. The single 4,850 lb thrust de Havilland Ghost 103 turbojet provides the Venom with a maximum speed of 640 mph and a range of up to 1,000 miles, while four 20 mm cannon are fitted in the nose and up to 2,000 lb of bombs or rockets can be carried on underwing strongpoints.

vertical take-off and landing: the term vertical take-off and landing, or V.T.O.L., generally refers to fixed-wing vertical take-off types, rather than to rotary-wing aircraft which can safely be assumed to possess such characteristics.

The first true V.T.O.L. machine, although lacking any aerodynamic surfaces, was the Rolls-Royce 'Flying Bedstead' test rig which made tethered flights during 1953, and free flights in 1954, using two Rolls-Royce Nene turbojets. A prototype transport, the Fairey Rotodyne (q.v.), proved to be successful in tests during the late 1950s, while the Hawker P.1127 vertical take-off fighter was developed into the Hawker Siddeley Harrier (q.v.) V./S.T.O.L. fighter, in service with the R.A.F. and U.S.M.C. today. Another vertical take-off transport design, the Dornier Do.31E (q.v.), also made successful test flights during the late 1960s, and Canadair have experimented successfully with their CL.214 design. Amongst other interested manufacturers can be included Bell, Westland, Sikorsky, Boeing, and M.B.B.

There have been almost as many different approaches to V.T.O.L. development as there have been aircraft manufacturers concerned with the concept. Four basic approaches can be identified: the use of vectored thrust from the engine used for forward flight, as on the Harrier; the use of separate lift jets, which are not used during forward cruising flight, as on the Dornier Do.31E; the use of turboprop or ducted fan engines tilting either on their own or with the whole mainplane assembly for the transition from vertical to horizontal flight, as with the Canadair, Westland and Bell designs; and the use of some form of rotor for take-off, sometimes folding away during cruising flight and sometimes augmenting the lift of the mainplane, as on the Rotodyne and the Sikorsky designs.

Most of these concepts have drawbacks – including excessive power for forward flight when vectored thrust is used, while the payload on take-off is restricted by the thrust available. Lift jets are unproductive weight in flight, and extra weight and complication is a drawback also of the tilting engine and tilting wing designs. Rotors, unless folded away, create drag at

273

higher speeds in flight, and a folding rotor is a costly complication. However, the benefits of vertical take-off are such that manufacturers have a considerable incentive to persevere with the concept, which offers to military users considerable mobility and air support without providing concrete runways as targets for enemy attack, and to civil users an end of the land waste which results from major airport development. The area subjected to aircraft noise on take-off and landing is also reduced.

Vertol: *see* Boeing-Vertol.

V.F.W.-Fokker: the German V.F.W. (Vereinigte Flugtechnische Werke) concern was formed in 1963 by the merger of Focke-Wulf (q.v.) and Weser Flugzeubau, joined in 1964 by Heinkel (q.v.). Much of V.F.W.'s early workload consisted of component and assembly production for the C-160 Transall, the Dornier Do.31E, and the Fokker F-28 Fellowship, in addition to work on the German components of the European Launcher Development Organisation's space projects. Development of V.F.W.'s own designs, the V.F.W. 614 (q.v.) airliner and the VAK. 191B experimental vertical take-off fighter, also continued.

The company took part in the first international aircraft industry merger when it amalgamated with the Dutch Fokker concern, the former owners having equal shares in the new V.F.W.-Fokker. Current work includes production of the Fokker F-27 (q.v.) Friendship and F-28 (q.v.) Fellowship and the V.F.W. 614 airliners, and subcontract work on the Panavia 200 (q.v.) Panther, the McDonnell Douglas F-4 Phantom II, and the European A.300B airbus, as well as continued space work.

V.F.W.-Fokker 614: the first German airliner design of the post-war period to enter production, the V.F.W.-designed 614 first flew in late 1971, and is a small short-haul jet aircraft with accommodation for up to forty-four passengers. An unusual feature is the mounting of the engines in pods above the upper surface of the wing, partly to reduce ingestion while ground running, but mainly to reduce the noise heard on the ground below the aircraft while in the air. Two 7,700 lb thrust Rolls-Royce/SNECMA M45H turbofans provide a maximum cruising speed of 457 mph and a range of 1,000 miles.

VH-: international civil registration index mark for Australia.

Viasa–Venezolana International de Aviacion: the Venezuelan airline Viasa was formed in 1961 to take over the international services of Avensa and L.A.V., leaving these two airlines to operate domestic services, and to hold part of the 45 per cent of Viasa's capital which is contributed by private enterprise – the remainder being held by the Government. Technical and operational assistance has been received from K.L.M. (q.v.), and Viasa now operates to major destinations in Latin America and the United States, with services to Europe in pool with Iberia and K.L.M. The fleet includes Douglas DC-10, DC-8 (including the 'Super-Sixty' series), and DC-9 aircraft.

Vickers: the aircraft manufacturing division of the armaments and engineering group, Vickers first noteworthy aircraft design was the 'Gun Bus' pusher-biplane fighter of 1915. By the end of World War I the company was building the Vimy (q.v.) bomber, which became the first aircraft to fly non-stop across the North Atlantic in 1919. During the early 1920s a number of airliner versions of the Vimy, with a new fuselage, were built. Developments of the Vimy during the 1920s were the Virginia bomber and the related Victoria military transport; these aircraft were still biplanes, and a further biplane design was the single-engined Vildebeest torpedo-bomber of 1930, which was an exceptionally heavy aircraft for a single-engined type.

An all-metal monoplane airliner and executive transport, the Viastra, first appeared in 1931, and normally used two engines, although single-engined and trimotor versions were also built. Vickers did not build any seaplane or flying-boat designs, possibly because of the position of Supermarine (q.v.), which Vickers had acquired during the early 1920s and kept in existence as a separate entity in this field.

The late 1930s saw the introduction of another famous Vickers aircraft, the Wellington (q.v.) medium bomber, into R.A.F. service in time for World War II; the aircraft continued in production and service throughout the war, operating on reconnaissance, maritime-reconnaissance, and navigational training duties as well as normal bombing missions. After the war the Wellington's aerodynamic surfaces were married to a new design of fuselage to produce the Viking (q.v.) airliner and the related Valetta military transport and Varsity navigational trainer.

The post-war period saw a number of achievements for Vickers, although technical and commercial success did not always come together, except in the case of the Viscount (q.v.), which was the world's first successful turboprop airliner. Vickers also produced the first British strategic nuclear bomber, the Valiant (q.v.), and another airliner, the Vanguard (q.v.), before building the first long-range airliner in the world to have tail-mounted engines, the VC.10 (q.v.) of the early 1960s, with its Super VC.10 development. Although Vickers and Supermarine became part of the British Aircraft Corporation (q.v.) in the mergers of 1960, the VC.10 and Super VC.10 continued to be built, as Vickers products, by B.A.C. The Vickers Group today holds 50 per cent of B.A.C.'s share capital.

Another activity of Vickers during the 1960s, which did not pass to B.A.C., was hovercraft development and manufacture, the company building one of the first passenger-carrying types. This activity passed to the British Hovercraft Corporation (q.v.), in which Vickers held a 25 per cent interest until they were bought out by Westland Aircraft (q.v.).

Victor, Handley Page: the third and last of Britain's V bomber force of strategic nuclear bombers, the Handley Page Victor first flew in December 1952, deliveries to the R.A.F. of the B.Mk.1, with four wing-mounted Armstrong-Siddeley Sapphire turbojets, starting in 1958. A developed version, the B.Mk.2, first flew in February 1959 using Rolls-Royce Conway turbojets, and having larger (but thinner) wings. The B.Mk.2 eventually replaced the B.Mk.1, before itself eventually being relegated from the nuclear bombing role (including delivery of the Blue Steel (q.v.) air-to-surface missile) to in-flight refuelling duties. A number of aircraft were however retained in the photographic-reconnaissance role until the early 1970s.

The B.Mk.2 uses four 20,600 lb thrust Rolls-Royce Conway 201 turbojets for a maximum speed of 640 mph and a range of up to 4,600 miles, and a heavy bomb load could be carried internally and on underwing strongpoints before the conversion to the inflight refuelling role, for which refuelling points are fitted on the wingtips and at the tail. An unusual design, with a characteristic crescent wing shape, the Victor has a high wing and a 'T' tailplane. Plans during the late 1950s for an airliner development did not advance far.

Viggen, SAAB-37: see SAAB-37 Viggen.

Vigilante, North American Rockwell A-5: see A-5 Vigilante, North American Rockwell.

Viking, Lockheed S-3A: see S-3A Viking, Lockheed.

Viking, Vickers: originating as the VC.1 (Vickers Commercial 1) design, the Vickers Viking was the first post-war British airliner design and was based on the Wellington wing and tail. The first flight of a prototype was in June 1945, and the aircraft was soon in service with B.E.A. and other airlines, including many independents. A military version, the Valetta, was produced for the R.A.F., which also received the Varsity navigational trainer, with a nosewheel instead of the tailwheel of the Viking and Valetta. Two 1,675 hp Bristol Hercules 634 radial engines produced a maximum cruising speed of 210 mph and a range of up to 1,150 miles with the maximum payload of thirty-two passengers.

Vimy, Vickers: a large bomber biplane for the R.A.F., with which it entered service

275

in 1919, the Vickers Vimy soon earned fame by becoming the first aircraft to fly non-stop across the Atlantic in June of that year. Two 360 hp Rolls-Royce Eagle engines powered the Vimy, which remained in R.A.F. service until 1931, while airliner versions with a ten-seat fuselage were also built during the early 1920s.

Viscount, Vickers: the first successful turboprop airliner in the world, the Vickers Viscount first flew in prototype form as the Viscount Srs.630, with Rolls-Royce Trent turboprops, during 1951. This version also did some airline flying. Initial production aircraft were designated as Srs.700 aircraft, with a larger fuselage than the prototype and the substitution of Rolls-Royce Dart turboprop engines. An up-rated 700 was the 700D, while a further power increase was provided for the Viscount Srs.800, which had a larger fuselage and first flew in 1957. Production of the Viscount ended in 1964 after 444 aircraft had been built, making it Britain's most successful airliner in terms of the numbers sold. The four 1,990 shp Rolls-Royce Dart turboprops of the Viscount Srs.800 provided a maximum cruising speed of 360 mph and a range of up to 1,760 miles, while up to eighty passengers could be carried. A large number of Viscounts remain in service.

Voisin, Charles (1882–1912) and Gabriel (1880–19 ?): the brothers Charles and Gabriel Voisin based their work on that of Hargrave (q.v.), Gabriel building two float-gliders in 1905, co-operating with Archdeacon and Blériot respectively. A feature of the Voisin designs, including the powered versions which first appeared in 1907, was the lack of lateral control and the rigid adherence to the boxkite structure. A series of successful flights with the gliders took place on the River Seine in Paris in 1905 and 1906, but by 1910 the Voisin designs were out-dated, although the factory established by the brothers continued to produce their designs until the outbreak of World War I.

Voodoo, McDonnell F-101: *see* F-101 Voodoo, McDonnell.

Vosper-Thorneycroft: a British shipbuilder

and ship-repairer, with a considerable reputation for building the lighter naval vessels, usually of patrol-boat size but also including frigates, Vosper-Thorneycroft entered hovercraft production during the late 1960s with the VT-1 (q.v.). There has been only a limited demand for this craft so far, although operationally it seems to have performed satisfactorily.

Vostok 1: the first manned satellite to enter earth orbit, the Vostok 1 was launched on 12 April 1961 from the U.S.S.R., carrying Flight Major Yuriy Gagarin (q.v.) on a single orbit, taking 108 minutes from take-off to landing.

VP-B: international civil registration index mark for the Bahamas.

VP-F: international civil registration index mark for the Falkland Islands.

VP-H: international civil registration index mark for British Honduras.

VP-L: international civil registration index mark for Antigua.

VP-P: international civil registration index mark for the Islands of the Western Pacific High Commission.

VP-V: international civil registration index mark for St Vincent.

VP-X: international civil registration index mark for Gambia.

VP-Y: international civil registration index mark for Rhodesia.

VQ-B: international civil registration index mark for Barbados.

VQ-F: international civil registration index mark for the Fiji Islands.

VQ-G: international civil registration index mark for Grenada.

VQ-H: international civil registration index mark for St Helena.

VQ-L: international civil registration index mark for St Lucia.

VQ-M: international civil registration index mark for Mauritius.

VQ-S: international civil registration index mark for the Seychelles.

VQ-ZE/ZH: international civil registration index mark for Botswana.

VQ-ZI: international civil registration index mark for Swaziland.

VR-A: international civil registration index mark for South Arabia.

VR-B: international civil registration index mark for Bermuda.

VR-G: international civil registration index mark for Gibraltar.

VR-H: international civil registration index mark for Hong Kong.

VR-N: international civil registration index mark for the Cameroons.

VR-O: international civil registration index mark for Sabah.

VR-U: international civil, registration index mark for Brunei.

VR-W: international civil registration index mark for Sarawak.

VS-300, Sikorsky: although not the first successful helicopter design, the Sikorsky VS-300 was nevertheless the first successful design produced outside of Germany, as well as being the first practical single-rotor design, introducing the now familiar stabilising tail-rotor. A single-seat helicopter, the VS-300 underwent tethered trials in 1939, with free flight trials during 1940–2, and led directly to the Sikorsky R-4, which was used for operational trials with the British and American armed forces.

VT-: international civil registration index mark for India.

VT-1, Vosper-Thorneycroft: a semi-amphibious air cushion vehicle, being a sidewall craft with beaching capability, the Vosper-Thorneycroft VT-1 uses two Avco Lycoming marine gas turbines to drive four centrifugal lift fans and one water propeller, giving a maximum speed of 40–50

knots and a capacity of up to 148 passengers and ten cars. In operation in the British Isles and Scandinavia.

V.T.O.L.: *see* vertical take-off and landing.

Vuia, Trajan (1872–1950): although Vuia's connection with aviation was brief and not altogether successful, it was nevertheless significant in that he designed and built in 1906 the first full-sized tractor-propeller monoplane, known as the Vuia No. 1, which made a number of hops during that year and in 1907. A further feature of this aircraft was the provision of the first pneumatically-tyred under-carriage. The Vuia No. 2 of 1907 was no more successful than the earlier aircraft.

Vulcan, Hawker Siddeley: undoubtedly the world's first delta-wing jet bomber, the Avro-designed Vulcan was the second British strategic nuclear bomber when it first flew in August 1952 as a prototype, using four Rolls-Royce Avon turbojets. Deliveries to the R.A.F. of the production B.Mk.1, with Bristol Siddeley Olympus turbojets, started in mid-1956. The first of the B.Mk.2s, with modified aerodynamic surfaces, flew in August 1958. Deliveries to the R.A.F. of the B.Mk.2 started in mid-1960, and during the 1960s the later aircraft replaced the remaining B.Mk.1s. Able to carry either Blue Steel (q.v.) air-to-surface missiles or up to twenty 1,000 lb bombs, the Vulcan has four 20,000 lb thrust Bristol Siddeley Olympus 301 turbojets giving the B.Mk.2 a maximum speed of 640 mph and a range of 5,000 miles.

Vultee: formed in 1931, the Vultee concern produced its first design the following year in the V-1, a single-engined, eight-passenger monoplane airliner. Later it developed the BT-13a Valiant trainer for the U.S.A.A.F., before the company's acquisition by Consolidated Aircraft (*see* Convair).

W

Wal, Dornier: originally designed and developed after World War I in Germany as a passenger and mail flying-boat, the Dornier Wal was also used for maritime-reconnaissance duties by various European and South American air arms, although not at that time by Germany itself. Lufthansa used the Wal during the mid-1920s on a number of pioneering and route-proving flights, notably to South America. A monoplane (although the wing was positioned well above the fuselage), the Wal used two engines, of B.M.W. or Napier manufacture, which were placed in line, one driving a pusher propeller and the other a tractor propeller.

Walleye: the Martin Walleye air-to-surface unpowered glide bomb has been in U.S.N. service since 1969. Television guidance is used, the missile being focused on to the target before being released. Eleven feet in length and fifteen inches in diameter, the Walleye is mainly used against hard targets such as bridges, and carries a high-explosive warhead.

Wallis, Sir Barnes: a British aircraft and weapons designer, Sir Barnes Wallis worked on the Wellington bomber before World War II, and during the war was responsible for the design of the 'bouncing bomb' used by the R.A.F. during the raid on the Ruhr Dams. He also worked on the large earthquake bombs later used by the R.A.F. After the war he took a keen interest in variable geometry (q.v.), and his work, although neglected in Britain, was followed closely elsewhere. Before retiring from the British Aircraft Corporation in 1972, he also worked on designs for commercial hypersonic flight.

Walrus, Supermarine: the Royal Air Force development of the Supermarine Seagull,

the prototype Supermarine Walrus first flew in June 1933, before entering service in 1936 and serving on air-sea rescue duties throughout World War II. A biplane with a single radial engine and both wings mounted above the fuselage, the Walrus gave dependable service. It was nicknamed the 'Shagbat'.

Warsaw Pact: a military and economic multilateral alliance formed in 1955, and more properly known as the Eastern European Mutual Assistance Treaty, the present members are the Soviet Union, Bulgaria, Czechoslovakia, East Germany, Hungary, Poland, and Rumania. China has never been a Pact member, and nor has Yugoslavia. Albania was a founder-member, but withdrew from membership in 1968 after eight years of non-participation – the only country to be able to do so, and doubtless this was only possible due to China's backing for the move, and the difficulty of mounting a successful military operation in some of the most difficult terrain in Europe. There has been no hesitation in using Warsaw Pact forces to enforce strict Communist control of Hungary and Czechoslovakia, even when there has been no intention of withdrawal, and internal security operations have been mounted in East Germany and Poland.

The predominant aspect of the Pact is the military alliance, which is controlled by its Joint High Command in Moscow. Operational commands are believed to be centred on three groups: one of these is Soviet-controlled and in East Germany, with complete control of all East German forces; another is the northern group based in Poland; and the third is the southern group, based in Hungary.

Wasp, Westland- *see under* Scout, Westland.

weather radar: in common with other forms of radar, weather radar provides a cathode-ray display of conditions ahead, in this case clouds or high ground. A problem associated with clear air turbulence (q.v.) is the lack of weather radar warning for an aircraft due to the absence of tell-tale cloud formations.

Wellington, Vickers: designed by Sir

Barnes Wallis, the twin-engined Vickers Wellington was the standard R.A.F. medium bomber after its introduction in 1937, and served on bombing and maritime-reconnaissance duties throughout World War II. Initially this mid-wing monoplane, with single-fin tail and nose and tail gun-turrets, used two 1,000 hp Bristol Pegasus radial engines, but later versions used 1,500 hp Bristol Hercules engines, giving a maximum speed of 250 mph, a range of 1,250 miles, and a bomb load of 4,500 lb. After the war Wellingtons were used on navigational trainer duties, and the Wellington aerodynamic surfaces were incorporated on the Vickers Viking (q.v.) airliner design.

Wenham, Francis Herbert (1824–1908): a marine engineer by profession, Francis Wenham, an Englishman, came into prominence in 1866 with the presentation of a paper to the Aeronautical Society supporting Sir George Cayley's (q.v.) views on the importance of cambered wings, and also stating that the front part of a cambered wing provided most of the lift. A number of successful experiments followed with a model, and later with a number of full-size gliders with multiplane configurations. Wenham also built the first wind tunnel in 1871.

Wessex, Westland: the Westland development of the Sikorsky S-58 (q.v.), but now out of production. The main difference between the two machines was the use of turbine engines in the British-built machine.

Western European Union: an organisation formed as the result of a treaty signed in 1955 between Belgium, Federal Germany, France, Italy, Luxembourg, the Netherlands, and the United Kingdom, the Western European Union operates within the NATO (q.v.) alliance but not as a part of it. The only significance of the W.E.U. is that British Army and R.A.F. units in Germany are there under a W.E.U. obligation, while the scope of the W.E.U. is considerably weakened by the attitude to it of France, and the absence of Norway, Denmark, and Portugal.

Westland Aircraft: the first Westland design of any note was the Wapiti, a general-purpose biplane for the R.A.F., intended to replace the D.H.9A while using the de Havilland aircraft's components as far as possible. A development of the Wapiti was the Wallace, which first entered R.A.F. service in 1933. However, the most interesting of Westland's designs during the late 1920s and early 1930s were G. T. R. Hill's 'Pterodactyl' series of tailless aircraft, including cabin monoplanes and fighters, all of a purely experimental nature.

The late 1930s saw the emergence of the famous Westland Lysander army co-operation monoplane for the R.A.F. and the R.C.A.F., which used the aircraft (nicknamed the 'Flying Carrot', because of the fuselage shape) during World War II. Also during the war Westland produced (in small numbers) the Whirlwind, a twin-engined long-range escort fighter for the R.A.F. The immediate post-war period saw Westland developing its last fixed-wing aircraft (if the Fairey-designed Gannet can be excluded) in the Wyvern, which was the only turboprop fighter aircraft to enter operational service, aboard the Royal Navy's aircraft carriers.

Westland's main interest after the war lay in licence-production of Sikorsky helicopter designs, in which Westland was given considerable freedom for further development of the designs and for marketing the resulting helicopters in Europe and the British Commonwealth. In return, Sikorsky has first refusal of licence-production of Westland designs, and a reciprocal agreement on technological development also exists. The first Sikorsky design to be produced by Westland became the Dragonfly, based on the S-51 (q.v.), and this was followed during the 1950s by the Whirlwind and the Wessex, based on the S-55 (q.v.) and the S-58 (q.v.) respectively. Also during this period, Westland developed its own Scout (q.v.) and Wasp helicopters, and built the Bell 47 Sioux under licence for the British Army.

During the 1960s, Westland produced the Black Knight rocket, which was used for space research. By this time the company had acquired Fairey Aviation, which had built the Gannet (q.v.) carrier-borne anti-submarine and airborne-early-warning aircraft and the Rotodyne (q.v.) V.T.O.L. airliner; the helicopter division of Bristol Aeroplane, with the Sycamore and the large Belvedere twin-rotor helicopters; and Saunders-Roe, which had built the Skeeter light helicopter and, more important, hovercraft. These acquisitions resulted from the 1960 reorganisation of Britain's aircraft industry. The hovercraft interest was soon vested in the British Hovercraft Corporation (q.v.), which was initially owned 65 per cent by Westland and 25 per cent by Vickers, although Westland has since acquired the Vickers interest.

Production in recent years has largely centred upon the Sikorsky S-61 (q.v.) Sea King, produced in anti-submarine, air-sea rescue, and assault (Westland Commando) versions; joint production with Aerospatiale of the S.A.330 (q.v.) Puma and S.A.341 (q.v.) Gazelle; and the Westland W.G.13 (q.v.) Lynx helicopter – the latter is also under offer by Sikorsky to the U.S. Navy. In addition, the company has studied a number of tilt-wing and tilt-engine V.T.O.L. airliner designs, but without production plans as yet.

W.E.U.: *see* Western European Union.

W.G.13 Lynx, Westland: first flown in March 1971, the Westland W.G.13 is the British component of the Anglo-French helicopter programme, and is being built in Britain by Westland and in France by Aerospatiale. Sikorsky in the United States will build the W.G.13 under licence in the event of a United States Navy order. Two Rolls-Royce BS.360.07 shaft turbines of 900 shp each drive a single rotor giving a maximum speed of 180 mph and a range of up to 480 miles, while up to eight passengers can be carried. The W.G.13 is available in military and naval versions, the latter having a wheeled undercarriage and provision for anti-submarine torpedoes.

Whirlwind, Westland: originally the name for a World War II long-range escort fighter using two Rolls-Royce Peregrine engines. The Whirlwind is today the name for Westland-built versions of the Sikorsky S-55 (q.v.), although also now out of production.

White, Major Edward Higgins, U.S.A.F.: the first American to float in space and the second man to do so (the first being the Russian Leonov (q.v.)), Major Edward Higgins White, U.S.A.F., was a member of the Gemini 4 crew on a flight lasting from 3 to 7 June 1965, establishing a record for a multi-man spacecraft of ninety-nine hours in space. The other crew member and the captain of the flight was Major James McDivitt, U.S.A.F.

Whitley, Armstrong-Whitworth: one of the early monoplane bombers for the Royal Air Force, the Armstrong-Whitworth Whitley medium bomber first appeared in 1936. Two 918 hp Bristol radial engines provided a maximum speed of 220 mph and a range of 1,000 miles, while 3,500 lb of bombs could be carried. Tail and nose gun turrets were also fitted. The aircraft was used operationally during the early stages of World War II, but was undistinguished in action.

Whittle, Air Commodore Sir Frank, R.A.F.: a Royal Air Force engineer officer, Frank Whittle became interested in the possibilities of the turbojet engine while a cadet at the R.A.F. College, Cranwell, possibly as a result of Dr A. A. Griffith's work during the period 1926–8, although Griffith was primarily interested in developing what would now be described as a turboprop engine. Whittle patented his basic design in 1930, afterwards helping to form a company, Power Jet, to build a prototype engine. This he was able to complete and test successfully on 12 April 1937, making the Whittle engine the first turbine in the world to be tested successfully.

After further development, the Whittle engine was fitted to the Gloster E.28/39 (Gloster Whittle) monoplane, which flew successfully on 15 May 1941. The engine was later fitted to the only Allied operational jet fighter of World War II, the Gloster Meteor (q.v.), and also to the Bell Airacomet. Whittle was knighted for his achievement, which gave the United

Kingdom many years of supremacy in turbojet development.

Whittle, Gloster E.28/39:   *see* E.28/39 Whittle, Gloster.

wind tunnel: the first wind tunnel was built in 1871 by Francis Wenham (q.v.), and today wind tunnels are considered vital in aircraft development, scale models of designs usually being subjected to exhaustive testing long before prototype aircraft are constructed, although these too can be subjected to wind tunnel tests. Smoke is sometimes filtered into the airstream in the wind tunnel so that photographic records of the behaviour of the air currents passing over the aerodynamic surfaces of an aircraft may be taken. The use of the wind tunnel is no longer restricted to aviation, since many road and rail vehicles are also subjected to the same tests.

wing: the usual term for the primary lifting aerofoil or mainplane of any aircraft, the term 'wing' applies to the complete structure, i.e. it is inaccurate to talk about an aircraft's 'wings' if it is a monoplane, since the aerodynamic surfaces on either side of the fuselage are regarded as being a single wing. Fixed wing aircraft are simply conventional aircraft, while rotary wing types are helicopters. *See also* variable geometry and variable incidence. For development of the wing *see* Sir George Cayley, Horatio Phillips, and F. A. Wenham.

wing loading: wing loading is the weight of an aircraft divided by the wing area. High wing loadings are normally found on high speed aircraft, while in aircraft requiring good low speed, take-off, and landing handling characteristics, a low wing loading is usual.

wingspan: the distance from wingtip to wingtip along the wing.

wing warping: a system of control in roll developed and perfected by the Wright brothers (q.v.), wing warping requires a twisting of the wing to alter the curvature and angle of incidence. Although many other early aircraft designers adopted the practice, it was out-dated by the start of

World War I, when ailerons were in general use.

World War I: both world wars played a considerable part in aviation development, both in technological terms and in making aviation a more practical proposition, although it is difficult to comment effectively on the idea sometimes mooted, that four years of war did more for aviation than fifty years of peace could.

In the field of military aviation, aircraft had only been used experimentally prior to the outbreak of World War I; they were as a rule unarmed, with the duty of reconnaissance and air observation, although sometimes also used for communications duties. Pilots were provided with revolvers or rifles at first, but during 1915 the fighter developed as a distinct type, to be followed by the bomber. Britain showed an early interest in the strategic bomber concept with the large Handley Page aircraft, followed immediately after the war by the Vickers Vimy, while in Russia, Sikorsky's Bolschoi and Ilya Mourometz pointed the way to the aircraft of the future. The first bombing raids were by the Royal Naval Air Service on Düsseldorf and Cologne in September 1914, followed by a German raid on Dover in December, and the first airship raids, on Great Yarmouth, in January 1915.

At sea, first the seaplane carrier and then the aircraft carrier (q.v.) evolved, while trials for launching aircraft from barges towed by destroyers were conducted, with less success.

The implications for civil aviation after the war included the ready availability of qualified pilots and mechanics, while many bomber designs, including the D.H.4 and D.H.9, were well-suited for conversion to transport duties, and airline versions of the Vickers Vimy, with a new fuselage, were also built.

World War II: the techniques of World War I continued into World War II, although with considerable refinements, including heavy strategic bombing and fighter formations. Aircraft carriers gained in importance, breaking into a more strategic role in the British attack on the

Italian Fleet at Taranto in November 1940, and the Japanese attack on the U.S. Fleet at Pearl Harbor in December 1941, while during the Battle of the Coral Sea, in May 1942, the American and Japanese fleets relied entirely on their aircraft rather than on guns.

The real progress since World War I lay in the development of missiles, including the German V-1 (q.v.) and V-2 (q.v.), and of rocket and turbojet propulsion. There was also the greater use of the aeroplane for transport, often over long distances, which led to a rapid post-war expansion in long-haul airline operations. The use of atomic weapons against Japan was also not without significance.

Wright, Wilbur (1867–1912) and Orville (1871–1948): the two Wright brothers, Wilbur and Orville, made the world's first true powered heavier-than-air flights on their Flyer I at the Kill Devil Hills, near Kittyhawk, North Carolina, on 17 December 1903. Four flights were made, the toss of a coin deciding that Orville should make the first flight (to be exact, it had decided that Wilbur should do so three days previously, but he stalled and damaged the aircraft, leaving the next turn to Orville on the 17 December), which lasted for twelve seconds and covered 120 feet at about 30 mph. The second flight was of 175 feet, the third of 200 feet, and the fourth of 852 feet, although it has generally been accepted that this was half a mile through the air against a headwind.

These were the first flights in history in which an aircraft had lifted itself and its pilot off the ground under its own power and without any downhill assistance, and sustained itself in flight, eventually landing on a point no lower than that from which it took off. However, the Wrights had already enjoyed considerable success with their gliders, the first of which had flown in 1900, and the third in 1902, incorporating the main features of their aeroplanes – wing warping for control in roll, rudders aft of the mainplane, and the elevators mounted in front, while the pilot lay flat on his chest in the middle of the mainplane. All Wright designs were biplanes.

The Flyer I was merely a powered Wright glider, with a single Wright engine driving two pusher propellers, take-off being along a single rail; one of the brothers piloted the aircraft while the other ran alongside to steady it before take-off. Further developments of the Flyer I led to the Flyer II of 1904, and to the Flyer III of 1905, which was the first practical aeroplane, first flying on 29 September 1905, for twelve miles, and being able to bank, turn, and fly a figure of eight. The Flyer III also had the distinction of making the first passenger-carrying flight, on 14 May 1908, with Orville as pilot, carrying a Mr C. W. Furness on a two-and-a-half-mile flight.

The next Wright aircraft was the 'A', which made the first public demonstration flights on 8 August 1908 in France, but also suffered the first fatal accident in aviation while on trials with the U.S. Army on 17 September 1908, when Lieutenant Selfridge, U.S. Army, was killed and the pilot, Orville Wright, injured. The 'A' was followed by the 'B', the first Wright machine to have a wheeled undercarriage, and then by the Model R, the Baby Wright, and the Baby Grand, all of which retained the pusher-propeller and Wright-built engines which characterised Wright designs.

Wyvern, Westland: the first and only operational turboprop fighter-bomber, the Westland Wyvern, a carrier-borne low-wing monoplane was developed during the late 1940s, before finally entering service during the early 1950s and replacing the Royal Navy's Blackburn Firebrands. A single-seat design, the Wyvern used a single 3,670 shp Armstrong-Siddeley Python A.S.P.3 turboprop for a maximum speed of 383 mph, and could carry a wide variety of underwing stores in addition to the four wing-mounted 20 mm cannon.

X

X-: United States designation for experimental aircraft, sometimes used as a prefix for other type designations, e.g. XC- is an experimental transport type. X- usually refers to an aircraft of a purely experimental nature rather than a prototype.

X-1 Skyrocket, Bell: the first aircraft in the world to break the sound barrier, the Bell X-1 (sometimes known as the XS-1, experimental supersonic-1) Skyrocket used a four-nozzle 6,000 lb thrust rocket engine, and reached a speed of 762 mph in October 1947. The achievement is not always recognised since the X-1 did not take off under its own power, but was air-launched from a Boeing B-29 Superfortress bomber.

X-2, Bell: an air-launched, rocket-powered aircraft, the Bell X-2 was built largely of stainless steel, and was used for exploration of high supersonic speeds and high-altitude flight, reaching 2,148 mph and 126,200 feet on 27 September 1956.

X-15 and X-15A-2, North American: starting in 1960, the United States Air Force conducted a series of extensive trials first with the North American X-15 air-launched rocket-powered aircraft, and then with its development, the X-15A-2, using a Boeing B-52 Stratofortress as launch aircraft. The first speed record reached by the X-15 was 2,196 mph on 4 August 1960, while the last, by the X-15A-2, was 4,534 mph on 3 October 1967. A 60,000 lb thrust rocket motor powered the aircraft.

X-22A, Bell: a vertical take-off research aircraft, the Bell X-22A used four 1,250 shp General Electric turboprop engines to power four ducted-propellers, two of which were tail-mounted, and the other two mounted on either side of the fuselage behind the flight deck; these were tilted for vertical take-off and landing. Experiments were conducted during the mid-1960s, and a maximum speed of 325 mph and a range of 540 miles were achieved.

XA-, XB-, XC-: international civil registration index marks for Mexico.

XC-142A, L.T.V.-Hiller-Ryan: a tilt-wing tactical transport for vertical take-off trials, the XC-142A is a joint project involving Ling-Temco-Vought, Hiller, and Ryan. Four 2,805 shp Lycoming turboprops provide a maximum speed of 430 mph and a range of 345 miles with a four-ton-payload.

XT-: international civil registration index mark for Upper Volta.

XU-: international civil registration index mark for Cambodia.

XV-: international civil registration index mark for South Vietnam.

XW-: international civil registration index mark for Laos.

XY-, XZ-: international civil registration index marks for Burma.

Y

YA-: international civil registration index mark for Afghanistan.

Yak-11 'Moose', Yakovlev: an intermediate trainer with tandem twin seats, the Yakovlev Yak-11, NATO code-named 'Moose', remains in service with some Communist Bloc air arms. A low-wing monoplane with retractable undercarriage, the Yak-11 is based on the wartime Yak-9 fighter, and uses a single 730 hp Shvetsov ASh-21 radial engine, giving a maximum speed of 286 mph and a range of up to 800 miles, while practice bombs and a machine gun can be fitted.

Yak-12 'Creek', Yakovlev: a high-wing four-seat liaison and A.O.P. monoplane, NATO code-named 'Creek', the Yakovlev Yak-12 has been in production and service for many years in Poland and China as well as the Soviet Union. Early versions were built of wood and metal, but the current aircraft are all-metal. A 240 hp AI-14R piston-engine provides a maximum speed of 112 mph and a range of 475 miles.

Yak-15, Yakovlev: the first Soviet jet fighter design, the Yakovlev Yak-15 was based on a modified Yak-9 airframe and a single 1,980 lb thrust RD-10 turbojet (based on a captured German Jumo 004B engine), giving a maximum speed of 495 mph. The Yak-15 was in limited service, and then only during the late 1940s and early 1950s, until replaced by Mikoyan-Gurevich MiG-15s.

Yak-18 'Max'/'Mouse', Yakovlev: the Yakovlev Yak-18, NATO code-named 'Max', has been the standard Soviet basic trainer since 1946, and is a low-wing monoplane with retractable undercarriage and tandem twin seats. Early versions had a tailwheel, but current versions have a tricycle undercarriage. A single-seat version, the Yak-18P, is NATO code-named 'Mouse'. A 260 hp AI-14R piston engine provides a maximum speed of 153 mph and a range of 630 miles.

Yak-25 'Flashlight', Yakovlev: in Soviet service since 1955, the Yakovlev Yak-25, NATO code-named 'Flashlight', is a now obsolete twin-seat all-weather fighter, reconnaissance, and tactical strike aircraft, also used for training duties, with two wing-mounted 5,500 lb thrust Type 37U turbojets. Excessive wing anhedral requires wing-tip wheels. Maximum speed is 630 mph. Trainer versions are code-named 'Mangrove'. There have been a number of developments of the basic aircraft.

Yak-28 'Firebar', Yakovlev: a development of the Yak-25 (q.v.), but with a high wing rather than the mid wing of the earlier aircraft, the Yakovlev Yak-28, NATO code-named 'Firebar', has been in service with the Soviet Union's armed forces at least since 1961, and is an all-weather fighter, reconnaissance, and tactical strike aircraft. A trainer version is code-named 'Maestro', and the tactical strike version is known as 'Brewer'. Two 9,500 lb thrust TDR Mk. R37F turbojets provide a maximum speed of 740 mph and a range of about 1,150 miles.

Yak-40 'Codling', Yakovlev: the first Soviet airliner design to be certificated for operations in Western countries, the Yakovlev Yak-40, NATO code-named 'Codling', first flew in October 1966, and was first on offer in the West in 1971 and 1972. Three tail-mounted Ivchenko AI-25 turbofans of 3,300 lb thrust each provide a maximum cruising speed of 470 mph and a range of up to 470 miles with a full load of forty passengers, although a lower seating capacity is usually specified.

Yak-? 'Freehand', Yakovlev: an as yet undesignated V.T.O.L. fighter, this Yakovlev aircraft has been NATO code-named 'Freehand', and has only made one public appearance, at the 1967 Domodedovo Air Display. A twin-engined aircraft with a single seat and a cranked mid-wing, vectored thrust is used for vertical take-off. Engines are estimated to be 8,000 lb thrust turbofans, giving a maximum speed of 600 mph.

Yakovlev, Alexander S.: originally an aero-engine designer, Alexander S. Yakovlev was persuaded to turn to airframes by the head of another Soviet design bureau, Andrei Tupolev (q.v.), and first came to the attention of the West with his Yak-3 fighter of World War II, which was followed by the Yak-9. After the war Yakovlev quickly developed the Yak-9 airframe to take captured German engines, and the resulting airframe was the Yak-15 (q.v.), the Soviet Union's first jet aircraft. His main activity during this period, however, was designing a number of training and liaison types, including the Yak-11 (q.v.), the Yak-12 (q.v.), and the Yak-18 (q.v.). He returned to combat aircraft during the 1950s with the Yak-25 (q.v.) and the Yak-28 (q.v.), while also building a number of prototypes of a large helicopter, the twin-rotor Yak-24. More recent designs from the Yakovlev Bureau have included the 'Freehand' V.T.O.L. fighter and the Yak-40 (q.v.) feeder-liner, the first Soviet design to be certificated for operations in the West.

YI-: international civil registration index mark for Iraq.

YJ-: international civil registration index mark for the New Hebrides.

YK-: international civil registration index mark for Syria.

YR-: international civil registration index mark for Rumania.

YS-: international civil registration index mark for El Salvador.

YS-11, N.A.M.C.: the first post-war Japanese transport aircraft to enter production, the YS-11 was developed and produced by a consortium of manufacturers trading as the Nihon Aircraft Manufacturing Company, the first flight taking place in August 1962. Deliveries to the airlines started early in 1965, and 182 aircraft had been built by the time production ceased. Two 3,060 shp Rolls-Royce Dart 542 turboprops give the low-wing YS-11 a maximum cruising speed of 297 mph and a range of up to 1,475 miles. Maximum payload is sixty passengers or 12,500 lb of freight.

YU-: international civil registration index mark for Yugoslavia.

Yukon, Canadair CC-106: *see* CC-106 Yukon, Canadair.

YV-: international civil registration index mark for Venezuela.

Z

ZA-: international civil registration index mark for Albania.

Zeppelin, Count Ferdinand von (1838–1917): the inventor of the giant Zeppelin airships, but not of the practical airship itself, since that distinction effectively belongs to the French Lebaudy (q.v.) brothers, Count Ferdinand von Zeppelin built his first airship in 1900. During the early years of the century Zeppelin's airships, with their potential for civil and military use, succeeded in attracting the attention of both the German public and of the Kaiser's ministers.

Perhaps one of the first air services in the world, a fleet of five Zeppelins carried some 35,000 passengers within Germany between 1910 and 1914 without mishap. During World War I Zeppelins were used on reconnaissance duties and on bombing raids on London and the English East Coast ports, although the highly inflammable hydrogen-filled airships were an easy target for the fighters of the R.F.C. and R.N.A.S. Interest in the Zeppelin remained strong during the post-war period. The most famous Zeppelin built, the Graf Zeppelin, flew around the world in 1929, taking twenty-one days and making the first trans-Pacific flight *en route*. One of the last Zeppelins was the ill-fated Hindenberg (q.v.) of 1937.

Count Ferdinand von Zeppelin was also considerably impressed by Sikorsky's (q.v.) aircraft designs, including the Bolschoi, before World War I, and took an interest in the possibilities of developing larger aircraft for passenger use, although he died before German designers started to put what was to be a considerable amount of effort into this aspect of aviation.

'Zero', Mitsubishi A6M: the most famous Japanese fighter of World War II, the Mitsubishi A6M derived its popular name from the use initially of the designation, Navy Type O. A low-wing, single-seat monoplane, the 'Zero' was operated from Japanese aircraft carriers and shore bases throughout the war, frequently being used as a fighter-bomber, and towards the end of the war also undertaking Kamikaze suicide operations. First flight of the 'Zero' prototype was in April 1939.

ZK-, ZL-, ZM-: international civil registration index marks for New Zealand.

ZP-: international civil registration index mark for Paraguay.

ZS-, ZT-, ZU-: international civil registration index marks for South Africa.